INNOVATION SYSTEMS IN THE SERVICE ECONOMY

MEASUREMENT AND CASE STUDY ANALYSIS

Economics of Science, Technology and Innovation

VOLUME 18

The titles published in this series are listed at the end of this volume.

INNOVATION SYSTEMS IN THE SERVICE ECONOMY

MEASUREMENT AND CASE STUDY ANALYSIS

edited by

J. Stanley Metcalfe
and
Ian Miles

ESRC Center for Research on Innovation and Competition
The University of Manchester, U.K.

KLUWER ACADEMIC PUBLISHERS
Boston / Dordrecht / London

Distributors for North, Central and South America:
Kluwer Academic Publishers
101 Philip Drive
Assinippi Park
Norwell, Massachusetts 02061 USA
Telephone (781) 871-6600
Fax (781) 871-6528
E-Mail <kluwer@wkap.com>

Distributors for all other countries:
Kluwer Academic Publishers Group
Distribution Centre
Post Office Box 322
3300 AH Dordrecht, THE NETHERLANDS
Telephone 31 78 6392 392
Fax 31 78 6546 474
E-Mail <services@wkap.nl>

Electronic Services <http://www.wkap.nl>

Library of Congress Cataloging-in-Publication Data

Innovation systems in the service economy : measurement and case study analysis / edited by J. Stanley Metcalfe and Ian Miles.
p.cm.-- (Economics of science, technology, and innovation ; v. 18)
Based on 2 workshops held at the ESRC Centre for Research on Innovation and Competition in 1998.
Includes bibliographical references and index.
ISBN 0-7923-7730-3 (alk. paper)
1. Service industries--Congresses. 2. Service industries--Government policy--Congresses. 3. Technological innovations--Economic aspects--Congresses. I. Metcalfe, J. S. (J. Stanley) II. Miles, Ian. III. Series.

HD9980.5 .157 1999
338.4--dc21 99-049240

Printed on acid-free paper.

Printed in the United States of America

Contents

List of Contributors

Dr Birgitte Andersen, ESRC Centre for Research on Innovation and Competition, University of Manchester, UK

Professor Cristiano Antonelli, Dipartimento di Economia, Universita di Torino, Italy

Professor Gilberto Antonelli, Department of Economics, University of Bologna and IDSE-CNR, Milan, Italy

Mr Giulio Cainelli, IDSE-CNR, Milan, Italy

Professor Rod Coombs, ESRC Centre for Research on Innovation and Competition, University of Manchester and School of Management, UMIST, UK

Dr Nicola De Liso, IDSE-CNR, Milan and Faculty of Law, University ofLecce, Italy

Dr Ken Ducatel, IPTS, Seville, Spain and PREST, University of Manchester, UK

Professor F. Gallouj, IFRESI, Université de Lille 1, France

Ms Christiane Hipp, Mannesmann Pilotentwicklung, Munich, Germany

Dr Jeremy Howells, ESRC Centre for Research on Innovation and Competition and PREST, University of Manchester, UK

Professor Alan Hughes, Margaret Thatcher Professor of Enterprise Studies, Judge Institute of Management Studies, and Director of the ESRC Centre for Business Research, University of Cambridge, UK

Professor Alfred Kleinknecht, Delft University of Technology, The Netherlands

Professor J Stanley Metcalfe, ESRC Centre for Research on Innovation and Competition and School of Economic Studies, University of Manchester, UK

Professor Ian Miles, ESRC Centre for Research on Innovation and Competition and PREST, University of Manchester, UK

Dr Paul Nightingale, Research Fellow, CoPS, SPRU, University of Sussex, Brighton, UK

Mr Rob Poll, Vice President, IT, Morgan Stanley Witter, London, UK

Dr Brigitte Preissl, German Institute for Economic Research (DIW), Berlin, Germany

Professor Jon Sundbo, Department of Social Sciences, Roskilde University, Denmark

Dr Bruce S Tether, ESRC Centre for Research on Innovation and Competition, University of Manchester, UK

Eric Wood, Research Officer, Manufacturing Roundtable, Graduate School of Business, University of Cape Town, South Africa and Research Associate, ESRC Centre for Business Research, University of Cambridge, UK

Mr Roberto Zoboli, IDSE-CNR, Milan, Italy

Preface

The essays contained in this volume were presented in their original form at two workshops held at the ESRC Centre for Research on Innovation and Competition in March and May 1998. The first workshop was organised around the theme "Innovation Systems and Services" and the second around the theme "Conceptualising and Measuring Service Innovation". The revised versions of most of the papers presented at these two workshops are collected here and they illustrate in many different ways the importance of innovation in services and services in the innovation process. We are grateful to all the participants for the material they provided and for contributing to an extremely interesting discussion. The workshops marked the first stage of CRIC's ongoing research programme in the service economy and we are grateful to the support of the ESRC in funding this programme. In putting this volume together we have also benefited from the support of the European Commission through a project on European Services and Innovation Systems co-ordinated by Professor F. Malerba, Director of CESPRI at Bocconi University. We are grateful to members of this project for helpful discussion on the theme of innovation in the service economy.

We take this opportunity of thanking all our CRIC colleagues for helping to develop our evolving research agenda and Sharon Hammond, Deborah Woodman and Sharon Dalton for arranging the workshops. We are particularly grateful to Sharon Dalton who has taken on the burdensome task of putting the essays together in their final form.

Chapter 1
Introduction, Overview and Reprise

J Stanley Metcalfe and Ian Miles

1.1 SERVICES AND THE ECONOMY

Modern studies of the relations between science, technology and innovation have frequently observed that the practical application of new ideas to the economy more often than not leads a theoretical understanding. So it has turned out in the study of the modern service economy. Long after services acquired economic and social significance, social scientists began tentatively to map this phenomenon and analyse its foundations. Despite the suggestions of Fisher (1933) and Clark (1957) it was not until the 1960s that scholars such as Fuchs (1965) began to chart the long-run growth of service employment in the USA, and to speculate on how this reflected particular interactions between productivity and growth and the evolution of the pattern of demand. Many scholars followed and we now have an increasing understanding of the role of services in the development of the division of labour in modern economies[1]. Even more significant for our purposes is the length of time it has taken to recognise the importance of services as loci of innovation in two senses. Firstly, service firms can be innovative in their own right, even though the process of innovation and the kinds of innovation may be different from those traditionally associated with manufacturing and other primary activities. Secondly, service firms and associated activities play an important role in the evolving division of creative labour which is constituted by modern innovation systems. It is to the elaboration and understanding of these two themes that this volume of essays is devoted.

No doubt part of the difficulty in coming to terms with service innovation has been the mental gymnastics entailed in defining and measuring services. Traditionally, scholars have felt comfortable with defining primary and secondary manufacturing activities in a positive sense, while allocating, in a negative sense, everything else to a tertiary, 'services' sector. In so doing they bundled together activities of great heterogeneity in application and production method, and added to the difficulties of understanding what turned out to be the most rapidly growing sector in modern economies.

The key point is that services are activities directed at creating changes, transformations in some entities. These transformations may be of form, place or time of availability, and the entities involved may be material objects, goods, people, the natural environment or symbolic representations, data, text, etc. As with all transformations, material and energy inputs are required and the resulting outputs may be classed as part of final economic demand or, particularly important in this case, as part of intermediate output. In so far as the activities transform people, the outcomes may be physical (a beauty treatment, a good meal, an appendectomy) or mental (a gain in knowledge, the sensory pleasure of a good concert). None of the classifications, which build on these principles, are watertight. It is often a matter of convention where the activities of a particular firm are classified, and often this convention is simply a question of the organization structure of an industry. Seen in these terms, services are endemic and industrial societies have always had a substantial service element. Perhaps this suggests the importance of looking at the service nature of manufacturing activities as a way of unravelling the division of labour in production <u>and</u> innovation. Thus, it is helpful to think of service *functions* as being performed throughout the economy. These service functions – such as design, R&D, delivery, after-sales, marketing, maintenance, etc. – may be performed by specialised service*firms*, and thus attributed to the service *sectors* of the economy. Service firms' output takes the form of their service *products* or service *commodities*. But such products can be generated and delivered (whether sold or free of charge) by firms in any sector, manufacturing included. For example, many companies listed as computer manufacturing firms are actually making the bulk of their profits from the retail of computer *services*. Furthermore, companies in all sectors of the economy may perform service functions internally for their own use, rather than contracting them out to specialised service firms. Thus service *occupations* – such as white-collar and other 'non-production' jobs like security, catering, cleaning – may also be found in all sectors. The white-collar share of industrial workforces and the service share of manufacturing firms' outputs are generally growing, along with the growth of specialised service firms. The widely used concept of 'the growth of the service sector' therefore has a multi-faceted meaning which goes well beyond a simple shift in the relative sizes of various industrial sectors. The 'service economy' is a matter of service being more prominent right across the economy, not merely the expansion of the specific service sectors.

1.2 SERVICES AND INNOVATION

A frequent complaint in the literature is that services have been largely overlooked by innovation researchers – and largely neglected by innovation and technology policymakers. Despite the diversity outlined above, their treatment in analyses of economic and technological change has, until recently, been very one-dimensional, when it has been attempted at all. Services' roles in technological change, in particular, were largely seen as so insubstantial as to be barely worth examination. They were, and still generally are, assumed to be innovative laggards – 'supplier-driven' industries. A very few services were always recognised as exceptions to this rule by the relatively few researchers who have examined the services economy. There were efforts to deal with these anomalies in some of the classic studies by defining technology-intensive services such as telecommunications and airports out of the sector. And the point that R&D and technology management activities are themselves services was rarely noted – something that is much more difficult today as specialised service firms carrying out such activities have become more prominent.

Given the unarguable growth in the importance of service sectors, increasing numbers of researchers and policymakers have taken a fresh look at service activities. This includes questioning received wisdom about the innovative capacity of these firms and sectors. The changes that have taken place in some services have made it evident that preconceptions about the sector as supplier-driven and relatively slow in the uptake of innovation are no longer valid – if they ever were. At the very least, these are certainly not generalisations applying to all or even most services.

Thus, it is apparent that services are major users of new Information (including Communication) Technology (IT). That IT use means a dramatic increasing in the technology-intensity of services. That services' technological expenditure is growing; their investment is changing focus from constructing plant to installing equipment (again, a process which previously took place in manufacturing sectors during the industrial revolution). As well as being users of new technology, some services are innovators – for their own use in new system configurations and applications, and for wider use when they sell on IT services such as software, databases, and telematics to other clients. Finally, it is apparent that other services also play a substantial role in helping diffuse technological knowledge, via marketing, training, and consultancy. 'Knowledge-Intensive Business Services' (KIBS) are major agents of change in the wider economy. KIBS are often both producers and agents of transfer of knowledge, and may well be involved in co-producing innovations with their clients. Though there are obvious examples connected with IT – computer and telematics services, web designers, etc. – there are also important roles for services in environmental, construction, biotechnological and other fields.

1.3 SERVICES AS AGENTS OF INNOVATION

Of particular interest to the study of innovation systems are Knowledge-Intensive Business Services (KIBS). KIBS are among the most rapidly growing sub-sectors of the service economy. These exemplify the general process of knowledge-intensification in industrialised economies, and play, we argue, an important role in innovation processes across the economy. Their growth reflects increased demands for certain types of knowledge in the economy, together with trends in the division of labour which lead to specialised services emerging and playing prominent roles in knowledge accumulation and transfer. Two main types of activity can be distinguished.

Traditional professional services help users deal with complex social, physical and psycho-biological systems. These include administrative rules and regulations (as in legal and accountancy services); social groups and interests (marketing and consultancy); physical systems (architecture and building services which may have considerable technical content); psychological and biological systems (medical and veterinary services, educational and clinical psychology and psychiatry, counselling – some of these also have high technical content). Their relation to new technology is typically more one of being users rather than as agents in development and diffusion.

New-technology-based KIBS' work focuses upon emerging technologies and technological challenges. IT, as a generic technology, is obviously particularly important; biotechnology and new materials, and less pervasive technologies with large and daunting knowledge-requirements (for example, radiology) are also the focus of much activity. It is also interesting to note the emergence of a large group of services whose activities concern problem-driven technological issues associated with environmental challenges (sustainable development and 'clean' technology).

These new-technology-based KIBS are of particular interest in the context of innovation processes. They rely heavily upon professional knowledge: accordingly, their employment structures are heavily weighted towards scientists, engineers, and other experts. Many supply information and knowledge to their users as their central function (for example, in the form of measurements, reports, training, and consultancy). Some others use their knowledge to produce intermediate inputs to their clients' own knowledge generating and information processing activities (for example, communication and computer services). And yet others use their specialised knowledge to provide other technical functions, such as pollution detection and remediation services.

KIBS thus span a broad range of activities, from the diagnosis and analysis of problems, to the identification and implementation of technological and other solutions. Some KIBS are highly standardised; their service products are largely supplied as packaged commodities. Others typically supply products that are highly customised to clients and whose production involves much interaction. Conventional economic sub-sectors conceal much of this diversity: the software sector, for instance, spans the complete range from firms producing standardised commodities to firms who only ever work to client specifications. One trajectory for much services' innovation is the move from client-intensive to commodity packages and combinations of modules.

KIBS have proliferated around new technologies, and around new technological problems (such as those created by environmental regulation and litigation). Considering services concerned with the development of IT, for example, there are specialised IT training and consultancy services; second-, third- and now fourth-party equipment (repair) services; sales services; management services; facilities management and outsourcing of computer and telematics services; software production and maintenance of various kinds; dataware services such as CD-ROM and other multimedia, as well as established on-line database services.

The growth of such KIBS results from a number of factors. Structural changes in industry (downsizing, concentration on core activities) have led to the externalisation of activities that were in the past provided in-house. Even activities which were usually regarded as a part of the core business – such as R&D – and which analysts normally assumed could not be delegated to outsiders, are being treated in this way. There is a strong element of firm strategy in this, with different firms taking different decisions – and with perspectives changing over time as experience accumulates on the part of both potential suppliers and potential clients for services.

A similar contracting-out of services from public sector bodies, usually under political imperatives, has also stimulated use of certain services. Some of the growth of private R&D-intensive services relates to the process of privatisation and 'marketisation' of government laboratories and similar facilities, as Howells argues in this volume.

But many services are too new to have been developed in-house and then externalised by many of their users. It is not a case of KIBS substituting for in-house activities, often the relationship is one of complementarity rather than substitution (Gallouj and Weinstein, 1997). New highly technical or complex services may be too costly to maintain in-house, and the knowledge may not even be present to establish these in-house functions. The requisite skills or equipment may be seen as experimental, as too different from core capabilities; as being only required occasionally, or as problematic in terms of achieving a minimum efficient scale. Some firms demand service inputs as a function of the need for co-operation with other members of a network. Recently, many professionals have been shed from firms that are 'downsizing' and are seeking to establish themselves as self-employed service suppliers – thus a growth in one-person consultancies.

The growth of Knowledge-Intensive Business Services is only one facet of the general theme of the interdependence of manufacturing and service activities in the operation of innovation systems. Recent work by Tomlinson (1997), for example, has shown that the growth of productivity in manufacturing is particularly correlated with the consumption of KIBS.

The growth of the 'services sector', as traditionally defined, to a dominant position in the share of output of the industrial economies is an important development in societies. It means that we can no longer treat the innovation which goes on in this sector as a residual problem, to be ignored or simply assumed to follow the patterns and be explained by the concepts derived from a manufacturing context. Furthermore, although statistical constraints often force us to use received notions of sector, we have to recognise that the understanding of innovation processes and consequences related to service activity forces us to go beyond these notions of sector.

1.4 OVERVIEW

The essays which follow explore the innovation and services theme from different perspectives. The chapters in Part I provide background discussion of the innovation systems perspective, and Part II contains a group of essays which either directly or by reflection draw upon new databases in relation to innovation and services. Finally, Part III contains a group of essays, each one a case study of different aspects of services and the innovation process.

Part I contains three chapters which consider conceptual frameworks in relation to innovation systems and innovation in services.

The nature of innovation systems is the theme of Chapter 2 by B. Andersen, S. Metcalfe and B. Tether. The focus of this survey of different approaches to the study of systems of innovation is the idea of distributed innovation processes and the division of labour in the production and application of practically useful knowledge. The discussion is located within the evolutionary tradition in which selection processes and development processes interact to shape patterns of innovation and economic change. Innovation systems are the sets of organizations and institutions which define the relevant developmental processes and which interpenetrate market and commercial systems. Particularly important in this chapter is the emphasis on understanding the dynamics of system genesis, growth, decline or demise, and in understanding the impact a system has on the timing and nature of specific innovations. A variety of contrasting approaches to the systems idea are summarised in terms of different ways of mapping the division of labour in distributed innovation processes. The chapter concludes with a number of questions to shape the next generation of innovation systems thinking in general and in relation to innovation in services in particular.

In Chapter 3, J. Sundbo and F. Gallouj continue the systems theme in their exploration of the notion of loosely coupled service innovation systems; a device they use to identify a number of common components to the innovation process across a wide range of service industries. An important element in their analysis is the generalisation of the idea of technological trajectories (frameworks of thought and practice which guide the innovation process) to encompass trajectories in management ideas and trajectories in professional practice which influence one another and interact at a higher level with institutional and social trajectories. This leads to a comparative analysis of different models of the innovation process (the interaction between management and organization, trajectories and external actors) in relation to service activities, the range of models being a consequence of the heterogeneity of service activities. They conclude that innovation processes in services are weakly institutionalized, they are not national in domain, and because of their fluid, informal, loosely coupled nature they provide an insubstantial basis for policy interventions of the traditional kind.

Chapter 4 by C. Antonelli has a twofold purpose. Firstly, he challenges the view, first expounded by Kenneth Arrow, that knowledge is freely available at zero cost to all potential users, that knowledge spillovers are automatic and 'in the atmosphere'. Per contra, the acquisition of information is a costly process and the combination of external information with internal knowledge in firms and other organizations involves a substantial commitment of resources. This recombinative process, Antonelli argues, is contingent upon the nature of the relevant innovation systems, it frequently involves market transactions, and it is greatly influenced by the development of modern communication technologies. The second objective of the chapter is to test for the economic significance of external knowledge flows at national level by using OECD data on the technological balance of payments as an indicator of market mediated transactions in information. Not only do such transactions represent a substantial proportion of national R&D outlays they are also growing more rapidly than those outlays. Antonelli finds that they have an output elasticity comparable to that for national R&D expenditures and that their inclusion in the statistical analysis leads to a downward reassessment of the national output-R&D elasticity.

All of the chapters in Part II are based on the analysis of new data sets in relation to services innovation. Individually and collectively they point to major challenges in data gathering which need to be faced and surmounted if we are to have a clear understanding of the modus operandi of innovation in the service economy.

In Chapter 5, R. Coombs and I. Miles begin with the troublesome observation that received concepts of innovation and the corresponding measurement techniques are rooted in the study of manufacturing innovation. They propose that the solution to this problem is neither to assimilate services into manufacturing nor to demarcate sharply innovation in services and manufacturing. Rather, they develop a synthesis transcending these distinctions in which consideration of services highlights previously neglected aspects of the innovation process. What they propose is no less than a thoroughgoing reappraisal of established measurement procedures in relation to services innovation, a reappraisal which captures the essential idea that service activities are distributed across all aspects of the modern economy. As they suggest, the industrialisation of service firms proceeds side by side with a greater service role in traditional manufacturing activities.

Chapter 6 by A. Hughes and E. Wood is based on an original and large-scale dataset of innovative activities in small and medium manufacturing firms together with a sub-set of service firms. This dataset was created by the ESRC Centre for Business Research at the University of Cambridge and it provides important microeconomic information on the similarities in and differences between innovation processes in services and manufacturing. Their framework is based upon Freeman's (1979) innovation taxonomy and they find that firms with a knowledge base in engineering, whether in manufacturing or business services, exhibit strong similarities in their innovation processes and substantial differences from those in non-engineering based firms. Important differences between engineering based firms in manufacturing and services are found in relation to the employment of R&D staff and to the presence of barriers to innovation. This chapter provides support for the view expressed by Coombs and Miles in Chapter 5 that a synthetic approach to the service innovation distinction is needed.

B. Preissl continues in Chapter 7 the theme that service activities play a major role as catalysts, carriers and facilitators of innovation in manufacturing as well as services themselves. Her work concentrates on the distinctive differences between service and manufacturing innovation using data gained from research in Germany. Among the conclusions are the need to revise innovation indicators because of the informal nature of R&D in services, the need to develop specific appropriability concepts relevant to services innovation, and the pervasive importance of active customer/client participation in the innovation process.

Chapter 8 by C. Hipp is also based on the elaboration and interpretation of German data. Starting from the proposition that the knowledge economy involves many other kinds of knowledge other than formal science and technology, she supports this claim through an investigation of the role of Knowledge-Intensive Business Services in national systems of innovation. The emphasis on boundary spanning activities identifies very well the role of bridging organizations and processes in the modern knowledge economy, highlighted in Chapter 2. Among her detailed empirical findings are the fact that KIBS are more R&D intensive than other services, that they make greater use of external sources of knowledge in the innovation process, and that KIBS' activities are concentrated in the production of scientific and technical services, software services and other business services. It is clear from this analysis that KIBS play key roles in the modern innovation system in relation to the production of and combination of tacit and explicit knowledge.

The problem of measuring innovation activity in services is also the theme of Chapter 9 by A. Kleinknecht, who draws on data for the Netherlands gathered in the Community Innovation Survey. After reviewing the strengths and limitations of different innovation indicators, a comparison is undertaken of new indicators in manufacturing and services. Among the principal findings, supported by the new innovation indicators, are that the greater R&D intensity of manufacturing does not map over into the importance of innovative products in the sales of each sector – the two sectors show an almost equal degree of innovativeness. A multivariate analysis, comparing three new innovative output indicators, finds the expected positive association with R&D inputs, the increasing size of a firm, and the rate of growth of the market. No association is found between these indicators and measures of R&D collaboration and technology transfer. It appears that indicators based on new product announcements have an important role in the future development of innovation statistics.

The final chapter in this section, by G. Antonelli and colleagues, is based on original and new data on service activity in the Italian economy. They too emphasise the interdependence between manufacturing and services in the innovative division of labour, but their data points to large regional differences in the nature of these relationships. In particular, the growth of business services is heavily concentrated in Northern Italy, that is to say, in the manufacturing heartland. They identify trends in the size of firms and establishments in different service activities, the changing composition of service employment, and they conduct some preliminary econometric investigation of employment growth in the service sector. They also develop a taxonomy of service activities based on criteria of tradability, productivity growth, and the extent of network interactions for each service activity. This leads them to a detailed elaboration of regional differences in service activities. It appears that regional differentiation must be considered side by side with sectoral differentiation.

The chapters in Part III are of a different nature, being in the form of a range of case studies of different aspects of innovation and the service economy.

Chapter 11 by K. Ducatel is an original analysis of service innovation in unwonted places, complementing the emphasis given by other authors to KIBS and financial services. Its focus is upon non-information services such as tourism, cleaning, distribution and household services and he shows convincingly how many of these activities are being transformed by the application of information technology. Ducatel develops his analysis with a sophisticated application of a characteristics approach to innovation in services, analyses some of the key innovations in these 'marginal' corners of the service economy, and concludes with a categorisation of these innovations. A central lesson of this chapter is the barriers which the client/producer interface places in the way of radical innovation in these non-knowledge based services. Instead, the emphasis is upon incremental and recombinant innovations, innovations which streamline internal production processes and integrate previously separate activities to meet client needs. Business imperatives drive this process, the new information and communications technologies are the means to these ends and are embedded at the key point, the client-provider interface.

We have drawn attention above to the significance of information technology investments in the provision of services. This is brought out very sharply in Chapter 12 in which P. Nightingale and R. Poll analyse the role of hardware and software developments in the trading activities of an investment bank. They argue that the IT system becomes the analogue to the manufacturing plant, with information rather than materials being the resource which is processed. This information system provides the basis for exploiting Chandlerian economies of scale and scope, and its development co-evolves with the control system of the organization. They highlight the significance of these investments for competitive advantage and their crucial role in the provision of information to manage risks on line. Their central finding is that the ability to generate an appropriate dynamics of information control systems is the key to understanding the relationship between innovation and competitive performance in this sector. They conclude with some interesting generalisations of their approach to the innovation process in other information and control intensive service activities.

J. Howells approaches the innovation systems and services question from a different direction in Chapter 13, that of the increasing importance of market mediated outsourcing of new technology and research services. This emerging division of labour is co-ordinated by market mechanisms, the R&T market with its evolving supply structure of primarily contract research and technology organizations. Howells traces the development of outsourcing over time and discusses the competing explanations of these trends. He argues that a proper understanding of outsourcing requires the investigator to abandon a single firm perspective on the innovation process, and embrace the systemic consequences of the division of labour in innovation; and to abandon a static perspective on innovation systems, and replace it with a more developmental perspective in which the set of acts and their interlinkages change over time. Howells suggests that we treat outsourcing as a 'phenomenon-based' system of innovation, within which service firms increasingly play a pro-active role.

The final chapter, by B. Tether and S. Metcalfe, is a study of 'soft' innovation in an important service sector which affects many people on a daily basis, namely air passenger transport. Despite the attention given to innovations in this sector in general, surprisingly little attention has been given to those procedural innovations which have greatly improved the efficiency with which aircraft are handled on and off the runway. This is a vital issue in relation to the overall productivity of an airport's facilities. The chapter focuses on the development of runway handling capacity at Heathrow Airport and analyses the substantial increase in runway productivity which has occurred as a result of the emergence through a trial and error process of new aircraft managing procedures. Collectively these innovations have led to the substantial increase in productivity which the chapter identifies. The innovations have arisen through a process of interaction between airlines, airline pilots, airports and air transport regulators. This process is not simply a question of agreeing procedures but of developing clear operating practices through which they are implemented. Significantly, few of these procedures require any significant capital investment, a feature which is reflected in the title of the chapter.

1.4 REPRISE

Taken together the essays in this volume demonstrate convincingly, and contrary to received opinion, that services have an important innovation dimension. They are important loci of innovations in their own right and they are an integral part of innovation systems in modern economies. These essays also provide several pointers to the nature of modern economies and the innovation processes they contain. The first is the distributed nature of these innovation processes in which there is an extended division of labour and a strong degree of interdependence between manufacturing and service activities. Secondly, this division of labour is itself evolving, with a shifting balance over time of specialization and internal vs. external modes of knowledge acquisition. Innovation systems are restless, in part because the accumulation of knowledge continually opens up new areas for recombination and specialization, in part because the economics of knowledge production and utilization put a premium on the exploitation of scale economies and create a changing pattern of comparative advantage between different suppliers of specialist knowledge. This connects to another important implication of these essays, that innovation systems and market systems interpenetrate one another, and that market knowledge of the client's needs is as significant as more formal, scientific and technological, knowledge in that innovation process. Consequently, the regulation of market processes may have considerable implications for the pattern and process of innovation in service activities, as the case studies of investment banking and airport operation seem to suggest. The third set of implications derive from the heterogeneity of service activities and functions which feature so strongly in these essays. Heterogeneity though places special demands on our understanding of services innovation, creates major benefits from having penetrating taxonomies of service activities, and requires considerable ingenuity in developing indicators of innovation activities. The final set of implications flow from the significance of knowledge accumulation processes in relation to services. It seems that the Arrovian tradition in the study of the economics of knowledge has reached definite limits. It is clear that the transmission of information, even public information, is not costless and that service activities play a crucial role in the communication process. Moreover, services are part of the market-mediated process supplying specialist knowledge to other innovating firms. The emphasis on market failure in the operation of knowledge-based economies seems to be seriously misplaced.

In their modest way, the chapters in this volume have made a good deal of progress in addressing these difficult to resolve issues. More than anything they illustrate the value of having access to comparative data from different national economies and the value of approaching the study of services innovation from a range of disciplinary backgrounds. Of course, a substantial research agenda remains but we hope that the concepts, statistical analyses and case studies provided in this volume will help others to develop further our understanding of a dynamic service economy.

References

Bryson, J.R. and Daniels, P.W. *Service Industries and the Global Economy*, London: Edward Elgar, 1998.

Clark, C. *The Conditions of Economic Progress*, Macmillan, 1957.

Fisher, A.G.B. 'Capital and the Growth of Knowledge',*Economic Journal*, 1933; Vol. 43, pp.374-389.

Freeman, C. 'Technical Innovation and the British Trade Performance' in Blackaby, F. (ed.), *De-industrialisation*, NIESR, London: Heinemann, 1979.

Fuchs, V.R. 'The Growing Importance of the Service Industries',*Journal of Business*, 1965; Vol. 38, pp. 344-373.

Gallouj, F. and Weinstein, O. 'Innovation in Services', *Research Policy*, 1997; Vol. 26; pp.537-556.

Tomlinson, M. 'The Contribution of Services to Manufacturing Industry: Beyond the De-Industrialised Debate', *CRIC Discussion Paper*, 1997; No. 5, ESRC Centre for Research on Innovation and Competition, University of Manchester.

---

[1]The valuable anthology edited by Bryson and Daniels (1998) collects together many of the important papers and illustrates a range of views and disagreements on the measurement and relevance of service activities.

PART I:
CONCEPTUAL FRAMEWORKS

Chapter 2
Distributed Innovation Systems and Instituted Economic Processes

Birgitte Andersen, J Stanley Metcalfe
and Bruce Tether

2.1 INTRODUCTION

There can be little doubt that the most significant recent development in the study of innovation has been the emergence of a 'systems of innovation' literature. Following Freeman's (1987) first statement of the concept, a rich body of investigation has developed, shaping in turn the further development of ideas on innovation, the nature of the firm and the role and methods of science and technology policy. Our purpose in this chapter is twofold: to draw out what is systemic about the innovation systems approaches in the literature and to review a number of the various ways in which innovation systems have been approached. Our way of summarising these diverse and rich literatures is to claim that they are all contained within the idea of distributed innovation processes and that they are a consequence of the division of labour in the generation and application of practically useful knowledge. Division of labour is in turn related to specialization in the generation of complementary forms of knowledge accumulated by different processes in different organizational contexts. Specialization entails co-ordination of different knowledge generating activities which are valued for their complementary contributions to the innovation process, and leads to outcomes which depend upon the interrelatedness of the different kinds of knowledge. As soon as the innovation problem is posed in this way a systemic dimension emerges naturally.

Now one of the interesting aspects of the innovation systems literature is that it has developed in a largely pragmatic fashion; only recently have attempts been made to pull together the various systemic threads within a coherent framework (Edquist and Johnson, 1996). There is, to our knowledge, no connection with systems thinking more generally in the social sciences or with the cross-disciplinary systems movement established by Von Bertalanffy (1968), Rapoport and others in the mid-1950s. From one perspective pragmatism has been a strength; it has contributed a fresh empirical foundation to innovation studies. From different perspectives, however, the empirical focus alone is unsatisfactory as it has left the concepts weaker in their theoretical foundations than they need be.

A central purpose of this chapter is to establish connections between the evolutionary and systems perspectives on innovation. In the evolutionary perspective, attention is focused upon populations of actors and the dynamic consequences which flow from their different behaviours. By taking the population itself as one unit of analysis, we can treat it as part of a wider system of institutions and relationships for the generation of innovation. This is the approach we will follow. However, populations are constituted by their members and here the principal contribution of the distributed innovation perspective is to raise difficult questions about what it is which constitutes a distinct unit of membership. It cannot simply be the firm and to say that it is a network is only to rephrase the problem, unless we have a clear understanding of the principles of network genesis, growth, decline and demise.

Our discussion is structured as follows. We begin with an outline of the role of innovation in evolutionary competition and link it to the innovation systems concept (Section 2.2). We then summarize some characteristics of a systems approach to innovation (Section 2.3) and then briefly review some of the principal contributions to the literature (Sections 2.4 and 2.5). We then discuss some approaches to systems thinking more generally (Section 2.6). This, we believe, has particular significance for the innovation in services debate and we conclude with a (very) brief outline of some aspects of services innovation systems (Section 2.7).

2.2 AN EVOLUTIONARY EXCURSION

Like many, but by no means all, scholars interested in innovation systems we locate our interest in terms of evolutionary approaches to innovation, competition and economic growth. We begin with a brief outline of the relevant evolutionary thinking in relation to innovation.

As is well known, an evolutionary theory of innovation-driven economic change is necessarily based on a three-stage scheme of ideas. First, there is variety of behaviour across economic actors, typically firms, within a population; secondly, there is a process within a specific environment to apply selection pressure to those differences in behaviour and resolve them into changes in the relative importance of the corresponding actors; and, thirdly, the consequences of selective change feedback to generate a new pattern of differences across the actors in the population. In most approaches the actors are firms, selection is provided by market mechanisms, and the crucial third step is defined by innovation mechanisms at the level of the firm. We say crucial because no evolutionary argument can be mounted without differential behaviour, one may speak of statistically representative behaviour in the population but one cannot treat this in terms of uniform behaviour. Moreover, the primary source of differential competitive behaviour is to be found in the differential innovative activities of the rival firms. Thus the canonical partial evolutionary model considers a population of firms which differ one from another in key competitive attributes (usually unit production costs or product range and characteristics) and who compete in specified market environments. Since one cannot mount an explanation of selection without a supporting account of the existence of variety it follows that any theory of selection needs to be based upon a supporting developmental account of variety generation. Thus the two central concepts in an evolutionary account of economic change are selective processes and developmental processes. These exist at different interacting levels and developmental processes themselves are often shaped by processes of selection. Where the one ends and the other begins becomes difficult to establish.

Competition involves interaction and interdependence, first to determine the profitability levels of the rival producers and second to simultaneously determine how the relative outputs of the different firms change over time. In the short term, this depends in part on the behaviour of the relevant populations of customers and in part on the disposition of a firm's profits to finance the growth of its production capacity. Over the long term the dynamics of market competition interact with the dynamics of differential innovation through the further allocation of profits to innovative activities, the dynamics of learning by doing and by using, and through good or bad fortune. Here we find the importance of market feedback and the foundations of competition as an evolutionary process. Market processes determine the distribution of profits, and thus the resources to innovate, across the firms, they reflect the interaction between firms, their customers and suppliers and the generation and selection of 'innovation' problems in that context. What this tells us is that market processes also shape the generation of new knowledge, and that the innovation systems literature provides a perspective on how markets interact with the wider institutional context. To this extent the interaction of market processes with innovation systems leads directly to the idea of endogenous innovation.

Now if we notionally suppress the innovation process and consider only populations of firms with fixed performance attributes, operating in a given selection environment, we find a remarkable evolutionary property. In the process of competitive market selection, economic variety is destroyed, and without some source of ongoing variety regeneration the evolutionary process comes to a halt. And, as the biologists put it, evolution consumes its own fuel. The problem is rather analogous to that of the uniform energy state in statistical physical systems, without variety there is nothing to induce change, the system sinks into stationarity. Hence, if change is to continue innovation must occur to redefine what constitutes 'best practice' in the population either via new entrants or through innovation by the incumbents. To continue the metaphor, a dynamic economic system must be kept far from equilibrium by the injection of innovations into the market. This is the developmental problem par excellence.

Now, this requirement places a particular premium on understanding the development process in evolutionary economics, and raises important questions as to the stimuli to and the boundaries constraining a firm's capabilities to innovate. In this deep sense all economic change is premised upon differential capabilities to innovate. This is where the innovation systems perspective connects with the argument. Firms do not normally innovate in isolation, in many cases their innovative activities are embedded in interrelations with a wider matrix of organizations and institutions which shape the way innovation develops in particular areas. Innovation involves interdependence as well as interaction, collaboration as well as competition. We suggest that a particularly helpful way to investigate innovation systems is to recognise the division of labour in the generation of innovative capabilities, a theme which is as old as Adam Smith. Firms must acquire knowledge together with the skills to articulate that knowledge in their activities. The resulting expertise defines what they can achieve, that is to say, the firm's capability. Now this capability perspective has led to a recent revival in the theory of the firm which provides important insights into the innovation process (Foss and Knudson, 1996; Montgomery, 1995). In particular, it focuses our attention on the development of capabilities in different institutional contexts, capabilities which, in part, must be externally acquired by the relevant firms and organizations. Firms are the node at which these distributed innovative activities are brought together but firms are not the only important source of capability. The manner in which firms gain access to distributed capabilities should be a central concern to any discussion of innovation systems (Coombs and Metcalfe, 1998). In particular, these systemic influences play a crucial defining role in the feedback mechanisms which link market selection to innovation.

This gives a system perspective two primary roles within the evolutionary tradition. The first is to help explain the way knowledge itself grows in an evolutionary fashion by means of blind variation and selective retention (Campbell, 1987; Vincenti, 1990). Innovation systems constitute a variety generating and selection environment for new practical knowledge, whether scientific and technological or managerial and operational; they are systems for keeping the economic system 'far from equilibrium' (Allen, 1988, 1999).

Secondly, the systems perspective helps explain the differential capacity of firms to innovate within and between different economic sectors. For if the systemic influences applied equally to all firms they could add nothing to the evolutionary argument and differential innovation performance would have to be explained by other means. Per contra the differential ability of firms to benefit from the relevant innovation systems is at the centre of how economic sectors develop and indeed how markets and innovation systems co-evolve. Indeed, what is important about markets and innovation systems is that they are both devices for adapting to new opportunities in terms of technologies, needs and resources; each supports the open-ended nature of economic change. How they support and constrain each other is, we believe, of crucial importance and an important element in the analysis of innovation systems. For example, if there is one thing about the innovation process that we do know, it is that innovation does not occur randomly, nor is it unbounded. The question of the limits to innovation is thus central to an evolutionary approach; innovation becomes guided variation along specific developmental pathways, it is cumulative in terms of the accumulation of knowledge and practice within particular sets of design principles. This is the important implication of the paradigm perspective on technological change (Dosi, 1982; De Liso and Metcalfe, 1996). Paradigms indicate fruitful areas of advance while simultaneously closing off other areas for investigation such that progress is made only by limiting perceptions of progress. Now such cognitive paradigms do not exist in the abstract; they are institutionally and organizationally specific, they reflect technological beliefs about what has worked in the past and expectations of what possible developments await in the future, and these beliefs and expectations are distinctively shaped by the social and institutional context in which they arise. Here there is a relation to be explored between the paradigmatic nature of technologies and the system contexts in which they are generated.

It is particularly important to make clear that a sharp separation between unit of selection and selection environment is not permissable in the economic and social sphere. Not only is the environment itself made up of other units of selection, but there are conscious attempts to mould the selective process in favour of particular units. Thus firms devote a good deal of time to influencing the regulatory environment or to creating barriers to selection in order to protect their position. Moreover, intentional, purposive behaviour plays a crucial role in the social science context for which there is no analogue in biology. However, this simply means that the sources of economic and social variety, the fuel with which innovation works, are far richer than they are in the natural world.

Some brief comment on our underlying perspective of the innovation process is necessary before we proceed. Broadly speaking we can distinguish four kinds of influence on innovation; namely, (1) the opportunity to innovate in terms of the acquisition of unexploited knowledge, (2) the distribution of incentives to innovate, (3) the distribution of resources available to invest in innovation, and, (4) the capabilities of the relevant organizations to form innovation strategies and manage the process of innovation. Each of these influences relates to the investment nature of innovation in terms of conjectured relations between current inputs to generate expected future outputs. But such innovation opportunity relationships are necessarily imprecise; they are conjectures not facts; there are no reliable metrics for inputs or outputs; the value of outputs often depends on unforeseeable complementary developments in markets and other technologies; and, there is genuine uncertainty so as to preclude the use of probability indicators in innovation decisions. This is why innovation is synonymous with entrepreneurship in the Schumpeterian sense of confronting the unknowable. All innovative activities are clouded by ignorance, they are acts of faith and thus particularly sensitive to changes in sentiment and expectation. As Loasby has so perceptively expressed the matter, innovations are the imagined deemed possible (Loasby, 1999).

Here then lies a further dimension of the innovation systems perspective, the way in which opportunities, incentives, resources and capabilities are open to systemic influence, the way in which innovation systems co-ordinate investments in the generation of knowledge and expertise. To the extent that they do, this will surely reflect on the search and selection mechanisms within firms, and other knowledge generating organizations, the stage, as Ulrich Witt puts it, of pre-revelation.

2.3 CHARACTERISTICS OF A SYSTEMS APPROACH TO INNOVATION

In Sections 2.4 and 2.5 we will provide a brief overview of several important but different perspectives upon innovation systems. In this section we begin the process by considering the possible meaning of a systemic perspective on innovation. This is a complicated but we argue important issue as scholars seek to develop the next generation of innovation studies based upon the systems method. Indeed our discussion ultimately connects to deep issues in the evolutionary tradition namely the relationship between reductionist individualism and holism as ways of analysing economic and social phenomena. Our position is that of a sophisticated methodological individualist, the parts are essential to the picture but not all the attributes of the system are reducible to the properties of the parts. Figure 2.1 provides a crude way of presenting what we think is implied in an innovation systems approach. Here we depict the three principal system elements:

- the parts, those distinct organizations (actors) which reflect the division of labour in the generation of knowledge, skills and expertise, in this case they are firms, users, universities and other kinds of capability generating organizations;

- the interrelationships and connectedness between the parts in terms of their different strengths and pattern, the mode as it were of their interdependence, the problem being to specify what makes for interconnectedness; and,

- the boundary relationships both in terms of the system as a whole and in terms of the definition of the autonomous component parts, it being recognized that boundaries are not barriers but permeable routes to interaction with the wider environment[1].

Figure 2.2 provides an alternative way of organising a systemic view by distinguishing different general levels at which knowledge is generated and applied, and by making the associated distinction between horizontal and vertical patterns of interaction. The different levels contain qualitatively different kinds of organization each with appropriate opportunities, incentives, resources and capabilities. Different linkages relate to the flow of particular kinds of knowledge which combine together to let a firm, say firm 1 or 2, innovate in a particular way. By distinguishing the horizontal from the vertical we can begin to ask how the connections are engineered, by what spontaneous process they may emerge, how stable they might be and how good they are as communication channels? We can also infer from even this limited approach that innovation systems are inexorably embedded in supply chains and market interrelationships.

Since the organizations at different levels are highly specialized products of a division of labour it is frequently necessary for intermediary 'bridging organizations' to be put in place to ensure effective communication of knowledge. These may be, for example, engineering consultancies or retail suppliers.

It is a chief determinant of a systems method that the nature of the whole cannot simply be deduced from the properties of the parts in isolation but depends on the nature of their interrelationship. To this degree the whole is more than the set of the parts and the part-whole relation is central to what the system achieves. In turn this relation depends on what is meant by interaction and interconnectedness of the system. For example, what 'flows' within the system and what flows across the boundary of the system are essential to understanding what the system does. To use an innovative systems approach analytically means that these flows, that is to say, interactions, must be identified with requisite precision. Another way to pose this question is to ask to what extent the behaviour of the component parts can be said to be co-ordinated? To answer this question would indicate how the division of labour operates in the generation of capability. Evidence of co-ordination may be found, for example, when decisions on the development of innovations are jointly determined or when the resources devoted to R&D in one organization respond to the R&D plans of another member of the system.

Within the innovation systems literature, the parts-whole problem is closely connected to the distinction between organizations and institutions (Edquist and Johnson, 1996). The parts are organizations, firms, their customers and suppliers, universities and technical colleges, public and private research establishments and whatever other knowledge accumulating bodies are deemed relevant. The transactions between the organizations, the manner of their interaction, are shaped by institutional factors some of which constitute a background to the system as a whole, for example, the general regime of intellectual property rights, while others reflect specific bilateral or multilateral relationships between the parts. What exactly these transactions are will occupy us further below. It follows from this that system change may be of three specific types, change in the actors, in the mode of their interconnection and in the boundary relationships.

Figure 2.1: An Innovation Systems Schema

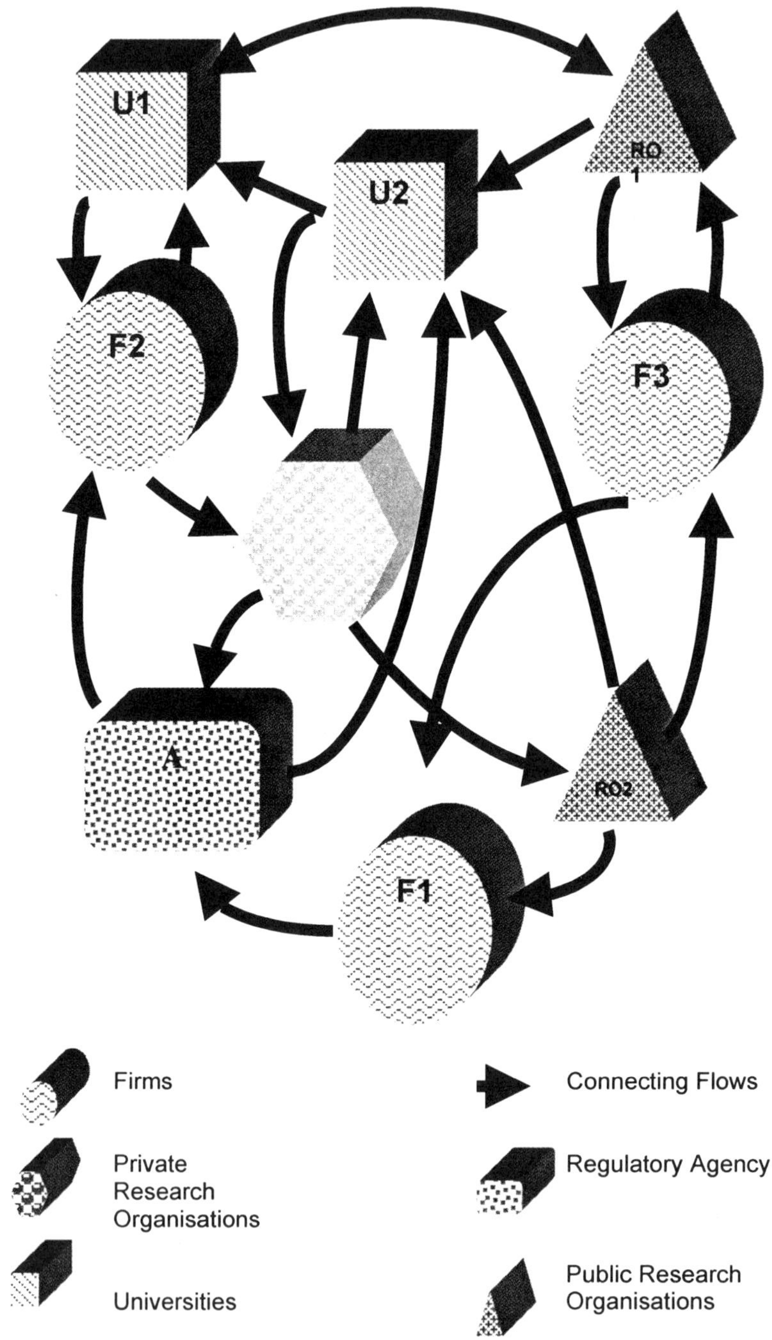

Figure 2.2

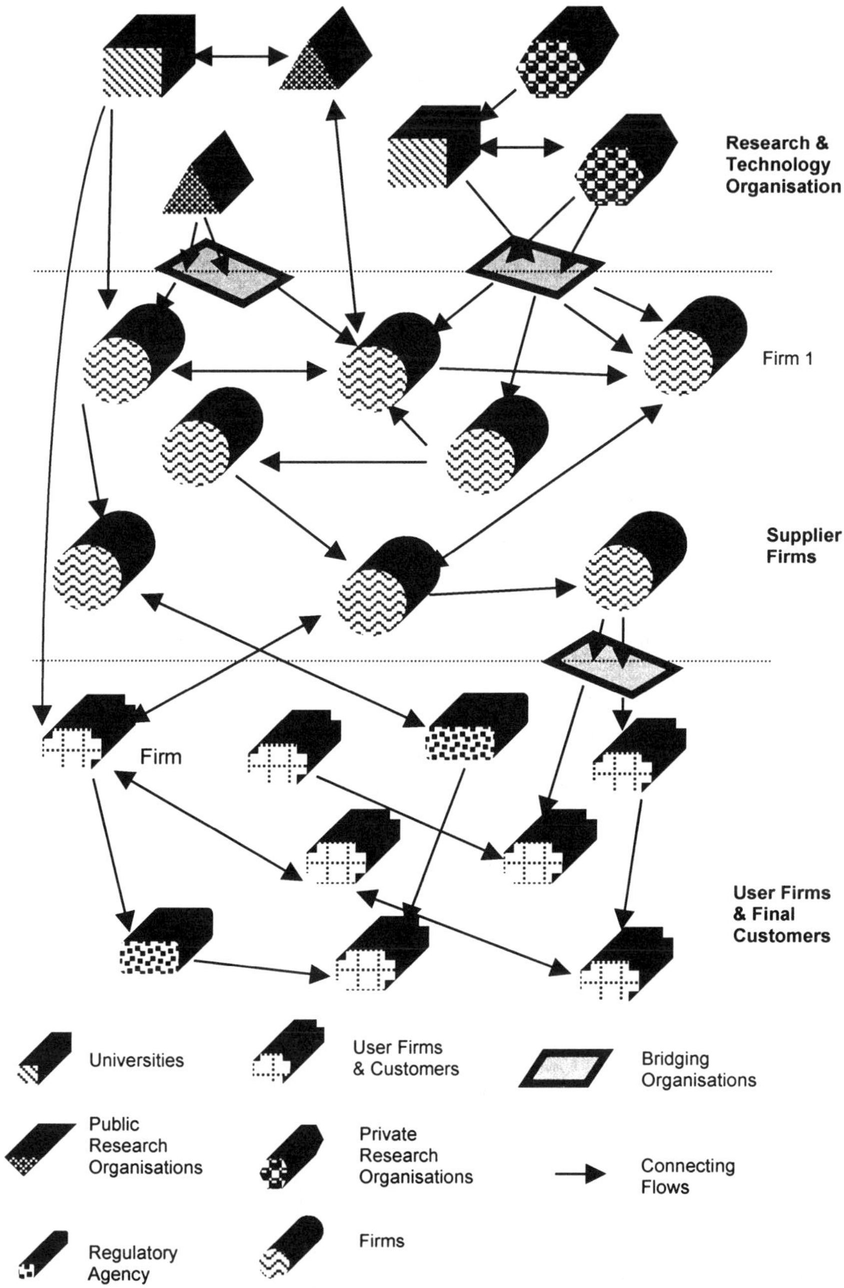
Research &
Technology
Organisation
Firm 1
Supplier
Firms
Firm
User Firms
& Final
Customers
Universities
User Firms
& Customers
Bridging
Organisations
Public
Research
Organisations
Private
Research
Organisations
Connecting
Flows
Regulatory
Agency
Firms

That innovation systems are based upon a division of labour in knowledge generation and utilisation has some important consequences, not least because specialization gives each organization an idiosyncratic set of properties. Each organization will have its own objectives defined over different timescales. Each organization will have its own internal structure, communication codes and social practices. Each organization, in developing its own knowledge, will develop a set of decision premises, its own attempt to define what it does, how it should be done and how the results should be judged. What will happen, one might ask, to the system's connectedness when decision premises are different and perhaps incompatible (Loasby, 1996). Each organization may have, in Bettis and Prahalad's terms (1995), a different dominant logic, a reflecting but distorted lens through which it sees its world. There is no reason to believe for a moment that the perceived world of knowledge and therefore innovation opportunities are the same across organizations. Nothing in this account requires the different organizations to be of equal status or significance in the generation or utilization of knowledge. Strong asymmetries are to be expected based upon capability, reputation, authority and scale of the various parties[2].

Moreover, not only do the different kinds of organization develop different kinds of knowledge, they accumulate it in different ways. The theory-based nature of experimental pure science is quite distinct from the trial and error processes by which more practical knowledge is acquired. They do so in response to different incentive structures, different tests of the validity of their new knowledge (science is judged by its 'truth' content, technical knowledge by its 'value') and on the basis of different structures of incentives. To summarise thus far, an innovation system has a number of facets. It is a pattern of resource use and a pattern of incentives to innovate in particular ways. It constitutes a system of specialized and distributed actors and organizations to generate and communicate particular capabilities. It constitutes a system for compound learning in which the knowledge generation activities of one part depend on the knowledge generation and dissemination activities in other parts. However, this can only be so to the extent that the parts are interconnected.

This brings us to the second main issue, the nature of what it is which turns the parts into a whole; by most accounts this lies in procedures to promote or limit the exchange of information (David and Foray, 1997). Rather more precisely we should say it consists of procedures for the exchange of messages and data as representations of knowledge. Here it is important to identify a serious weakness in the commonly expressed view that, because knowledge has the property of a ubiquitous non-rival public good, so the relatively low cost of transmitting information implies that knowledge is a free good available to those who do not produce it. Nothing could be further from the truth. Information flow does not map into an equivalent flow of changes in knowledge of the recipients. Given the division of labour it is by no means obvious that either the transmitter of information or the recipient of information will hold beliefs such that the information flow corresponds to some equivalent item of knowledge. Depending on the parties and the context, information may have multiple meanings. There are also considerable differences in the method of interconnection. This may involve unfocused broadcasting, as it does with scientific and technical publication, or it may involve focused, private bilateral exchange. The interchange may be accidental or intended, it may be informal (Von Hippel, 1991) or it may involve a formal commitment which at the extreme may require setting up a joint organization to manage the interaction. Whatever the methods there will be costs of communication, of learning the communication codes of different organizations (Arrow, 1974). Intellectual property issues may limit attempts to communicate, and the social organization and incentive structures of the respective organizations may be expected to greatly affect patterns of communication. Consequently, knowledge is sticky, viscous, localised and the channels through which information flows must be actively created and supported by background investments in understanding (Antonelli, this volume).

In further exploring the nature of the interactions between system parts it is useful here to draw upon one of the key ideas in economic sociology, that of the embeddedness of instituted economic processes (Granovetter, 1985; Polanyi, 1958). Embeddedness conveys the idea that the behaviour of particular actors is conditioned by ongoing social relationships. Action is neither wholly atomistic nor wholly dictated by external factors but rather individual action is socially guided in a blend of choice, command and custom. Action within systems reflects constrained voluntarism, and because there is scope for independent behaviour the system becomes a framework within which novelty can be generated and variety expressed - within limits. Within a voluntaristic framework individuals are seen as exercising a considerable degree of freedom in the way they pursue their objectives and interests but their choices are constrained by the framework of norms and values that are shared by the people with whom they interact. Consequently their actions cannot be reduced only to their material circumstances and their rational self-interest, nor can they be totally irrational expressions of arbitrary behaviour (Thorelli, 1986; Scott, 1995). Trust is an important example of what Granovetter means by social conditioning, as indeed are authority and reputation, two particularly important features of the institutions of knowledge accumulation. Closely related to this perspective are those which emphasise systems or networks of intercommunicating agents and organizations (Green et al, 1998 for survey). However, there are a number of different ways in which network analysis is approached depending on whether the focus is upon the network structure or the network's activity. In particular, the broad thrust of the French School (for example, Callon, 1992) is upon networks as co-ordination devices and the context in which networks develop. Who interacts with whom, on what terms, with what authority and to what effect is one way to summarize the nature of the information interlinkage.

Even this limited discussion suggests a number of questions relevant to the appraisal of innovation systems. Since they are based on a division of labour in knowledge accumulation it is not obvious that their parts will develop branches of knowledge which are readily communicable within the system. The more specialized the process of accumulation the greater the potential problems of interchange. Consequently duplication of effort is likely to be the necessary prerequisite for any conversation to take place at all (Rosenberg, 1990). This further suggests that the ability to participate in an innovation system is unlikely to be equally open to all firms and will normally require prior investment in capability before it can be achieved. Network formation may be spontaneous but it is not costless. In the light of the differential knowledge creating capabilities of the different organizations, one would not expect their capabilities to contribute to a given field of innovation to increase at the same rate. Consequently, bottlenecks occur in the innovation process, in the sense that there are critical shortages of capability which are organizationally specific. Thus imbalances, reverse salients, structural tensions, to deploy just three of the accepted ways of describing the naturally uneven pattern of technological evolution, have a specific location within any given innovation system (see Section 2.4 below). In this way an innovation system will exhibit its own dynamic of development in which organizations which fail to overcome the respective bottlenecks are eliminated from that specific system.

A further observation is relevant here, namely that innovation networks are bound to overlap with and share the characteristics of broader commercial networks. As Richardson (1972) has forcefully reminded us, the anonymous market transaction and centralized co-ordination within an organization are only the limiting cases of a wide spectrum of co-operative interrelationships which define a market economy. From joint ventures to interlinking directorships, firms devise ways of co-ordinating those qualitatively rich and often poorly defined transactions that depend for their success on the continued interchange of information and intention. Innovation systems are simply one dimension of this institutional complexity and one should expect that they interpenetrate and partially define the corresponding commercial systems and vice versa. Indeed, this is the focus of Dahmén's notion of development blocks which are built on sets of input-output relations and that promote the transformation of the detailed microeconomic basis of industry and trade (Dahmén, 1988, 1989).

Of particular importance is the fact that, there is no reason to see a system as representing a steady state of affairs. Changes in the organization and in the pattern and strength of the interconnections between them are surely to be expected. Indeed one should expect that attributes associated with interaction such as authority, trust and reputation are generated over time in the development of the system and do not exist independently of the mode of interaction. Indeed, because the activity of the network generates new knowledge it is certain to change the beliefs of the various participants in relation to the opportunities they perceive and the objectives they pursue. Knowledge networks are inherently restless, and at times may change kaleidoscopically when radical new knowledge is discovered. Thus the durability of the network relationships is contingent; like the partners in a dance, when the music stops the option to change can always be invoked. To pursue this further one should recognise that the way in which firms adapt to changes in market pressure will be contingent upon the innovation system in which they are embedded and their abilities to draw upon that system in solving innovation related problems. In turn their success at innovation will contribute to how the system develops over time. The interdependence between market processes and innovation system processes appears crucial from this viewpoint and suggests interesting research programmes which focus on the emergence, growth and ultimate decline of market-innovation system complexes. Moreover, much systemic change is dependent upon happenstance, and since modes of interaction are socially self-reinforcing it follows that innovation systems evolve in a path-dependent way. Consequently, small events will not be massaged away nor will the evolution of systems be predictable. Surely here there is important scope for comparative and historical work on similar sectoral systems operating in different national contexts. Moreover, it cannot be assumed that innovation systems, at whatever level, always promote innovation. Since they necessarily acquire a mind set of their own and promote innovations consistent with this world view, they may as a consequence prove particularly resistant to more radical forms of innovation. That they are distributed and involve the decisions of multiple actors is likely to compound the problem of radical change. From our evolutionary perspective the genesis of new systems and the demise of old ones are critical features of the broader competitive dynamic.

Finally let us turn to the question of the domains on which systems can be studied. As we explain below, scholars have distinguished national, sectoral and regional systems of innovation which should not be seen as exclusive approaches but rather as complementary levels of investigation, posing important questions as to how the different levels interact. An alternative take on the levels problem is provided by approaches related to the concept of institutioned economic processes, techno-economic networks (Green et al, 1998), communities of practitioners (Constant, 1980) and industrial districts. It is also important to distinguish innovation systems from the narrow concept of technology support systems. Innovation is dependent upon much more than R&D and the accumulation of technology: it involves the identification and development of markets, the supply of finance and skilled human inputs, and the creation and operation of regulatory frameworks. Innovation systems proper have technology (support) systems, essentially the R&D and technological learning and transmitting infrastructure, as only one component. We must also distinguish technology support systems from technology systems in the sense of Hughes (1983) and other like-minded historians of technology. They are concerned with the systems characteristics of productive activities such as power generation and railways where particular sub-system activities have to be co-ordinated for the system as a whole to work. At any one time the level of development of the relevant sub-systems creates constraints on the operation as a whole and the associated bottlenecks Rosenberg (1976) or reverse salients (Hughes, 1983) provide the incentive structure to shape innovation effort. At a micro scale the same themes of interrelatedness are explored by Henderson and Clark (1990) where they distinguish between the architecture of a technology and its component parts.

2.4 SOME CATEGORIES OF SYSTEMS THEORY

To further our discussion of these issues we now touch upon three different, complementary, types of systems perspectives that reflect the contemporary systems movement.

First, there are the early approaches which have been termed hard system thinking. They are based on a positivist (mostly quantitative) functionalist view, and include models based on biological analogies as well as mechanical models of society. These approaches are based on a certain criteria of equilibrium (steady state or dynamic) or the conditions for the preservation of given structures. They have not really been found in the innovation system literature, as it is this type of functionalist thinking that the innovation system literature has avoided (Langlois, 1983). Within this category, autopoietic systems are ones defined by the property that their components and interactions are self-reproducing such that the network of relationships becomes a unity with a self-defining boundary. In these systems it is the internal pattern of interaction which establishes the response to changes in the external environment, which influences the systems sustainability, and which controls the development of the system (Zeleny, 1996). Related to this category are the autogenetic systems described by Csanyi (1996), the chief distinguishing characteristic of which is that the components are not give, a priori, but develop as a result of creative acts within the system. Clearly such system perspectives transcend the extended and spontaneous orders associated with Hayek's work.
Secondly, there are the soft systems approaches based upon qualitative interpretative thinking, in which the systemic perspective is guided by a criteria of consensus, common understanding and shared meaning, all arising from social processes of interaction. In this approach any asymmetry bringing instability into the system remains outside the interpretative paradigm's agenda. This kind of approach might be compared to Freeman and Perez's socio-economic paradigm model (Freeman and Perez, 1988; Perez, 1983), in which each paradigm reflects a common understanding of an 'ideal type' of production which is most suitable for the technological regime in question.

Thirdly, there is the new wave of critical system thinking, which is based on a specific approach to system problem solving. This approach suggests that organizations as well as social and institutional systems are generally characterised by power asymmetries, unequal distributions of wealth, structural conflicts and contradictions. Hence, systemic stability and common consensus are not a natural fixture but could be imposed by a certain type of social control or power relationships or by social discipline (Oliga, 1990).

These three analytical systems approaches are not meant to substitute one for another. The last of the three, i.e. critical system thinking, developed as a complementary addition, in part because system methods from the natural sciences (i.e. the functionalist models) could not transfer easily to the domain of the social sciences and soft methods could not be used where technological, economic and social structures gave rise to coercive constraints or conflicts. Hence critical system thinking treats conflicts and change as endogenous phenomena in which change in the system arises as a consequence of the architecture of the system and modifies that architecture.

Complex, Complicated, or Both?

Much systems discussion is bedevilled by confusion between the notions of complicatedness and complexity. By any standards, Figures 2.1 and 2.2 are complicated; there are multiple parts and multiple connections which directly and indirectly connect the various kinds of organization. In Figure 2.1, a change in firm 3 (F3), for example, affects public research organization 1 (RO1) which feeds back on firm 3 again indirectly via university 2 (U2), private research organization C and public research organization 2 (RO2). Students of input:output systems in economics or trophic webs in ecology would find these kinds of complication perfectly normal, we doubt if they would mean that they were complex. For complexity is a function of the lack of pattern in the arrangements between the parts, it is patternless systems that are complex. It is only when we cannot trace all the important linkages and their relative strengths that we are entitled to talk about complexity (McShea, 1991). If we do mean that innovation systems are complex then they are to a degree unobservable and hence the patterns of interaction are likely to have unpredictable outcomes. It is this element of indeterminacy which is the distinguishing characteristic of complex systems proper (Allen, 1999). This is not what students of innovation systems have in mind in their analogies. Complexity leads us into a different issue namely how innovation systems develop over time through essentially trial and error processes. This seems to be a considerably under researched area, as is the idea that systems have development pathways which constrain the kinds of interaction which take place. May it not be that established innovation systems can create lock-in to particular technologies and may suppress viable challengers? Is there a possibility of competition between different systems of innovation within the same sector? What are the processes of system emergence, growth and decline? Dynamic questions of this kind will be important for the next generation of innovation studies.

2.5 MAPPING THE DIVISION OF LABOUR

In this section we provide a brief overview of the principal approaches to research on innovation systems, each of which has provided a distinctive contribution to the understanding of a distributed innovation systems perspective.

Innovation Institutionalism

The study by Nelson and colleagues (1992), which covers fifteen advanced nations, stands as the only large scale attempt to put the innovation systems concepts within a comparative framework of study. It is very much within a Marshallian tradition. By an innovation system is meant the 'institutions and mechanisms supporting technological outcomes', the 'set of institutions whose interaction determine the innovation performance of national firms'. One of the considerable strengths of the work is its breadth of view, that there is much more to innovation than public and private R&D and that the national focus becomes problematic if it is applied too rigorously. Yes there are national dimensions, which vary from country to country, but other dimensions spill across natural boundaries. Science, whether pure or applied, has 'always' been an international enterprise and transnational companies have a modest but increasing influence on national R&D efforts. Yet other aspects of the system fit more happily in a sectoral and possibly regional context.

Perhaps the best way to interpret the empirical side of this work is in terms of a comparative analysis of the division of labour in knowledge generation in relation to innovation. The various contributions trace the public and private institutions involved, their profit or not for profit basis together with the broader cultural and national characteristics which shape innovation activity. Thus while firms naturally appear as the primary innovating institutions they are supported by the complementary activities of universities and other research organizations. Universities, for example, provide general support in terms of advances in generic knowledge and experimental methods, and in terms of the supply of trained minds. However, they also interact more specifically with firms in more or less tightly specified bilateral relationships to jointly develop and share knowledge. So significant have the latter become in some areas (for example, biomedicine) that the traditional openness of academic science has begun to be questioned by several observers (David and Foray, 1997).

To summarise, the Nelson study bases its theory of innovation systems on the interaction between institutions and on the mapping of institutions across countries to explain different national systems of innovation, different patterns of specialisation, and the impact upon national comparative competitive performance. Hence, the study makes much of the considerable international differences at the micro level in terms of the contributing institutions (Nelson, 1992).

From Knowledge Institutions to Learning Institutions

Whereas Nelson, and also Freeman in his work on Japan (Freeman, 1987), take as their starting point the existence of national systems of innovation, the IKE group in Denmark attempts to present a theory as to why this might be the case (Lundvall, 1992).

The distinctive feature of their innovation system approach is to investigate the process of knowledge organisation or process of learning. By using a notion of innovation processes they question the traditional separation between invention, innovation and diffusion, and its corollary the linear model of innovation. Although they argue that interactive processes of innovation can be studied at different levels of aggregation, Lundvall in his principal contributions (Lundvall, 1986, 1992) focused especially on the vertical division of labour between what he termed users and producers and from there he develops a model of national systems of innovation (see Figure 2.2).

When investigating the microeconomics of interaction, Lundvall focuses upon the systemic interdependence between what in standard economic textbooks are formally independent, atomistic economic subjects. In the IKE group's model of interactive learning, the accumulation of less formal knowledge is embedded in the social relations between users and producers and other relations. Hence, compound learning is important.

In a historical perspective, Lundvall (1986, p.357) writes that the rigidity of user-producer relationships arising from the sunk investments in trust, co-operation and shared mental frameworks, might be regarded as one important aspect of the Freeman and Perez (1988) mismatch problem. When technological paradigms change, new user-producer networks may take considerable time to become established. This is an important point, systems constrain patterns of change as much as they influence the rate and direction of change.

When moving from Lundvall's user-producer interaction to national innovation systems, the IKE group's emphasis is on those differences in geographical and cultural distance, language, financial infrastructure as well as government imposed standards and regulations etc, which may impede the interaction between users, producers and other organizations. From here they explain why different national systems display different patterns of development.

Communities of Practitioners

A distinctively different approach to the study of innovation systems has been developed in recent years by historians of technology, building primarily on the work of Constant (1980) (Laudan, 1984; Constant, 1984). The starting point for the approach is the claim that technological knowledge is different from other kinds of knowledge. It is neither pure science nor science applied; in that it is tested not by its truth properties but by its economic and social viability; in that it is holistic requiring the integration across boundaries of expertise; and, in that its accumulation depends crucially on its pattern of application. Moreover, while both science and technology are based upon communities of practitioners those working in technology are primarily located in firms governed by market forces. Technological communities embody traditions of practice (for design, construction, testing and use) which provide a common currency for exchanges between the members. The accepted state of the art defines what Constant calls normal technology and this is the basis from which (incremental) improvements are identified, tried and tested. Radical improvements, particularly those located in new sets of design principles typically require the creation of new communities and new traditions (Constant's example is the invention and development of aero-jet engineering) embodying new people, new professional expertise and often new firms. The analogue with Schumpeter's parallel emphasis on 'new-men' is telling. Traditions in this scheme have a three-fold role. They are the embodiment of the historical growth of knowledge, they are the embodiment of relationships between the current members of the community, that is to say, the framework within which they communicate, and they are the embodiment of their expectations as to the development of the technology. This rich network of ideas, which stimulated a great deal of subsequent work (Staudenmier, 1985), has from our viewpoint two very interesting implications. First, the concept of a community cuts across the idea of the division of labour in different organizations. Members of the community are located in different parts of the system but they are integrated by their common background and allegiance to the tradition. Thus, as with Von Hippel, informal knowledge trading takes place via networks defined by the community, with questions posed and answers formulated within the tradition. Secondly, the tradition is a framework for variation within that set of designs it subsumes, it becomes the developmental framework for guided variation, for the purposeful but limited construction of new technological knowledge. One might say that the tradition is the institution in which technological practice and exploitation is embedded.

Generic Technological Systems

One of the most comprehensive attempts to investigate the systemic aspects of innovation at a microeconomic level is provided by Carlsson and colleagues (1995, 1997). A central feature of their studies is the shift in emphasis from innovation to technological systems. Among the several important contributions made is the significance of the (unequal) distribution of what they term economic competence, the ability to develop and exploit new business opportunities. They also point out that economic competence within a sector is often built on the integration of very diverse technologies, and that systems can act as barriers as well as stimuli to knowledge accumulation. They also make the important step of studying the systems supporting the development of generic technologies as they evolve through several decades. In a series of detailed empirical cases in the Swedish economy they identify four factors as being of key importance in shaping the systems aspects of innovation. Namely, the nature of the relevant knowledge and the associated spillover mechanisms, the capabilities of the information receivers, the connectivity of the relevant organizations and the vigour of the relevant variety creation and selection mechanisms. They draw attention to the importance of competent suppliers and customers in relation to problem solving activity, the movement and distribution of trained engineers, and to the role of selected 'prime movers' in forming patterns of interconnection. They also make the important points that systems can fail through lack of connectivity, that they may become locked in a particular view of the technological opportunities, and that a system is conditioned by market forces that determine which firms survive to remain effective members of the relevant networks. In this way it becomes impossible to separate the generation of technology from the diffusion and use of that technology.

Putting together these abstractions, Carlsson (1997) and Eliasson (1996) use the idea of competence blocks to aid their understanding of the linkages within technological systems and to provide a basis for developing Dahmén's (1988) concept of development blocks. The essential idea of a development bloc is that the uneven development of different technological areas within it creates tensions which shape the ongoing pattern of innovation. Pressures and incentives to innovate are thus systemic. "While development blocks represent production systems achieving economies of scale through interdependence, a competence block functions as a Marshallian positive externality" (Eliasson, 1996, p.126). Hence, their notion of technological systems is broader than that proposed by Hughes' (1992) who also attempts to understand economic changes in relation to technological development.

Technological Infrastructures

An alternative way of emphasising the systemic aspects of the innovation process is provided by the notion of technological infrastructures (Teubal, 1996; Tassey, 1992). Teubal and colleagues (1996) have been particularly persuasive in developing an understanding of the role of technology infrastructure in relation to economic growth and the development of firm specific technological capabilities. They argue persuasively for the crucial role of infrastructure policy, to shape the micro-development and distribution of knowledge in terms of mechanisms for technology transfer, the support of bridging or transfer sciences in universities, the support of sectoral technology organizations and the support of generic technologies. One of the points they make with particular effect is that technology infrastructures are themselves increasingly dependent on market relations for the development and diffusion of knowledge so that we find an intriguing combination of market failure and market process shaping the creation and operation of technology infrastructure. Infrastructure is public capital on which firms can draw for their innovating activities either deliberately or through involuntary spillovers. At the most general level the infrastructure is constituted by shared knowledge of generic technologies, together with the infratechnologies of measurement and testing methods and public regulatory standards (Tassey, 1992). More specifically it relates to a variety of collaborative methods for generating shared generic knowledge to act as a linking agent for R&D, production and market development.

Social Shaping of Technology

Studies on the social shaping of technology (SST), which first emerged in the 1980s (for example, Pinch and Bijker, 1984; MacKenzie and Wajcman, 1985), have their intellectual origins in four broad academic traditions: the sociology of scientific knowledge (SSK), the sociology of industrial organisation, technology policy studies, and certain approaches within the economics of technological change (Williams and Edge, 1996). SST emerged as a critique of technological determinism, challenging the 'linear model' and problematising the innovation process. SST studies stress that technologies do not develop according to their inner technical logic, or innate performance characteristics, but are instead social products, patterned by the conditions of their creation and use (Williams and Edge, 1996). The essential feature of this approach is that technological designs are defined by form and function and the latter is necessarily a matter of economic and social acceptability. This points to the importance of social 'choices' in the development of technologies. Although these choices are normally constrained by path dependency and lock-in, the existence of choices leads to notions of 'interpretative flexibility' and the 'negotiability' of technology, whereby different social groups attach different meanings to technologies, and may try to shape the development of technologies for their own interests. As Bijker explains: "A central adage for this research is that one should never take the meaning of a technical artefact or technological system as residing in the technology itself. Instead, one must study how technologies are shaped and acquire their meanings in the heterogeneity of social interactions" (Bijker, 1995a, p.6)[3].

Artefacts are, so to speak, described through the eyes of the members of relevant social groups. Thus an artefact does not suddenly appear as the result of a singular heroic act of innovation[4], it is instead gradually constructed in social interactions between and within relevant social groups. 'Closure' is achieved when the meanings attributed to an artefact become less ambiguous and essentially consensual and stabilised. The process of closure is generally, although not universally, irreversible.

Bijker (1995a) introduced the concept of the 'technological frame' as a tool for understanding the social construction of technologies. A technological frame exists between actors of a relevant social group and structures their interactions. It is the interpretative framework by which artefacts are understood by relevant social groups and consists of all the elements that influence the interactions within relevant social groups and which lead to the attribution of meanings to technologies. These include their key problems or goals, their current theories and problem solving strategies, their tacit knowledge and testing procedures, exemplary artefacts and design criteria, as well as perceived function and users' practice. The idea of a technological frame is related to Kuhn's (1970) concept of a disciplinary matrix, but, Bijker (1995a) argues, it is more heterogeneous, including social, cognitive and material elements and is also applicable to all social groups, not just scientists and engineers.

Because the power of different social groups varies, so does their influence over the development of technologies. Moreover the social effects of technologies do not result from technology alone, but from the ways impacts are actively sought or avoided by actors. (Rip et al, 1995). The balance of power can also change over time, which can lead to a change in the perception of the problem, as happened with the Dutch plans to improve their coastal defences against flooding (Bijker, 1995b). From such studies it can be shown that, a priori and independent of context, it is never clear whether a problem should be treated as technical or social, and therefore where the solutions should be sought (Bijker, 1995b).

2.6 A NEW GENERATION OF SYSTEMS RESEARCH AND STUDIES OF INNOVATION?

A useful place to begin this section is with Edquist (1997) who has argued in a recent survey, that the relation between the various entities (institutions, organizations and markets) is not clearly articulated in the systems literature. One could go further and argue that it is not merely the relations within and between the entities and systems which are not well articulated, but that we have not yet a coherently and clearly defined view as to what difference the systems perspective makes to innovations in general and in particular. In a dynamic context we need to know what the connecting flows, linkages and interfaces are in order to understand the evolution of the system. We need to address their quality and effectiveness and we need to understand how innovation systems co-develop with the relevant market systems.

Addressing these difficult questions will help build on the contribution of existing innovation systems ideas while linking them more closely with the dynamics of competition and development in market contexts.

Hence, we suggest that the next generation of systems thinking in innovation studies should address issues concerning (i) the system perspective rationale, (ii) the system's modus operandi, and (iii) the quality and efficiency of a system. Clearly the response to these issues could carry substantial consequences for the content and implementation of science and technology policy.

Relevant questions to raise in relation to the above issues include the following:

(i) **The rationale of the systems perspective:** What difference does a systems perspective make to the innovative performance of firms, and to what extent can we identify the differences which the system made to the comparative innovativeness of firms, sectors and nations? Can we say that the features or timing of a particular innovation are consequences of the system in which that innovation process is embedded?

(ii) **Systems operation:** In coming to a view on the rationale for a systems approach it will be necessary to identify the consequences which flow from interrelationships between the parts. There are several ways this may be approached. How is a given system configured and what are its boundaries? How does a system emerge, and develop over time? What are the 'properties' of the connecting flows, interfaces, linkages and spillovers within, between and across systems and how, as well as why, do they evolve over time? Does interaction change the way organizations search for new knowledge, does it lead them to abandon some lines in favour of others? Does it lead to measurable changes in R&D programmes? Does it lead to explicit forms of co-ordination in the generation of new knowledge? Are there explicit forms of co-ordination in the generation of new knowledge? Of particular interest here is the work of McElvey (1997) who in a detailed study of the biotechnology industry has begun to address these questions. This she does in the context of an explicitly evolutionary framework which identifies the systemic effects precisely in terms of their consequences for variety generation and selection in knowledge accumulation. An evolutionary historical perspective on the patterns of system development would also seem to be very pertinent to this research agenda.

(iii) **The quality and efficiency of a system:** We also need to address the quality of a system, and whether criteria can be formulated to assess system performance. This can also be approached from different perspectives. As suggested above, it is quite possible that systems provide constraints to change rather than stimuli to development. Indeed, they may act as a barrier to innovation in a broad sense. An example could be that of compound learning in user-producer relationships, in which if the users are 'smart' the producers benefit and vice versa. As Lundvall argues, conservative users with weak technical competence might put their suppliers at a competitive disadvantage and thus distort the innovation process (Lundvall, 1986, p.356). Closely related to this theme is the prospect of developing the notion of system failure as an alternative to the market failure perspective on innovative policy. If systems are deemed to fail it could be because particular organisational components are missing or because the available components do not interact or communicate in an appropriate fashion. Either way there is scope for public policy intervention, creating appropriate incentives to interact, stimulating new organisational forms and relocating the boundaries of the identified system.

2.7 SYSTEMS AND SERVICES

As many of the chapters in this volume indicate the innovation systems literature has developed almost exclusively around the problem of manufacturing innovation, which, given the importance of services in the economy, is something of a paradox. No doubt part of the problem lies in the difficulty of defining and measuring service sector activities. It is certainly not accurate to define services negatively, that is, as all those activities which do not entail physical transformations of matter and energy. Many so-called services do in fact generate physical changes either in individuals or firms. Other services transfer matter, data and energy over space as with logistics and telecommunication services. Yet other services provide transformations over time of which retailing and financial services provide important examples. Given the importance of information handling in these activities it is not difficult to see why many of the chapters in this volume give a central importance to developments in IT in relation to services innovation.

From a systems perspective on service innovation, four themes appear to be relevant to further investigation in this area. First, it is no longer satisfactory to treat services and non-services as independent creators of innovations. It is their mutual interdependence which is important in the creation of new service delivery possibilities. The chapter by Ducatel in this volume gives many important examples of this interdependence. Second, service activities are themselves an important element in the innovation systems in manufacturing industry, a point which is explicitly acknowledged in the literature on knowledge intensive business services. It is likely that service innovation systems are closely interwoven with conventional supply chains, the food industry provides many important examples of this. Third, it seems clear that service innovation systems will call upon different kinds of knowledge and knowledge-generating institutions from those emphasised in relation to the manufacturing sector. The knowledge base for telecommunication services is manifestly different from that in air transport logistics which in turn is different from that in retailing logistics. In the latter, the role of operations research and operations management knowledge are of particular important. To establish the nature of the knowledge base in different service activities and the appropriate generating systems offers an exciting, comparative research agenda. Finally, the intellectual property regime is for historical reasons less well-defined in services than it is for manufacturing and agricultural activities. As Andersen and Howells (1998) point out, the regime of copyrights for written works, performing arts, visual arts and software involves significant differences from traditional patent regimes.

2.8 CONCLUSION

To summarize briefly, our purpose in this chapter has been to lead up to the case for the next generation of innovation system studies and to take particular account of service activities in their application. We have identified a number of gaps in the innovation systems literature but we do not doubt for one moment that the concept of innovation systems will remain central to any study of innovation and, indeed, the competitive performance of nations.

References

Allen, P. 'Evolution, Innovation and Economics' in Dosi, G. et al (eds.), *Technical Change in Economic Theory*, London: Pinter, 1988.

Allen, P. 'Knowledge, Ignorance and the Evolution of Complex Systems' forthcoming in Metcalfe, S. and Foster, J. (eds.), *Self-Organization, Evolutionary Economics and Innovation*, Edward Elgar, 1999.

Andersen, H.B. and Howells, J. 'Innovation Dynamics in Services. Intellectual Property Right as Indicators and Shaping Systems in Innovation', *CRIC Discussion Paper Series*, No. 8, University of Manchester and UMIST, 1998.

Antonelli, C. 'Recombination and the Production of Technological Knowledge: Some International Evidence', *This Volume.*

Arrow, K. *The Limits of Organization*, New York: Norton, 1974.

Bettis, R. and Prahalad, C.K. 'The Dominant Logic: Retrospective and Extensive', *Strategic Management Journal*, 1995; Vol. 16, pp. 5-14.

Bijker, W.E. *Of Bicycles, Bakelites, and Bulbs: Towards a Theory of Sociotechnical Change*, Cambridge, Massachusetts: The MIT Press, 1995a.

Bijker, W.E. 'Sociohistorical Technology Studies' in Jasanoff, S. et al (eds.), *Handbook of Science and Technology Studies*, New York: Sage Publications, 1995b.

Callon, M. 'Some Elements of a Sociology of Translation: Domestication of the Scallops and Fishermen of St Brieux Bay' in Law, J. (ed.), *Power, Action and Belief: A New Sociology of Knowledge?*, London: Routledge and Kegan Paul, 1986.

Callon, M. 'Society in the Making: The Study of Technology as a Tool for Sociological Analysis' in Bijker, W.E. et al (eds.), *The Social Construction of Technological Systems: New Directions in the Sociology and History of Technology*, Cambridge, Massachusetts: The MIT Press, 1987.

Callon, M. 'The Dynamics of Techno-Economic Networks' in Coombs, R., Saviotti, P. and Walsh, V. (eds.), *Technological Change and Company Strategies: Technological Change and Company Strategies*, Edward Elgar, 1992.

Campbell, D.T. 'Blind Variation and Selective Retention in Creative Thought as in other Knowledge Processes' in Radnitzky, G. and Bartley, W. (eds.), *Evolutionary Epistemology, Theory of Rationality and the Sociology of Knowledge*, New York: Open Court, 1987.

Carlsson, B. (ed.). *Technological Systems and Economic Performance*, Dordrecht: Kluwer, 1995.

Carlsson, B. (ed.). *Technological Systems and Industrial Dynamics*, Dordrecht: Kluwer, 1997.

Carlsson, B., Eliasson. G. and Taymaz, E. 'Micro-Macro Simulation of Technology Systems: Economic Effects of Spillovers' in Carlsson, B. (ed.) *Technological Systems and Industrial Dynamics*, Kluwer, 1997, pp.255-278.

Constant, E.W. *The Origins of the Turbo-Jet Revolution*, John Hopkins University Press, 1980.

Constant, E.W. 'Communities and Hierarchies: Structure in the Practice of Science and Technology' in Laudan, R., *op cit*, 1984.

Coombs, R. and Metcalfe, J.S. 'Distributed Capabilities and the Governance of the Firm', *CRIC mimeo*, University of Manchester, 1998.

Csanyi, V. 'Organization, Function and Creativity in Biological and Social Systems' in Khalik, E. and Boulding, K. (eds.), *Evolution, Order and Complexity*, London: Routledge, 1996.

Dahmén, E. '"Development Blocks" in Industrial Economics', *Scandinavian Economic History Review*, 1988; Vol. 36, pp.3-14.

Dahmén, E. 'Development Blocks in Industrial Economics' in Carlsson, Bo (ed.), *Industrial Dynamics*, Boston: Kluwer Academic Publishers, 1989.

David, P. and Foray, D. 'Information Distribution and the Growth of Economically Valuable Knowledge: A Rationale for Technological Infrastructure Policies' in Teubal et al, *op cit*, 1997.

De Liso, N. and Metcalfe, J.S. 'On Technological Systems and Technological Paradigms: Some Recent Developments in the Understanding of Technological Change' in Helmstader, E. and Perlman, M. (eds.), *Behavioural Norms, Technical Progress, and Economic Dynamics*, Michigan University Press, 1996.

Dosi, G. 'Technological Paradigms and Technological Trajectories - A Suggested Interpretation of the Determinants and Deviations of Technological Change', *Research Policy*, 1982; Vol. 11.

Edquist, C. *Systems of Innovations: Technology, Institutions and Organizations*, London: Pinter, 1997.

Edquist, C. and Johnson, B. 'Institutions and Organizations in Systems of Innovation' in Edquist, C. (ed.), *op cit*, 1996.

Eliasson, G. 'Spillovers, Integrated Production and The Theory of The Firm', *Journal of Evolutionary Economics*, 1996; 6, pp.125-140.

Foss, N.J. and Knudson, C. *Towards a Competence Theory of the Firm*, London: Routledge, 1996.

Freeman, C. *Technology Policy and Economic Performance: Lessons from Japan*, London, Pinter: 1987.

Freeman, C. and Perez, C. 'Structural Crises of Adjustment: Business Cycles and Investment Behaviour' in Dosi, G., Freeman, C., Nelson, R., Silverberg, G. and Soete, L. (eds.), *Technical Change and Economic Theory*, London: Pinter Publishers, 1988, pp.38-66.

Granovetter, M. 'Economic Action and Social Structure: The Problem of Embeddedness', *American Journal of Sociology*, 1985; Vol. 91, pp.481-510.

Green, K., Hull, R., McMeekin, A. and Walsh, V. 'The Construction of Techno-Economic Networks vs Paradigms', *CRIC Discussion Paper*, No. 17, University of Manchester, 1998.

Henderson, R. and Clark, K. 'Architectural Innovation', *Administrative Science Quarterly*, 1990.

Hughes, T.P. *Networks of Power: Electrification in Western Society, 1880 – 1930*, Baltimore: John Hopkins University Press, 1983.

Hughes, T.P. 'The Dynamics of Technological Change: Salients, Critical Problems, and Industrial Revolutions' in Dosi, G., Renato, G. and Tonielli, P.A. (eds.), *Technological Change and Enterprise in a Historical Perspective*, Oxford: Clarendon Press, 1992, pp.97-118.

Kuhn, T. *The Structure of Scientific Revolutions*, 2nd Edition, Chicago: Chicago University Press, 1970.

Langlois, R.N. 'Systems Theory, Knowledge and the Social Sciences' in Machlup, F. and Mansfield, U. (eds.), *The Study of Information*, New York: Wiley, 1983.

Latour, B. *Science in Action: How to Follow Scientists and Engineers Through Society*, Cambridge, Massachusetts: Harvard University Press, 1987.

Laudan, R. *Models of Scientific Change Relevant*, Dordrecht: Reidel, 1984.

Law, J. 'Technology and Heterogeneous Engineering: The Case of Portuguese Expansion' in Bijker, W.E. et al (eds.), *The Social Construction of Technological Systems: New Directions in the Sociology and History of Technology*, Cambridge, Massachusetts: The MIT Press, 1987.

Loasby, B. 'The Organization of Industry and the Growth of Knowledge',*mimeo*, Department of Economics, University of Stirling, 1996.

Loasby, B. *Knowledge, Institutions and Evolution in Economics*, Routledge, 1999.

Lundvall, B-A. *Production Innovation and User-Producer Interaction*, Aalborg: Aalborg University Press, 1986.

Lundvall, B-A. *National Systems of Innovation: Towards a Theory of Innovation and Interactive Learning*, London: Pinter, 1992.

McElvey, M. 'Using Evolutionary Theory to Define Systems of Innovation' in Edquist, C. (ed.) *Systems of Innovation: Technology, Institutions and Organizations*, London: Cassell, 1997.

MacKenzie, D. and Wajcman, J. (eds.) *The Social Shaping of Technology: How the Refrigerator got its Hum*, Milton Keynes: Open University Press, 1985.

McShea, D.W. 'Complexity and Evolution: What Everybody Knows',*Biology and Philosophy*, 1991; Vol. 6, pp.303-324.

Montgomery, C. *Resource-based and Evolutionary Theories of the Firm*, Dordrecht: Kluwer Academic, 1995.

Nelson, R. *National Systems of Innovation*, Oxford University Press, 1992.

Oliga, J.C. 'Power-Ideology Matrix in Social Systems Control' (*Systems Practice*, 1990; Vol. 3). Reprint - Flood, R.L. and Jackson, M.C. *Critical System Thinking*, Chichester: John Wiley and Sons, 1991.

Perez, C. 'Structural Change and the Assimilation of New Technologies in the Economic and Social System', *Futures*, 1983; Vol. 15, pp.357-375.

Pinch, T. and Bijker, W. 'The Social Construction of Facts and Artefacts, or How the Sociology of Science and the Sociology of Technology Might Benefit Each Other',*Social Studies of Science*, 1984; 14, pp.399-441.

Polanyi, K. 'The Economy as Instituted Process' in Polanyi, K., Arunsburg, C. and Pearson, H.W. (eds.) *Trade and Markets in the Early Empire*, Glencoe: Free Press, 1958.

Richardson, G.B. 'The Organization of Industry', *Economic Journal*, 1972; Vol. 72, pp.883-896.

Rip, A., Misa, T. and Schot, J. *Managing Technology in Society: The Approach of Constructivist Technological Assessment*, New York and London: Pinter Publishers, 1995.

Rosenberg, N. *Perspectives on Technology*, Cambridge University Press, 1976.

Rosenberg, N. 'Why Do Firms Do Basic Research (With Their Own Money)?',*Research Policy*, 1990; Vol. 19, pp.165-174.

Scott, J. *Sociological Theory – Current Debates*, Cheltenham: Edward Elgar, 1995.

Staudenmaier, J. *Technology's Story Tellers*, MIT, 1985.

Storper, M. and Harrison, B. 'Structure of Industrial-Production Systems and Their Forms of Governments in the 1990s', *Research Policy*, 1991; Vol. 20, pp.407-422.

Tassey, G. *Technology Infrastructure and Competitive Position*, Dordrecht: Kluwer, 1992.

Teubal, M., Foray, D., Justman, M. and Juscovitch, E. *Technological Infrastructure Policy: An International Perspective*, Dordrecht: Kluwer, 1996.

Thorelli, H.B. 'Networks – Between Markets and Hierarchies', *Strategic Management Journal*, 1986; Vol. 7, pp.37-51.

Vincenti, W.G. *What Engineers Know and How They Know It*, Johns Hopkins University Press, 1990.

Von Bertalanffy, L. *General System Theory. Foundation, Development Applications*, London: Allen Lane Penguin Press, 1968.

Von Hippel, E. *The Source of Innovation*, Oxford University Press, 1991.

Williams, R. and Edge, D. 'The Social Shaping of Technology', *Research Policy*, 1996; 25, pp.865–899.

Zeleny, M. 'On the Social Nature of Autopoietic Systems' in Khalil, E. and Boulding, K. (eds.), *Evolution, Order and Complexity*, London: Routledge, 1996.

---

[1] In a recent paper Carlsson has formulated these distinctions in terms of components, connections and consequences.

[2] See Storper and Harrison (1991) for an illuminating account of the roles of power and influence in the conduct of innovation agglomerations.

[3] Within this broad research tradition there are several strands. Bijker (1995a) identifies three strands in what he labels 'constructivist studies of technology' which analyse 'sociotechnical ensembles':

- the technological systems approach – exemplified by Hughes' (1983) work on 'Networks of Power';
- the actor network approach – pioneered by Callon (1986, 1987), Latour (1987) and Law (1987), which emphasises the 'seamless web' of society and technology;
- and the social construction of technology (SCOT) approach – in which the work of Bijker and Pinch is prominent.

[4] Even when the success of a technology or artefact does appear to have been the result of the ingenuity and emotional commitment of 'heroic' individuals, social constructionists argue that the characteristics of these individuals are also a product of social shaping (Bijker, 1995a).

Chapter 3
Innovation as a Loosely Coupled System in Services

Jon Sundbo and Faïz Gallouj

3.1 INTRODUCTION

In this chapter we will put forward a model of the innovation system in services. The model is theoretical, but based on a review of a large amount of existing empirical research on innovation in services, including a recent EU financed project, called the SI4S (Services in Innovation and Innovation in Services) project which we have participated in, and which has been the occasion for us to develop these thoughts[1] (Sundbo and Gallouj, 1998).

We find that the empirical results give us a basis for establishing a general model and that it is a fair generalisation of these results. Even if the service industries are different from one another, the empirical research indicates that there are some common characteristics of the innovation processes due to the specific nature of service production that is common to all service industries.

The main focus of the model is the innovation processes from the perspective of the service sector: How do service industries view their innovation activity? What are the driving forces behind the single service enterprise? More generally, how do these driving forces combine to constitute a system? And how should we characterise this possible system?

The material is divided in five sections. Section 3.2 is devoted to an attempt to present some general characteristics of services and of service innovations. On the basis of empirical work, Section 3.3 presents some typical innovation patterns in services. This presentation leads us to what we mean by innovation-system and whether an innovation system may exist in services is discussed in Section 3.4. In Section 3.5 we conclude about the issue of innovation system and discuss how the analysis might contribute to developing evolutionary theory of innovation.

3.2 SERVICE FIRMS DO INNOVATION

Services have been considered in most of the literature as an appendix to manufacturing (for example, Cohen and Zysman, 1987; Miles et al, 1994), a residual sector, or at least as a sector lagging behind the manufacturing sector in form of low productivity, low capital intensity, weak qualification levels and low innovation activity. This is not true. Service firms do innovate, but the innovations often take other forms than those in manufacturing and they may be organised differently.

The different forms of service innovations are to some degree related to the specific form of service production, and we will therefore start by briefly stating these characteristics.

By looking at the special forms that service innovations take, we may learn more about innovation processes which might even be valid for manufacturing innovations as well.

3.2.1 General Service Characteristics

The particular innovation pattern in services, for example compared to manufacturing, must be explained by the specific characteristics of service production. We will not detail these here as they have been treated intensively elsewhere (for example, Gadrey, 1992; Illeris, 1996; Normann, 1991), but only briefly repeat the facts that are most relevant to innovation.

In service industries, the product is not always perfectly 'formatted' and codified, i.e. precisely determined a priori. Each service transaction may be considered as unique as long as it is produced on demand (tailor-made) in interaction with the client or as a response to a specific, not standardisable problem, and in different environments.

Client participation (in various forms) in the production of the service may be the most basic characteristic of service activities, particularly knowledge-intensive ones. Various concepts have been developed in order to account for this client involvement (co-production, servuction, service relationship, the moment of truth, prosumption). At the interface between the service provider and its client different types of interaction are occurring. Different types of elements are being exchanged: information and knowledge, emotions, verbal and gesture signals of civility. This interaction also expresses power struggle, domination and reciprocal influence relationships.

The service industries are also under pressure to reduce the costs and that leads to a standardisation tendency. This standardisation means that service production is not unique in the single delivery situation. The service firms attempt to combine this with individual customer care in a modularisation system (Sundbo, 1994).

Though analytically useful (albeit difficult to use), a distinction between product and process is widely accepted in the case of manufacturing goods. The same is not true of services where the product mostly cannot be separated from the process. Here, the term 'product' frequently includes a process: a service package, a set of procedures and protocols, an 'act'.

In the case of services, and particularly those in which the intangible and relational aspects are important, the correspondence between competencies and other means brought to bear by the service provider and the 'product' are generally much hazier and much more difficult to codify; they are, to a large extent, tacit and subject to the difficulties caused by informational asymmetry. The emphasis on quality and trust is therefore an important dimension of service activities.

3.2.2 Service Innovation as an Interaction Process

The innovation process in services is, to a large degree, an interaction process, both externally and internally.

It is an interaction with external actors, particularly with customers. The customer's satisfaction with the total encounter (not only the core service delivered, but also the circumstances of the delivery) has been crucial in service production. Customer satisfaction has been more important than the issue of a new core service. Customer satisfaction, in terms of service quality, has thus been more important to service firms than innovation. There have been innovations, but mainly as delivery or process innovations. However, the standardisation or modulisation tendency has made it more relevant to emphasize product innovations and innovation generally.

Nevertheless, the tradition has led service firms to be focused on the customer encounter also in terms of their innovation activities. These emphasize the client interaction (sociologically: primary interaction) as an important parameter in the innovation process (Edvardsson, Thomasson and Ovretveit, 1994). It is a crucial factor in the process of getting the innovation accepted on the market. The primary interaction is often forgotten in the theories on manufacturing innovations that more or less implicitly presuppose that the marketing of a new product is a mass process (sociologically: secondary interaction), and it is so for mass goods.

The innovation process is also an interactive process at the internal level. Innovation is generally an unsystematic, collective process in which employees and managers participate in different interaction patterns at both formal and informal level. The organization of innovation is differentiated and various patterns can be observed in different types of firms, as we will see in Section 3.3. Service firms have not been good at organising the innovation process in a formalised and systematic way or learning from the process. This is even valid for external interaction with customers (Edvardsson, Haglund and Mattsson, 1995).

The contemporary tendencies in the service sector, however, are towards a more systematic innovation process, often based on certain trajectories. They are often service professional trajectories (ideas and logics within a service professional such as law, accountancy etc.), but may also be technological trajectories. Service firms still maintain a great deal of flexibility in innovation activities, which involve several actors and trajectories.

3.2.3 Innovation and Non-Reproduced Small Changes

Innovations in services are a mix of reproduced (although incremental) innovations and 'small', non-reproduced changes to solve single customers' problems (what we will also call ad hoc innovation). The latter is particularly a result of the customer interaction process.

This means that we must understand the development of service business by combining traditional innovation theory and a theory of continuous change as accumulated, reproducible, ad hoc innovations. We cannot combine all these individual changes in one theory, but we can understand the firms' attempt to guide this process through a combination of two theoretical elements: 1. On organizational creativity, 2. On organizational learning.

3.2.4 Three Approaches to Understanding Innovation in Services

A useful distinction can be made between approaches that might be described as 'technologist', which focus on an analysis of the introduction of equipment and technical systems, 'service-oriented' approaches that emphasise divergence by highlighting the specific characteristics of service activities, and 'integrative' approaches that emphasise convergence by advancing analyses that can be applied to both goods and services. All these works are more deeply analysed in F. Gallouj (1994) and C. and F. Gallouj (1996).

1) Innovation in services and technology: one referent, a multiplicity of relationships

Many studies in this first category (technologist approaches) concentrate exclusively on innovations that are both technological and adopted, usually at the expense of ignoring non-technological innovation and technological innovation produced by service firms themselves.

This technologist approach can be interpreted in various ways, empirical and theoretical. The first consists of recording the extent to which technologies have been diffused within services, as shown by statistical studies. Service industries are now the main users of information technology in all the developed economies (Miles, 1993). The second, linked to the first, is based on investigation of the nature of the effects produced by the adoption of these technologies on economic variables such as productivity, employment, skills, trade, etc. The use, implicit or explicit, of standard neo-classical economic theory (through the production function) constitutes a second line of interpretation. This theory has, in fact, contributed to the development of a 'mechanistic' approach to production and to a somewhat reductionistic, 'technologist' view of innovation that focuses on process innovations embodied in capital goods. Although they adopt a much wider definition of innovation, the new neo-Schumpeterian and evolutionary approaches to technical change are not immune to this technologist bias, and have even contributed to it to a certain extent (Gallouj, 1997). Their primary objective is usually to analyse the ways in which service firms and industries adopt or, in a few cases, produce technologies. For the moment, their theoretical horizons are limited to the application to the service sector of concepts and methodologies developed with reference to the manufacturing industry, such as the natural technological trajectory, the technological paradigm, sectoral taxonomies of forms of technological change, etc. (cf. in particular Soete and Miozzo, 1990).

Barras' work (1986, 1990) is particularly interesting because of its theoretical ambition. In certain services (banking, insurance, accounting, administration), Barras has observed a product life cycle that is the converse of the traditional industrial cycle. The basic element of this so-called 'reverse product cycle' theory is the adoption of an item of computer equipment by a service activity that triggers what might be called a 'natural technological trajectory'. This leads, in the first instance, to the emergence of incremental process innovations, the purpose of which is to improve the efficiency of the service being provided, secondly to an improvement in service quality through more radical process innovations and finally, in the last phase of cycle, to the emergence of product innovations. Thus innovation is not viewed in isolation from the technological potentialities, and Barras' model is less a theory of innovation in services than a theory of the diffusion within the service sector of technological innovations derived from the manufacturing industry.

2) The specificities of innovation in the service sector as a priority for analysis

Without ignoring the technological dimension, the 'service-oriented' approaches focus on non-technological forms of innovation; in this respect they follow Schumpeter's well-known, broad and open definition of innovation.

Studies based on this approach often take the 'purer' services as their field of investigation, i.e. those in which the criteria of intangibility and the co-production of output are assumed to be most evident. Consultancy services, for example, are an interesting area for empirical analysis of service-oriented innovation. In his study of consultancy firms, Gallouj (1994) highlights the existence of ad hoc forms of innovation that are not immediately reproducible and of institutional 'formalisation' trajectories (i.e. the search for a certain degree of formalisation, though not necessarily, or even predominantly, in tangible form).

The studies by Van der Aa and Elfring (1993), Gadrey et al (1995) and Sundbo (1997) also take a broad, Schumpeterian view of innovation. According to Sundbo (1996), innovations in services do not follow a technological trajectory (in Dosi's sense (Dosi, 1982)) but rather service-professional trajectories (for example, a certain number of ideas on management, banking, etc.) in which technologies are only one vector among several others.

3) The search for convergence and the desire for analytical integration

The notion of adopting a similar approach to the economic analysis of both goods and services is based on the observation that the boundary between goods and services is becoming increasingly less clear. Certain services are being 'industrialised' and, conversely, the production of certain goods is being 'tertiarised'. These converging tendencies are often described in terms of the goods-services continuum and functions.

The study by Barcet, Bonamy and Mayère (1987) adopts this approach and results in a classification of the forms of innovation that applies to both goods and services. These authors categorise innovations according to whether they relate to function, specification or the production process. The first category encompasses the emergence of new, undifferentiated, abstract functions, such as the storage of picture and sound in the case of video recorders, or the identification of a new risk to be covered in the insurance industry. The second category involves the concrete realisation and differentiation of the functional innovation while the third corresponds to a cost-cutting activity (as a result of standardisation, the use of new technical instruments, etc.).

The characteristics approach developed by Gallouj and Weinstein (1997) is also integrative. Following and extending Saviotti and Metcalfe's (1984) representation of the product, it shows that an approach to products in terms of competencies, service, technical and process characteristics offers a stimulating starting point for the study of innovation in both goods and services without sacrificing any of the specific aspects of innovation in services. Various modes of innovation are highlighted (radical innovation, innovation based on improvement, innovation involving the addition of new characteristics, ad hoc innovation, re-combinative innovation, innovation through formalisation) and interpreted in terms of a characteristic dynamic. This may take different forms: addition, elimination, improvement, bundling, unbundling, shaping of characteristics.

3.3 PATTERNS IN SERVICE INNOVATIONS

3.3.1 Internal and external driving forces: The components of innovation patterns

As a basis for presenting the innovation patterns, we will start by setting up the driving forces of innovation in services in a scheme. These are the determining elements in the innovation process and a possible system is composed by a combination of the driving forces.

The formulation of these driving forces are the result of an investigation of most of the existing research on innovation in service which has been part of the SI4S project[2]. It is also based upon the survey of other existing empirical literature.

Taking the firm as a focus, there are external and internal driving forces.

3.3.2 Internal driving forces

The innovation process in service firms is mainly driven by internal forces (however, this may be said about manufacturing firms as well).

There are three internal forces.

Figure 3.1: Driving forces behind service innovation

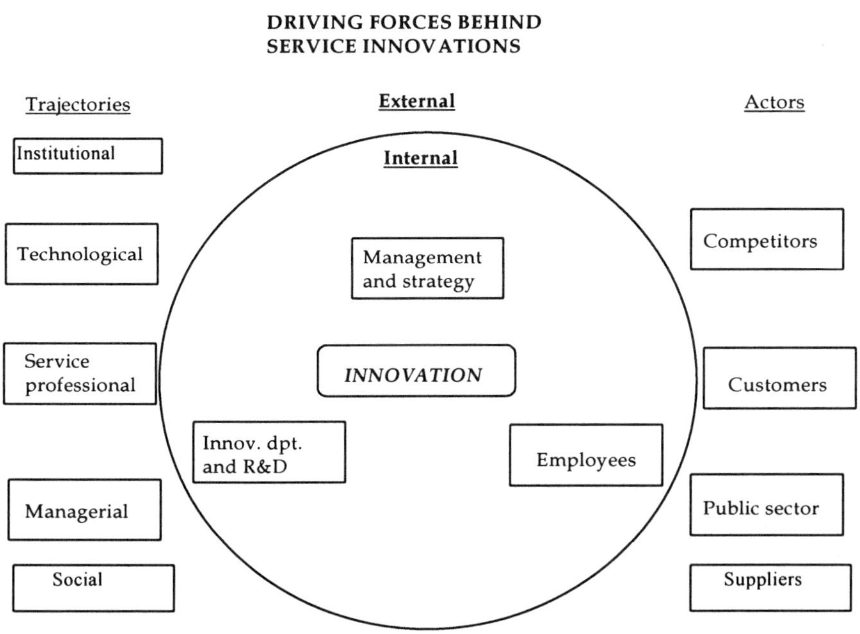

From: Sundbo and Gallouj 1998

The first is the management and strategy of the firm. Management could be the top manager, but is often the management of the marketing department, since service innovations very often are market driven, and the marketing department, which has the direct customer contact and market knowledge, is the leading actor in innovation activities.

The innovation process in services is mostly a loosely coupled process in which the employees (including managers at all levels) are involved, or they just function as corporate entrepreneurs and start the process. The employees are, therefore, an important driving force. They are the second internal driving force.

A third driving force is formalised R&D departments or other types of formalised department, which has the responsibility for ensuring that innovations will appear. The latter is a kind of communication department, that exist in some service firms and which has the task to induce innovation ideas among the employees and managers and to collect these ideas; the innovation department does sometimes innovate itself, but this is not always the case. Since service innovations rarely are science-based, it is very rare to find R&D departments in service firms.

3.3.3 External driving forces

The external forces can be divided into trajectories and actors. Trajectories are considered to be external driving forces even though the innovation activities of the single firm can contribute to the reinforcement of a given trajectory.

Trajectories are ideas and logics that are diffused through the social system (being a nation, an international network, professional networks, etc.) (cf. Dosi, 1982). They are often diffused through many and difficult identifiable actors. The important factor is, however, not the actors, but the ideas and the logic behind the ideas. We can identify five types of trajectories.

The most important factor is service professional trajectories by which we mean methods, general knowledge and behavioural rules (for example, ethics) that exist within the different service professions (for example, lawyers, nurses, catering (how to cook)).

Another type of trajectory is general management ideas or ideas for new organizational forms such as motivational systems, BPR, service management etc. These two first types of trajectories may highly overlap as far as knowledge intensive business services (KIBS) are concerned.

The third type of trajectory is technology trajectories in the traditional economic sense. New logics for using technology that generally influences service products and production processes. Examples are the ICT wave, and more specifically, the Internet, and the freezer and microwave oven, which together have created a new distribution system within catering. Some service fields (such as software, financial services, technical services, etc.) have contributed more to the ICT development than manufacturing (Miles et al, 1994).

The institutional trajectory describes the general trend of the evolution of regulations and political institutions (for example: the European construction, the European research programs).

The social trajectory displays the evolutions of general social rules and conventions (for example: ecological and environmental consciousness).

These different trajectories are not independent of each other, they may, in many situations, be intertwined.

Actors are persons, firms or organizations whose behaviour has importance for the service firms' possibilities for selling services and, therefore, for their innovation activities. The actors define the market possibilities and sometimes they are involved in the development of the innovations.

Customers are, of course, actors of major importance. They may be sources of information but they can also contribute more actively to the innovation process. In certain situations, the interface between the service provider and its client can be considered as a laboratory where innovation is co-produced.

Competitors are also important for the innovation activities. Service firms may imitate competitors' innovations, and since service industries generally have not been characterised by offensive innovation strategies, a condition for starting an innovation activity has often been that the competitors should be moving first.

Suppliers and especially knowledge business service suppliers are important sources of innovation as well. To complement the two well-known Schumpeterian models of innovation (Schumpeter Mark I and Schumpeter Mark II), it is possible to define what could be called 'a consultant-aided model of innovation' (cf. Gallouj, 1994). Therefore, knowledge business service suppliers or some of them may be considered as a new locus of the 'Schumpeterian enterprise spirit'. Technology suppliers (including software suppliers) are also sometimes important partners in the innovation process, much knowledge business service suppliers and software has been developed in co-operation between service firms that needed a new technology and technology suppliers.

The public sector is the least important actor, but nevertheless an actor of some importance. The public sector demands services, and it delivers research and education necessary to innovation activities, but the public sector is rarely a direct actor in service innovation processes, neither as change agent nor as deliverer of knowledge (since service innovations rarely are science based). Furthermore, the public sector has regulated the service sector, a function that in fact has led to many innovations, but has also impeded innovation. Many financial innovations are due to changes of tax laws. The contemporary tendency to deregulation makes this function of the public sector less important.

3.4 TYPICAL INNOVATION PATTERNS

Theoretically one can identify many patterns by combining these driving forces. We will mention here the patterns that have been found in empirical research until now.

3.4.1 The classic R&D pattern (the industrial pattern of innovation) and its evolution

This pattern is the less frequent in service industries (cf. Barcet, Bonamy, Mayère, 1987). Nevertheless, it can be found in large size firms specialising in the production of standardised operational services dealing with material or information. For example, large firms specialising in information mass processing, building maintenance or teleguarding. This is a copy of the traditional manufacturing R-D pattern which makes a clear dissociation between R-D and production. Compared, for example, to knowledge business service firms one can say that in these types of firms there is generally a dissociation between service production and its delivery. It is therefore possible to create an R-D department devoted to the improvements of the «products» that are to be delivered or to the design of new « products ».

This industrial pattern of innovation, as it is defined (notably by Barcet, Bonamy, Mayere, 1987), seems to us ambiguous. It refers to the old industrial pattern, which has changed a lot itself. The new industrial model, which has substituted a flexibility logic for the old standardisation logic, is far closer to the functioning of service activities (which are often interactive by definition).

One may distinguish between two variants of the classic R&D pattern: a traditional or fordistic variant, and a neo-industrial one.

A traditional or fordistic variant

This pattern (Figure 3.2) is defined as above. It is rare in services and it tends to be rarer in manufacturing as well. The main lever of innovation is the pursuit of the technological trajectory (technological and process innovations). One or several departments specialising in innovation do exist. They develop linear relationships (without any real feedback) with the other departments (traditional linear model of innovation). These innovation departments are generally production technical departments or information technology departments. The client is present in this pattern but only as a passive source of information.

Figure 3.2: The industrial pattern: traditional or fordistic variant

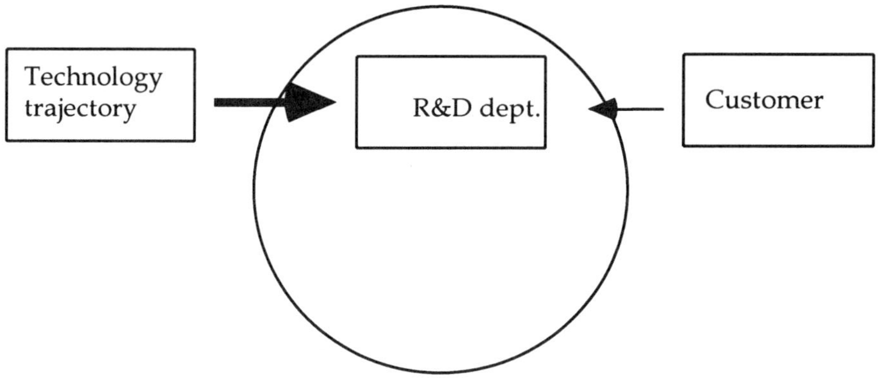

A neo-industrial variant

Firms run according to the traditional industrial pattern tend to move towards this new one. The neo-industrial pattern refers to certain evolutions that are occurring in mass informational services that traditionally were following a fordistic pattern and that are now facing important competition pressures (banks, insurance companies, postal services). In these firms, innovation is produced by interacting sources or actors. The innovation process does rarely follow the linear model, but often a more complex pattern (cf. Kline and Rosenberg, 1986). These interactions are « technical » unavoidable relations, no matter their quality and efficiency. In insurance companies the actors are, for example, information technology departments, the various actuarial departments, the marketing department, possibly a genuine research laboratory (cf. Gallouj and Gallouj, 1996). Transversal project groups are favoured and multiplied with more or less success. In this pattern, the levers of innovation are both the technological trajectory and the service-professional trajectory. The main actors participating in each trajectory and its corresponding forms of innovation interact. Management and strategy play an important role as well.

This model is more customer-oriented. More innovations come from a pull effect, namely, the expected future needs of the customers.

Figure 3.3: The neo-industrial pattern

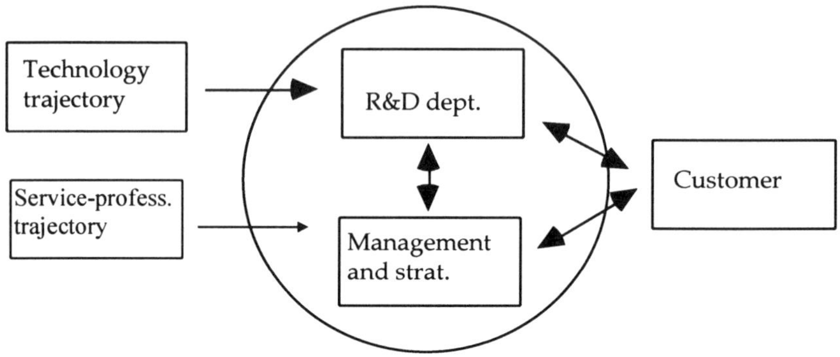

One can distinguish between two sub-variants of this neo-industrial variant (Gallouj, Djellal and Gallouj, 1997):

1) The variant in which the technological trajectory and the service trajectory are in an imbalanced interaction;

2) The variant in which they are in balanced interaction.

In the first case, if we define a product as a set of service, technical and process characteristics in correspondence (Saviotti and Metcalfe, 1984), one can say that the service characteristics change much more rapidly than the technical and process characteristics. There is, to a certain extent, proliferation of new services characteristics (or functions) while the technical and process characteristics remain unchanged. The incremental model of innovation (innovation by adding characteristics) plays a very important role here.

In the second case (when the technological trajectory and the service trajectory are in a balanced interaction), certain members of the organization in charge have to produce technical and process characteristics and others service characteristics. There is a share of tasks and a certain balance of power, which does not mean a lack of conflict.

3.4.2 The service professional pattern

This model which has also been described by Barcet, Bonamy and Mayère (1987) characterises the professional knowledge service firms. They are generally medium size firms devoted to knowledge intensive business services. These firms do not really sell product-services, but competencies, abilities to solve problems in different expertise areas (consultancy and engineering).

In this pattern, formalised structures dedicated to innovation do not exist. The innovation trajectory is of the service-professional type.

Figure 3.4: The service professional pattern

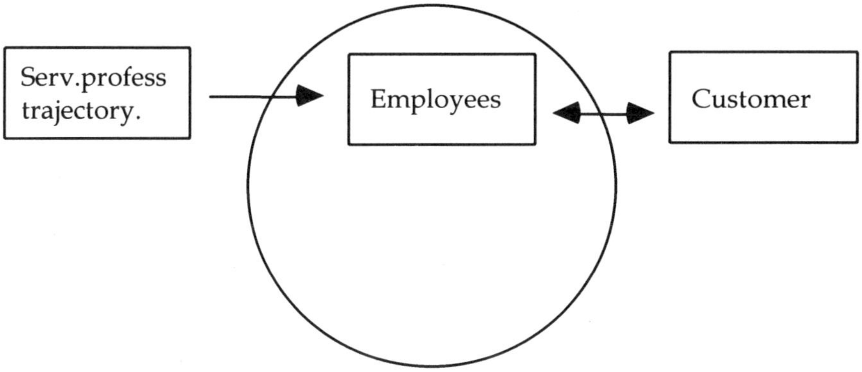

The innovation process is a collective process in which all the professionals are supposed to participate. It will often be a more disciplined, and less 'wild' and radical, process than in the entrepreneur model (see below). The professional often follows certain professional norms and methods in their innovation.

Extrapolating Martin and Horne's analysis (1993) one can describe this pattern as bottom-bottom or top-top. Because of this the service professional pattern has a certain number of advantages: it is flexible, able to answer quickly to market signals, able to cross synergically the individual ideas of its members. Conversely, as far as it depends heavily upon its individual components, it also has a number of disadvantages among which are the risk that the innovation process is not completed, and the absence of enterprise projects.

In the service professional pattern, the main driving force or lever of innovation is individual expertise and competencies which correspond to the service-professional trajectory. An important locus of innovation is the interface with the customer. Thus the client who is present here plays a much more active role than in the previous pattern. The ad hoc mode of innovation, without being exclusive, plays an important role here.

The analysis of the different steps of such an ad hoc mode of innovation shows that the steps of production, selling and innovation take place simultaneously or are merged. The client's problem (in its concrete sense) is the starting point of the innovation process. An important point here is that this service production process, which a posteriori becomes an innovation process, ends with a formalisation step. This formalisation step is achieved without the client's participation. It aims at going through the problem and the innovating final solution again and at formalising and modifying them in order to re-appropriate some of their components and to capitalise them in the memory of the firm (for example paper, software, IT-files audio-visual, etc.).

As a product of the customer interface, ad hoc innovations, particularly in consultancy firms, depend upon the nature and components of this. Thus sparring type interfaces conversely to jobbing type (Gadrey and Gallouj, 1998) are most propitious for the creation and success of this form of innovation, because they help to assure a better understanding and acceptance (legitimacy) of the innovation. Moreover, problems of a strategic nature, themselves potential sources of innovation, are most often the object of a sparring type interface. However, one must not conclude from this that only creative problems (as Kubr (1988) calls them) - those where one seeks to develop a completely new situation - are carriers of *ad hoc* innovation. Corrective problems, in which the consultant plays the role of therapist, and progressive problems, in which the consultant are expected to improve a given situation that is feared to be deteriorating, are also ad hoc innovation carriers. Furthermore, the opportunities for ad hoc innovation appear to increase with the size of the service provider and that of the client. The effective implementation of ad hoc innovations also depends upon the quality of the professionals in the client organization participating in the interface.

Some of the professional knowledge service firms currently have a tendency to move towards the third pattern, the organised strategic innovation pattern. The innovation process becomes guided by the top management and the firm's strategy and less anarchic and free for the professionals.

3.4.3 The organised strategic innovation pattern

The organised strategic innovation pattern - or the managerial model of organising the innovation - is the most typical within the service sector. It is definitively so for large service firms, but even small service firms are moving towards this model.

This pattern corresponds to the real existence of a policy, a strategy or a function of innovation in the firm, but to the absence of an R-D department. There is no permanent R-D innovation department. Research and more precisely new ideas research is everyone's task, but development, which requires much time, is done by ad hoc project teams. The logic which is favoured is that of designing products which are as reproducible as possible. Whether we can conclude that a trend of industrialisation exists in service is difficult to say. Empirical research suggests different conclusions. In some research, an industrialisation tendency has been found (Sundbo 1994, 1997). In other research, it was found that a professional rationalisation logic (combining technology, service-professional, managerial strategies) prevails rather than an industrial rationalisation logic (Gadrey, 1994).

Figure 3.5: The organised strategic innovation pattern

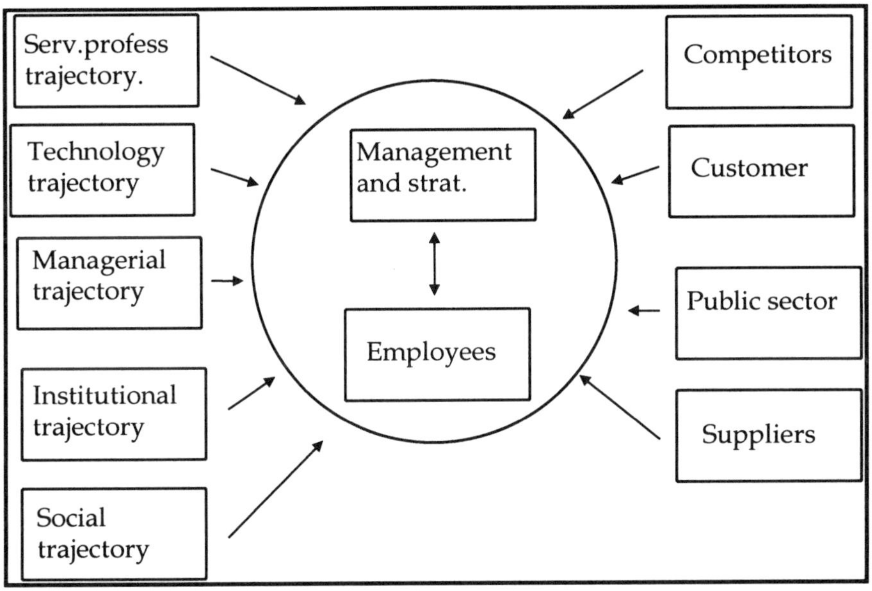

In this model, innovation policy is supplemented with two important actions: 1) knowledge accumulation in order to facilitate its reproducibility and the share of individual knowledge among the firm; and, 2) quality control as a means of checking adherence to service standards, but also as an indicator of the evolution of the nature of clients' demand. The innovation process becomes a process of balanced entrepreneurship (Sundbo, 1992, 1996). The employees act as corporate entrepreneurs, but the management attempts to regulate and control the corporate entrepreneurial process. The framework for the management's regulation is most often the strategy, which contents the policy for innovation (which types, for which market segment, etc.). The strategy can also function as an inspiration for innovative ideas.

The innovation process is often organised in different steps, starting with a free corporate entrepreneurial idea phase, which turns into a more guided development phase, often organised as team work, and finally ending as a test and marketing activity in which the marketing and production departments have the main role. The marketing department when it exists is often the strongest department in the innovation process.

The innovation policy in this pattern is very broad, which means that the firm is looking for many fields in which it could make innovations, for example, in marketing, production, organization, etc., and since the innovation policy is determined by the strategy, the management focuses more on what is going on in society. This will namely determine the future customer needs. Thus, all the trajectories are relevant here as are all the actors.

In previous works devoted to the consultancy sector, Gallouj (1995) shows that this pattern (for example the design of a formatted method or product-service) may follow the standard scheme of industrial R-D (be more or less formalised): as in the R-D pattern, innovation, production and selling may, at least in theory, be separated. A certain formalisation of the ideas' genesis is possible through internal gathering procedures of ideas; gathering procedures of customer ideas and dissatisfactions (user groups for example).

3.4.4 The entrepreneurial pattern

This pattern corresponds to the creation of service firms on the basis of a radical innovation. These firms are small and do not have any R-D department. Their main activity is to sell the initial radical innovation. The innovation processes that might follow later are generally focused on the improvement of the latter: it is generally incremental, and improvement of the first radical innovation. The appearance of IT services, of repairing services, etc. may be interpreted in these terms.

Numerous firms set up by university researchers often correspond to this entrepreneurial pattern as well, but the pattern can be found in all service industries. Some of the service industries, and in particular retail, are characterised by many new firms. This, however, does not represent large-scale entrepreneurial activity because most of the new firms are not based on any innovative idea.

Because of the radical character of the basic innovation that lies at the foundation of this pattern, one can say that none of the different innovation trajectories can be excluded from the entrepreneurial pattern.

3.4.5 The artisanal pattern

This pattern describes small firms involved in operational services (cleaning, guard services, hotels, restaurants, etc.). These firms do not have any innovation strategy. They do not have any R-D department nor information technology department.

Generally, these firms are not innovative, and, if they are, the renewals are normally small, non-reproduced changes. The firms are generally conservative, not-change and trajectory oriented and therefore not oriented towards external trajectories. The external innovation drivers are the actors. If innovation is present, it is through improvement models and learning processes.

Figure 3.6: The entrepreneurial pattern

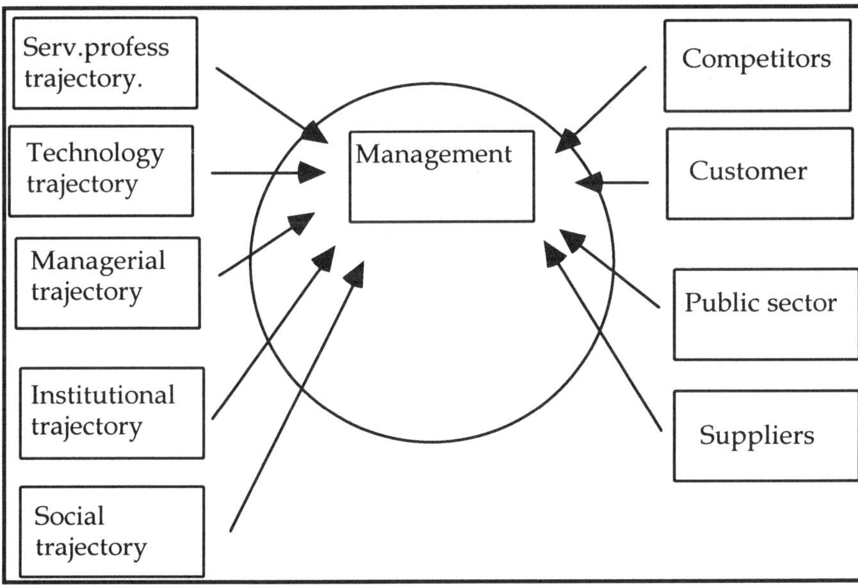

Figure 3.7: The artisanal pattern

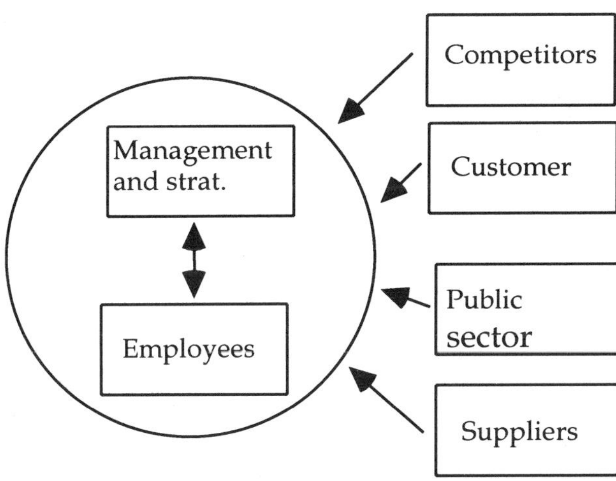

3.4.6 Network pattern

One pattern is a situation where a number of service firms have created a common network firm that has the purpose to innovate on behalf of the member firms or induce innovations in these. This pattern is found in tourism (Sundbo, 1997) and in certain financial groups (Djellal et al, 1998). This is a situation where the service firms put the innovation activity and the relation to actors and trajectories in the hands of a professional organization for innovation outside the firms.

The network firm could, in principle, have an R&D department, but this is not reported in any empirical research.

The network firm may be supposed to be not very customer or supplier oriented because it does not have the contact with these actors, who interact directly with the member firms. The clients are the member firms.

Figure 3.8: The Network Pattern

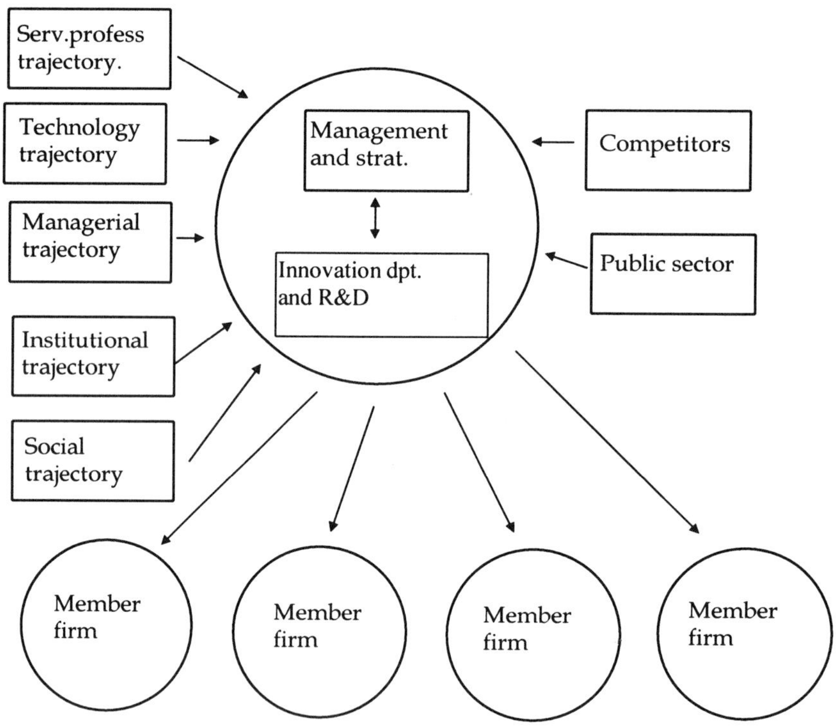

3.5 INNOVATION SYSTEM AND SERVICES

3.5.1 The notion of innovation systems

Before we can discuss whether an innovation system exists in service, we have to define that concept.

By innovation system we mean a general pattern that can describe the innovation activities in a sector, in this case, the service sector. That a pattern exists means that certain elements are determining the innovations and the development of new ideas and innovations and their diffusion follows certain ways. If there is some repeated common characteristics of the pattern, we may call it a system. An innovation system can be institutional or loosely coupled.

An institutional innovation system is a coherent system with a series of relationships between different actors through which knowledge and ideas for innovations are diffused. The actors interact and through this system of interaction innovations are developed. It may also include a 'snow ball' effect - one new idea leads to the generation of another in the system. One can follow the diffusion process because it follows certain patterns. The system can be said to be institutionalised because the interrelationship between the actors often follows a certain pattern with long lasting relations and co-operation and often the relationship is formalised through contracts or well-known norms for co-operation. There are often fixed positions in the system and generally accepted norms for the relationships and interactions.

A loosely coupled innovation system is composed of certain actors, certain trajectories of development within important fields like technology, management, etc. and some forms of behaviour that are common to the sector in case. The constellation between the actors, the trajectories and the behavioural forms is not very fixed; it may take various forms. It is not an institutionalised system in the way that there are no fixed norms for behaviour and relations that everybody in the sector know and follow. The firms co-operate less with external actors than supposed in the institutional innovation system, but they relate to the external actors, i.e. these actors are important providers of input or purchasers of output from the innovation process or are competitors. The interaction may be as large, or even larger, than in the institutional system, it is just more competition oriented and not institutionalised. Furthermore, the output purchasers are more important than the input providers in this system compared to the institutional system. The diffusion process does not follow a straight line, but is complex with many informal, and often unobservable, elements, like intuitive ideas generated by one person, whose identity has been forgotten by everybody if you ask the actors some time later. The loosely coupled innovation system cannot be theoretically understood from a coherent, explanatory model as the institutionalised system can because of the loose coupling of all elements and non-fixed behavioural patterns and traditions. The actors, trajectories and major behavioural and interactional elements can nevertheless be described and some scientific rules or laws of the average behaviour and relationships formulated. Strategy is a social behaviour and the actors are social beings thus the innovation process and the interaction system follow sociological laws as do all other human groups.

The loosely coupled innovation system may be supposed to characterise a situation with hard market competition and weak common push elements such as a common scientific or technological basis of the production (a trajectory). This calls for more strategic game approach towards the market of the single firm and little co-operation with other actors outside the firm.

The institutionalised innovation system may be supposed to characterise a situation in which a sector has gone through a long history where the independently determined behavioural patterns of the single actors have been common and fixed, general norms have been established with a sanction system to ensure that all actors follow the institutionalised norms and rules. Although this may look like a description of an inefficient conservation system, this does not need to be the case. One can argue that an institution could be oriented towards change and creativity. However, there could be a tendency to a routinisation of the creativity thus changes follow certain paths or trajectories that can not be broken. Radical innovations might not appear in an institutionalised system.

3.5.2 Is there evidence of the existence of an innovation system in services?

Do the different patterns that we have described in section 3.4 constitute a system? To describe them as a system demands that two conditions should be fulfilled: 1. There is at least some coherence in these patterns; and, 2. There is some repetition in the patterns thus one or a few patterns are general.

The possible system could be institutionalised if the coherency is strong and there is only one pattern that is repeated very much. If there are several patterns and it cannot be predicted which pattern will appear in which situations and if the coherence in the patterns is weak, it is a loosely coupled system.

Thus, we have three possibilities concerning the innovation system in services: A. There is an institutionalised system; B. There is a loosely coupled system; and, C. There is no system at all.

Lack of coherence

The innovation activities are only coherent to a small degree in services and we would state as a hypothesis that it would be difficult to find a route of imitation where different actors have a mutual relation and the diffusion of new ideas and concrete innovations could be followed through several links. Even the trajectories are often not coherent systems, service firms are still not very scientific-professional based, so the service professional trajectory is often weak; innovations are still often quick, practical ideas.

The technology trajectory is also often weak if we discuss it as a coherent system that leads to a wave of innovations that generate each other. The innovation process and the introduction of new technology are still often unsystematic and are a result of internal trial-and-error decisions and not a consequence of any external system.

However, this is a matter of degree, the service professional and, to a less degree, the technology trajectories have been found to be the strongest patterns in services and those that have been most institutionalised. Furthermore, these are general statements about the total service sector; there are differences between different service industries. Thus, a generalisation to all service industries might be doubtful, but the service industries have some common characteristics (that we have briefly emphasized in this chapter) thus we propose overall models at the general level of this theoretical discussion. In empirical studies one must investigate the possible innovation systems that each industry or maybe firm is involved in.

The lack of coherence means that we can find only a loosely coupled system of service innovation and even that may, in some service fields, be weak. The service innovation process as a societal activity could be characterised as an anarchic market-based process. However, this is an exaggeration because more systematic relations already exist and there are developments towards formation of innovation systems that can already be observed as we shall demonstrate below.

Some repetition of a few patterns

As argued above, there are several patterns of innovation in the service sector, and not only one. They are not always repeated in the same firm or industry; a new pattern may be selected for a new innovation. Service firms are only moderately aware of innovation as a means for developing the firm - to get a better competition position on the market, and to grow in turnover and profit. This means that they are not very clear about how they want to organise and manage their innovation activities. The form of organization, which actors and/or trajectories they choose, and how the relationship with actors should be, becomes often a coincidental decision, determined by the actual situation.

Nevertheless, some patterns are more common than others. There is also a tendency for one pattern to become dominant in most types of services and firms. This is the case with regard to the organised strategic innovation pattern and, within specific service areas, to the service professional and artisanal patterns, although the domination of these patterns is less clear. This means that there is at least some repetition of one, or maybe two or three, innovation patterns.

3.5.3 Conclusion concerning innovation system

Our conclusion is that since there is some coherence and some repetition in the innovations patterns, and since a limited number of patterns are repeated and seem to characterise the main part of the innovation situations, we can say that there is a system.

Since there is only some coherence, some repetitions and there are several patterns, it is only a loosely coupled system.

3.6 CONCLUSION AND PERSPECTIVES

3.6.1 Conclusion on the innovation system

We can conclude that there is a system of innovation in services but it is a loosely coupled system and there is a variation of patterns within the system. The system is not a national system, and the varied and loosely coupled character of the system makes it difficult to use it as a basis for political regulation and stimulation.

Whether the service innovation system in the future will be more institutionalised is difficult to say. Since it will still be characterised by a large variety of relations between trajectories and actors, and by many widespread interaction situations between actors, it is mostly likely that it will remain at least less institutionalised than the manufacturing system. This may not, however, be a disadvantage to the service sector. On the contrary, it may create a more dynamic innovation system that even manufacturing could learn from (and which, according to our hypothesis stated below, it will).

3.6.2 Convergence between service and manufacturing innovation systems

The service innovation system is different from the manufacturing innovation system, but one can assume that the service and manufacturing systems will converge in the future. There is some empirical evidence for stating such a hypothesis.

In the current post-fordistic period, manufacturing innovations get traits from the service innovation system: a heavy customer and market orientation, less standardised and more flexible products and production organization, mainly dominated by incremental innovations. The employees may be supposed to get a more central role as corporate entrepreneurs even in manufacturing.

Service innovations are moving in the direction of the manufacturing system in some ways; it seems to become more systematic, more technology is involved, and the service innovations may be supposed to be more push determined through R&D, although the sciences on which the service innovations will be based, are more social sciences than natural and technical sciences.

However, the service innovation system keeps some of its own elements:

- The customer encounter as core driver
- Many small, non-reproduced changes
- Person-to-person contact (non-technological) will remain a core characteristic for much service innovation
- A relatively loosely-coupled organization system, characterised by less R&D, more corporate entrepreneurship, strategic guidance, and service professional trajectories.

3.6.3 Contribution to evolutionary theory

Evolutionary economics, which is particularly interested in interaction-intensive economic phenomenon, is obviously at the heart of our analysis. It offers promising ideas to cope with innovation in services. Indeed there is, for example, a kind of proximity or similarity between the nature of the service and the way one could say that evolutionary economics defines innovation, or should define it (since it is difficult to make general statements on evolutionary economics).

Both are a process, an act and not merely a result. Both are interactive and both have some difficulties to follow optimising principles. We will now discuss how the attempt to define and find an innovation system in services could contribute to develop evolutionary economic theory

The models above (Section 3.3.3) are a proposal for a model of the organised and complex innovation systems that follows after the breakdown of the entrepreneur model. The most dominant pattern in the service innovation system is the organised strategic pattern. It emphasises the dualism of corporate entrepreneurship of the employees and the management who induce and control the innovation process within the framework of the strategy. This model is within the theoretical framework of the strategic innovation theory (Sundbo, 1998), which points to this dual innovation organization as important, and the firm's strategy as the framework for the management's decision and inducement.

This model might be more dynamic and more valid than the technological R&D model that has been discussed as the proposal for a Schumpeter II model. Further research will be necessary to show that.

The above structuring of a service innovation system can thus contribute to develop evolutionary theory in a way that offers a version of the determining and structuring mechanisms in the innovation process and thus in the evolution of the production system of the economy. This version may include elements, and constellation of these, that have not until now been stressed in evolutionary theory. Compared to Pavitt's general sectoral taxonomy (1984) or to Soete and Miozzo's service innovation taxonomy (1990), our model does not focus on technological trajectories alone, but it also takes into account several other trajectories: service-professional, managerial, social, and institutional.

REFERENCES

Barcet, A., Bonamy, J. and Mayère, A. *Modernisation et Innovation dans les Services aux Entreprises (Modernisation and Innovation in Business Services)*, Report for Commissariat Général du Plan, 1987.

Barras, R. 'Interactive Innovation in Financial and Business Services: The Vanguard of the Service Revolution', *Research Policy*, 1990; 19, No. 3, pp.215-37.

Barras, R. 'Towards a Theory of Innovation in Services', *Research Policy*, 1986; 15.

Cohen, S. and Zysman, J. *Manufacturing Matters*, New York: Basic Books, 1987.

Dosi, G. 'Technological Paradigms and Technological Trajectories: A Suggested Interpretation of the Determinants and Directions of Technical Change', *Research Policy*, 1982; 11.

Edvardsson, B., Haglund, L. and Mattsson, J. 'Analysis, Planning, Improvisation and Control in the Development of New Services', *International Journal of Service Industry Management*, 1995; 6, No. 2, pp.24-35.

Edvardsson, B., Thomasson, B. and Ovretveit, J. *Quality of Service*, London, 1994.

Gadrey, J. *L'Économie des Services (The Service Economy)*, La Découverte, 1992.

Gadrey, J. 'La Modernisation des Services Professionnels: Rationalisation Industrielle ou Rationalisation Professionnelle (The Modernisation of Professional Services: Industrial Rationalisation or Rrofessional Rationalisation)', *Revue française de sociologie*, 1994; XXXV, pp.163-195.

Gadrey, J. and Gallouj, F. 'The Provider-Customer Interface in Business and Professional Services', *The Service Industries Journal*, 1998; April.

Gadrey, J., Gallouj, F. and Weinstein, O. 'New Modes of Innovation: How Services Benefit Industry', *International Journal of Service Industry Management*, 1995; Vol. 6, No. 3, pp.4-16.

Gallouj, C. and Gallouj, F. *L'Innovation dans les Services (Innovation in Services)*, Paris: Editions Economica, 1996.

Gallouj, F. *Economie de l'Innovation dans les Services (Economics of Innovation in Services)*, Paris: Editions L'Harmattan, Logiques économiques, 1994.

Gallouj, F. 'Le Processus de Production de l'Innovation dans les Services de Conseil (The Process of Innovation in Consultancy Services)', *Revue Française de Gestion*, 1995.

Gallouj, F. 'Towards a Neoschumpeterian Theory of Innovation in Services',*Science and Public Policy*, 1997; Vol. 24, No. 6, December.

Gallouj, F., Djellal, F. and Gallouj, C. *Vers une Théorie de l'Innovation dans les Services, (Towards a Theory of Innovation in Services)*, Report for Commissariat General du Plan, July, 1997.

Gallouj, F. and Weinstein, O. 'Innovation in Services', *Research Policy*, 1997; 26, pp.537-556.

Illeris, S. *The Service Economy: A Geographical Approach*, Chichester , 1996.

Kline, S. and Rosenberg, N. 'An Overview of Innovation', in Landau, R. and Rosenberg, N. (eds), *The Positive Sum Strategy: Harnessing Technology for Economic Growth*, Washington DC: National Academic Press, 1986.

Kubr, M. *Management Consulting: a Guide to the Profession*, Geneva: ILO, 1988.

Martin, C.R. and Horne, D.A. 'Service Innovation: Successful Versus Unsuccessful Firms', *International Journal of Service Industry Management*, 1995; 4, No. 1, pp.49-65.

Miles, I. 'Services in the New Industrial Economy',*Futures*, 1993; Vol. 25, July/August, pp.653-72.

Miles, I., Kastrinos, N., Flanagan, K., Bilderbek, R., den Hertog, P., Huntink, W. and Bouman, M. *Knowledge-Intensive Business Services: Their Role as Users, Carriers and Sources of Innovation*, PREST, University of Manchester, 1994.

Normann, R. *Service Management*, London, 1991.

Pavitt, K. 'Sectoral Patterns of Technical Change: Towards a Taxonomy and a Theory',*Research Policy*, 1984; 13.

Saviotti, P.P. and Metcalfe, J.S. 'A Theoretical Approach to the Construction of Technological Output Indicators', *Research Policy*, 1984; 13, pp.141-151.

Soete, L., and Miozzo, M. *Trade and Development in Services: a Technological Perspective*, MERIT, The Netherlands, 1990.

Sundbo, J. 'The Tied Entrepreneur', *Creativity and Innovation Management*, 1992; Vol. 1, No. 3, pp.109-120.

Sundbo, J. 'Modulization of Service Production and a Thesis of Convergence Between Service and Manufacturing Organizations', *Scandinavian Journal of Management*, 1994; Vol. 10, No. 3, pp.245-266.

Sundbo, J. 'Balancing Empowerment', *Technovation*, 1996; 16, No. 8, pp.397-409.

Sundbo, J. 'Management of Innovation in Services',*The Service Industries Journal*, 1997; 17, No. 3, pp.432-55.

Sundbo, J. *Innovation Theory: Three Paradigms*, Cheltenham, Glous, 1998.

Sundbo, J. and Gallouj, F. *Innovation in Services: The Results of Work Package 3-4 of the SI4S Project*, 1998.

Van Der Aa, W. and Elfring, T. *A Typology of Innovations in Service Firms*, Conference on 'Managing Innovations in Services', Cardiff Business School, 5-7 April 1993.

---

[1] In the SI4S project national reports on innovation in services have been produced by the following research teams: Deutsches Institut für Wissenschaftforschung, Berlin, Germany (Brigitte Preissl); TNO, Centre for Technology and Policy Studies, The Netherlands (Pim den Hertog and Rob Bilderbeek); STEP group, Oslo, Norway; NUTEK Analys, Stockholm, Sweden; PREST, University of Manchester, UK (Paul Windrum, Kieron Flanagan and Mark Tomlinson); Roskilde University, Centre of Service Studies, Denmark (Jon Sundbo); and, IFRESI, Université de Lille 1, France (Faïz Gallouj, Faridah Djellal and Camal Gallouj).

[2] The basis for the driving force scheme has been the reports from the national teams of the SI4S projects on the service innovation situations in different countries. The scheme thus is a result of the SI4S work and we owe thanks to the national teams

Chapter 4
Recombination and the Production of Technological Knowledge: Some International Evidence[1]

Cristiano Antonelli

4.1 INTRODUCTION

Acquisition of external knowledge in disembodied technology markets and its recombination with internal research and development activities and competencies is emerging as an effective way to increase the total amount of technological knowledge each company and country can generate. Transactions in technological knowledge take place among industrial firms as well as between industrial firms and Knowledge-Intensive Business Services specializing in the generation of technological knowledge to be delivered and sold to third parties.

The growing trade in disembodied technology is increasingly horizontal in that it reflects the growing specialization of firms within countries and countries at large in limited technological fields. Much advance in technological and scientific knowledge is possible nowadays only by relying upon the complementarity among bits of technological knowledge generated by specialists and eventually implemented or recombined by knowledge-assemblers. The cumulative character of the scientific and technological knowledge generation process makes technological specialization and outsourcing even more necessary (Archibugi and Pianta, 1992; Archibugi and Michie, 1997, 1998).

Recombination, assembling, specialization and division of labour are now spreading into the organization of the production of knowledge after much experience in the production of manufactured goods. Increasing shares of R&D activities are conducted within each company but find their economic application elsewhere either in plants operated abroad by multinational companies, in other firms with which arm's length transaction schemes have been implemented, or by independent firms which buy external knowledge (Gibbons, 1994). A large body of empirical evidence at the company level confirms that external knowledge is becoming an important source of technology, and that it is a key input in the process generating technological change and productivity growth (Arora and Gambardella, 1990, 1994).

This dynamics can be considered a part of a broader process where the generation of technological knowledge is itself becoming closer to the production of goods. Exchange of technological knowledge takes place because the conditions for appropriability are now far better than currently assumed by a large traditional literature. This evidence requires a theoretical assessment and provides the background to appreciate empirically the role of external knowledge in growth accounting.

The international dimension of disembodied technological trade is gaining momentum. International specialization of countries and international technological trade are two facets of the same coin, one where an international market for technological knowledge is being implemented and developed, partly by means of new remote applications of information and communication technologies (Arora, 1995).

The chapter is structured as follows. Section 4.2 provides the general theoretical analysis about the nature of technological knowledge whether a fully public good or a semi-private good and relates this discussion to the basic distinction between pecuniary technological externalities and knowledge technological externalities. This discussion is finalized to specify the basic hypotheses. Section 4.3 gathers the empirical evidence; it provides the descriptive evidence about the technology balance of payments and focuses attention on the role of technology payments with respect to domestic research and development expenditures. The econometric tests of a technology production function, which has been re-specified so as to take into account explicitly the role of external knowledge, are also presented in this section. In the conclusion the general results are summarized and put in perspective.

4.2 THEORETICAL CONTEXT OF THE ANALYSIS AND THE HYPOTHESES

Standard Arrovian microeconomics suggests that technological knowledge, like information should be considered a public good, in that its use is non-excludable and non-rival and its production and use are characterized by high levels of indivisibility; it can also be easily transferred and learnt at little cost. Because it cannot be properly appropriated by innovators, it cannot be traded without disclosure; hence its value is difficult to assess. Conversely, however, the actual flow of technological information is considered to be a spontaneous aspect of economic systems. Intellectual property rights can increase appropriability but reduce the scope for the socialization of innovation benefits (Arrow, 1962, 1969; David, 1992; Geroski, 1995).

A recent surge of empirical and theoretical interest in technological spillovers at both the domestic and the international levels builds upon this line of analysis (Griliches, 1992).

According to this approach, technological knowledge generated by each firm and country cannot be fully appropriated by the innovator but 'spills' into the environment and generates important externalities. Other innovators can take advantage of the knowledge available in the economic environment without any dedicated market transaction: no exchange takes place and technology transfer is realized without any contractual interaction and no prices are actually paid. Other factors, such as proximity, complementarity, social interaction, matter. Clearly the analyses of such knowledge externalities are consistent and coherent with the Arrovian tradition of analysis of technological knowledge.

This approach provides the theoretical framework underpinning the empirical analyses, recently expanded to the international aspects, of technological spillovers. Such empirical analyses try and measure the effects of the technological knowledge generated elsewhere on the rates of technological change and productivity growth experienced by each country. Traditionally this approach consists in some growth accountancy where the stock and flow of research and development expenditures conducted in the rest of the world are related to the output growth of country i together with the domestic research and development expenditures conducted in country i. Along this line of empirical analysis some attention has been paid to the proper 'filter' to apply in order to catch the correct amount of technological knowledge spilling. Flows of trade have been considered a reliable indicator of the exposure and commercial proximity of each country to the technological opportunities generated by each other country (Park, 1995; Coe and Helpman, 1995; Engelbrecht, 1997; Frantzen, 1998)[2]. A more technology-aware effort has recently been made by Verspagen (1997) which attempts to estimate the actual effects of technological spillovers using a notion of technological proximity based upon technology flows matrices.

Within the class of technological externalities, that is complementarities among agents due to imperfect divisibilities among technologies to be developed, it seems useful to stress again the traditional distinction between pecuniary and technical externalities. This distinction has been further elaborated by Griliches (1992) between rent or pecuniary technological externalities and knowledge technological externalities. The empirical approach so far considered is clearly referred to knowledge technological externalities. The implicit assumption is in fact that knowledge technological externalities apply because of the public good nature of technological knowledge, its low levels of appropriability and non-excludability.

This traditional view of technology, as information, is being increasingly challenged by recent developments of the Schumpeterian approach which stress the distinction between information and knowledge; for information is an input in the production of knowledge. A growing number of authors, working in this tradition, have challenged the traditional view of technology as a pure public good, arguing that technological knowledge especially, as distinct from scientific knowledge, has a strong proprietary character. In fact it is largely excludable and its use if partly rivalrous. In particular, technological knowledge tends to be localized in well-defined national, technical, institutional, regional and industrial situations: it is specific to each country, industry, region, firm and individual or team of individuals and consequently costly to use elsewhere.

Specifically the notion of localized technological knowledge contrasts with the Arrovian notion of generic knowledge and elaborates an approach where technological knowledge is considered a quasi-private good with high levels of appropriability and excludability, generated by a process characterized by cumulativeness and path-dependence. The localized character of technological knowledge increases its appropriability but reduces its spontaneous circulation in the economic system. In fact, each unit of technological knowledge can be created, used and exchanged only by means of specific competencies acquired by firms, now viewed as learning organizations, within a larger framework which includes a whole array of complementary and inter-related units of knowledge generated by other firms, universities and research institutions (Antonelli, 1999).

The capability to innovate appears to be strongly conditioned both by access to available technological information and learning opportunities, and by the accumulation of tacit knowledge both internal and external to each firm. Technological information, the competence of each firm and the technological communication conditions within technological and regional innovation systems, represent the three basic inputs in the process of the creation of new knowledge. New technological knowledge is the result of a complex process of creation of new information building upon the mix of competence acquired by means of learning processes, the acquisition and recombination of external technological and formal internal R&D activities (Antonelli, 1995; David, 1975; Atkinson and Stiglitz, 1969).

External knowledge plays a major role in this context. External knowledge contributes to the internal production of knowledge by means of the recombination of bits of technological information which are re-organized and applied to different contexts than those originally conceived, and often implemented with formal co-operation between firms with their own R&D laboratories or between firms and universities.

Rarely can external knowledge, even when it consists mainly of the results of scientific 'top-down' undertakings, be reduced to a simple set of instructions which can be acquired without dedicated interaction with the firm producers. Both codified and tacit technological knowledge can be acquired only by means of direct and intimate relationships between researchers. Since technological knowledge tends to be highly specific and is embedded into the background and experience of each innovator and hence highly idiosyncratic, it is costly, if not impossible, to use elsewhere. In this context, communication conditions become very important. The transfer and adaptation of localized technological knowledge from one country, industry, region or firm to another involves specific costs which are affected by the quality and effectiveness of the communication channels within the innovation system.

External knowledge, in this approach, in order to be successfully assimilated and implemented, cannot flow in the air and cannot spill. External technological knowledge, without dedicated user-producer relations, is very expensive in order to be actually implemented: re-engineering costs may amount to a substantial share of the original research costs (Cohen and Levinthal, 1989, 1990).

In this approach, however, because of its semi-private character, technological knowledge can be traded and exchanged in market places, especially when implemented by an institutional set-up which makes the exchange more reliable for both parties. Arm's length transactions implemented by long-term contracts, technological clubs, joint-ventures, equity-agreements, direct foreign investments and interlocking directorates can help the establishment of international markets for technological knowledge (Williamson, 1985; Archibugi and Michie, 1997).

In this context, new information and communication technologies, specifically a variety of on-line control systems by means of international telematic links, make it possible to better enforce the application and use of the knowledge sold and effectively reduce the appropriability limitations of technological knowledge so that a market for it can gradually emerge. The direct telematic assistance of the sellers to the customers and users of external technological knowledge, within a contractual transaction framework, is a further necessary condition for the exchange to take place successfully. Customers can better acquire the external knowledge and sellers can secure and further enforce the appropriability conditions setting prices which are partly related to the effective eventual use of the knowledge which has been sold (Antonelli, 1998; Gibbons, 1994).

This analysis has important implications for assessing (from both a social and private viewpoint) the organization of knowledge production and its economic effects. In fact, the strong complementarity between internal and external sources of knowledge stresses the importance of the conditions of trade, communication and access to external knowledge.

The new conditions to accessing and making a direct and assisted use of the disembodied technological knowledge generated by third parties enhance the general productivity of internal research and development activities. Recombination with complementary sources of codified knowledge already elaborated in other contexts becomes an important source of additional knowledge. Internal R&D activities can be better focused on the implementation and recombination of scientific information offered by contiguous fields which, when properly combined with the competencies and technological capabilities accumulated by each firm, may translate into new opportunities to introduce innovations directly relevant to the products currently manufactured (David, 1992; Gibbons, 1994).

The systemic access to and use of external codified knowledge eventually integrated with both tacit knowledge, accumulated by means of learning processes, and R&D activities is likely to affect strongly the efficiency of internal research and development activities (Nelson, 1992).

Access to external tacit and codified knowledge depends on the extent to which effective communication among innovators is available through the system, influencing both the research process itself and its outcome. The properties of innovation systems, perceived as network structures through which technological communication flows, are therefore relevant in explaining the capability to generate new technological knowledge.

Technological information, in fact, does not flow automatically within an economic system but requires the understanding and application of specific knowledge. In order to acquire external knowledge, intentional efforts and actual transactions between parties are necessary. The active participation of customers is required; the assistance of sellers is also necessary. The assimilation of external knowledge demands specific actions which depend on the individual context in which firms operate in each country.

According to this line of analysis, it is clear that pecuniary technological externalities matter rather than knowledge technological externalities. The purchase of technological knowledge from third parties and the related technological outsourcing and technological specialization which technological trade makes possible, have clearly positive effects in terms of the general efficiency of the innovative process. Pecuniary technological externalities can be considered to exert a direct positive effect upon firms which can concentrate on the limited technological fields where they have relevant competence and rely upon external knowledge for complementary technologies. Reduction of redundancies and duplications are also positive outcomes of the new opportunities for technological specialization engendered by the new international trade in technological knowledge.

According to our hypotheses, the production of localized technological knowledge by firms in country i can be formalized as the result of the interaction between internal research and learning activities and the creative access to external technological knowledge made available by vendors within specific contractual obligations among which is a price. These arguments enable us to specify a production function for knowledge as follows:

$$(1)\quad LTK_i = f(R\&D\&L_i, EXTERNAL\ KNOWLEDGE_{i,})$$

where LTK_i is the localized technological knowledge produced by the firms in country i; R&D&L are the resources invested in learning and R&D activities by the firms in the same country; EXTERNAL $KNOWLEDGE_i$, are the resources invested by the firms in each country in acquiring external knowledge. The innovative output LTK is generated by the interaction between intramural resources dedicated to research and learning and the external knowledge flows.

Equation (1) can now be used to better specify the standard technological production function[3] which makes explicit the role of external technological knowledge into the production of goods:

(3) Y = (K, L, R&D&L, EXTERNAL KNOWLEDGE)

where Y is the aggregate output of each country, K is the real stock of physical capital, L is the labour inputs, R&D&L is the real stock of resources invested in domestic and internal research and development and learning activities and EXTERNAL KNOWLEDGE is the stock of the resources invested in acquiring on international technological markets the technological knowledge generated in other countries. The payments of the technological balance of payments seem to be a reliable indicator for the actual amount of EXTERNAL KNOWLEDGE.

According to our hypotheses and related re-specification of the technology production function, the actual levels of output a country is able to generate, under the standard control of the levels of the stocks of physical capital and labour, depends upon both the cumulated levels of both the domestic expenditures in research, development and learning activities and the imports of disembodied technological knowledge acquired on international technology markets.

4.3 THE EVIDENCE AND THE TESTS

Recent statistical work by OECD and other national statistical institutes has made available an interesting and reliable body of data about international transactions in disembodied technological knowledge. The technological balance of payments is built upon the records about international technological transactions in terms of technology payments and technology receipts among a large number of advanced countries[4].

According to such data, international markets for disembodied technology are growing very fast. Actually, through the 1990s, international transactions in technological knowledge have been growing faster than domestic expenditures on research and development activities.

Data show that the technology payments (TP) represent a significant share of total expenses in research and development activities (GERD) in most countries. TP have an order of magnitude very close to that of the research activities funded by the business sector in the main OECD countries (BERD) and BERD shows that most countries are converging towards a common level in the vicinity of 30% of BERD.

Traditional R&D-intensive countries, such as United Kingdom, Germany and United States, exhibit a clear trend towards an increasing use of external knowledge. The ratio of TP to BERD has been steadily increasing through the 1980s and 1990s in Germany from less than 15% to 34%. In the United Kingdom it grew from less than 15% in the early 1980s to 22% in 1994. In the same time interval in the United States it has increased from 1% to almost 5%. France remains stable around a 15% ratio. Countries with lower levels of BERD/GDP intensity, such as Belgium, show an increasing trend reaching 125% in 1994 from the 72% levels of 1981. Italy remains around the 30% level for all the period considered. Canada, in the vicinity of 25% in 1981, shrinks to 16% in 1994. Countries like Spain are instead reducing their dependence upon external knowledge: the ratio of TP to BERD shrinks from a ratio of 150-160% towards the 50% levels. Finally, Japan, once a strong importer of foreign technology, reduces the ratio of TP to BERD from 7% in 1981 to 4% in 1994[5].

The integration of the payments for the procurement of external knowledge, at the firm as well as the aggregate levels, into the total amount of resources invested in the search for new technological knowledge could help reduce the substantial gap in the growth accountancy of technological change. The econometric evidence in fact suggests that a major inconsistency exists between inputs and outputs, that is, funds invested in the generation of technological knowledge, traditionally limited to research and development expenditures, and productivity growth. For the latter exhibits an output elasticity in the vicinity of 30%, an order of magnitude larger than the ratio of R&D expenses to GDP.

The econometric equation, in logarithmic form, reads as follows:

$$(4)\quad Y = aL + bK + cGERD + dTPK + e$$

The econometric test has been conducted after transformation of flow data on GERD (and BERD) and TP into stocks with the standard capitalization procedure based upon a 3 year sum assuming standard depreciation rates[6]. Hence GERD (and BERD) and TP respectively are our proxies for the capitalized stock of research and development activities and the external knowledge for which the technological balance of payments has a record. Data for physical capital and labour are also drawn from the OECD data bank. Y is the standard output in terms of real gross domestic product. All variables enter the econometric equation in logarithmic form.

Available data have been organized in a panel framework based upon 10 countries[7] in cross sections on a 12 year time series length. Preliminary inspections have detected first-order autoregressive error structure with contemporaneous correlations between cross-sections in the data set. Hence the data gathered tested equation (4) with the Parks Method Estimation, executed by the SAS System. The results follow (t of Student between parenthesis):

$$(5)\quad Y = \underset{(28.103)}{0.401\, L} + \underset{(20.552)}{0.277\, K} + \underset{(11.470)}{0.259\, GERD} + \underset{(22.509)}{0.269\, TP}$$

The variance component estimates are: SSE = 118.8642; MSE = 1.0246; RSQ = 0.999; DFE = 116; Root MSE = 1.0122. The RHO of the First Order Autoregressive Parameter Estimates run from 0.997786 for State 1 to 0.891419 for State 10.

Equation (4) has been tested also with capitalized BERD figures instead of GERD ones. The results read as follows (t of Student between parentheses):

(6) $Y = 0.411\ L + 0.287\ K + 0.245\ BERD + 0.278\ TP$
(28.983) (20.762) (12.670) (21.009)

The variance component estimates are: SSE = 116.4665; MSE = 1.0398; RSQ = 0.998; DFE = 116; Root RSE = 1.0176. The RHO of the First Order Autoregressive Parameter Estimates run from 0.995676 for State 1 to 0.899829 for State 10.

Finally, equation (4) has been tested without the proxy for EXTERNAL KNOWLEDGE, that is without the capital figure for TP, in order to obtain a benchmark. The result follows (t of Student between parentheses):

(7) $Y = 0.411\ L + 0.287\ K + 0.363\ GERD$
(28.983) (20.762) (12.670)

The variance component estimates are: SSE = 113.2265; MSE = 1.0668; RSQ = 0.997; DFE = 116; Root MSE = 1.0165. The RHO of the First Order Autoregressive Parameter Estimates run from 0.993246 for State 1 to 0.892229 for State 10.

These results are highly satisfactory on many counts. First, the stock of TP is statistically significant: the estimated parameter enters equations (4) and (5) with a probability of 0.0001. Second, and most important, the size and magnitude of the estimated parameter is as large as that of the traditional R&D capital stock figures (i.e. both capitalized GERD and BERD). Thirdly, the estimates of equation (4) without the capital figure for TP, our proxy for EXTERNAL KNOWLEDGE, show a far larger output elasticity of the traditional research and development expenditures (capitalized BERD in this case). This confirms that the inclusion of a proper variable for an important input in the knowledge-generation process, such as the payments to acquire external knowledge, reduces to more reliable values the estimated output elasticity of domestic research and development expenditures.

In sum, our results confirm that external knowledge, acquired on international technological markets, plays a major role in explaining output growth together with traditional domestic research and development expenditures[8]. The flows of external knowledge purchases on international markets and measured by the technological balance of payments exhibit a significant output elasticity on their own count and contribute to output growth with an effect close to that of traditional research and development activities conducted internally by the business sector and at the country level at large including the public sector.

4.4 CONCLUSIONS

The distinction between pecuniary technological externalities and knowledge technological externalities seems important and deserved much attention. Technological knowledge is far more embedded and localized than currently assumed in the Arrovian tradition. The free mobility and circulation of technological knowledge in the 'air' is far less easy than currently assumed in the spillover literature. Much recent empirical evidence has focused attention on international R&D spillovers as if countries could benefit from foreign R&D activities without actually purchasing technological knowledge and without any actual participation of sellers in the technological transfer.

The growing attention to knowledge technological externalities is also justified by the need to elaborate an interpretative framework for the huge inconsistency between the output elasticity of research and development expenditures, usually estimated in the order of 0,30 (Griliches, 1992), and the share cost of R&D on GDP, usually lower than 3%. Beyond the limitations of equilibrium analysis applied to a domain where disequilibrium processes are clearly relevant, such inconsistency can also be related to the limitations of the traditional focus on R&D expenses, one of the activities directed towards the generation of new technological knowledge, but by no means the single source of technological knowledge. The lack of reliable indicators for the actual amount of learning activities and related competence involved in such a process still limits our capability to properly establish the basic growth accountancy of technological knowledge (Nelson, 1992; Carlsson and Eliasson, 1994). So far the explicit inclusion of foreign and external knowledge seems a useful step in the attempt to enlarge our understanding and measures of the actual inputs in the generation of new technological knowledge.

The notion of localized technological knowledge suggests in fact that the transfer of technological knowledge is not a spontaneous 'atmospheric' process. It can take place successfully only if both parties, vendors and customers, are actively involved and an actual transaction takes place. International (as well as domestic) markets for embodied technological knowledge are also emerging for the diffusion of new information and communication technologies and the related changes in the organization of the production of knowledge. Much evidence gathered on international (knowledge) technological spillovers risks to reflect the actual flows of technological transactions in international marketplaces. In other words, a major mis-specification problem may arise in that approach because of the omission of a key variable with evident effects of double-counting.

The analysis conducted in this chapter has highlighted the important role of external technological knowledge purchased by each country on international markets. Firms in each country have been able to take advantage of the opportunity offered by new emerging international markets for technological knowledge to complement their internal research and development expenditures and recombine external knowledge with internal competence and formal research activities. The supply of external knowledge clearly makes it possible to acquire inputs at lower costs than those generated internally: pecuniary externalities are at play. Technological outsourcing from firms specializing in the provision of technological knowledge as well as from other 'industrial' firms willing to sell disembodied knowledge can become an important source of technological knowledge and thus adds to the general efficiency of the generation of new knowledge.

A new framework with important poicy implications becomes evident in conclusion. Until now, the economic importance of formal intramuros R&D conducted by firms and scientific activities conducted by universities, as the unique and single input into the generation of technological knowledge has been exaggerated. As a consequence, too much emphasis has been put upon R&D policies and, more generally, science policies as the basic tools to sustain the rates of accumulation of new knowledge. In the generation of new technological innovations, firms rely on external knowledge acquired by means of formal interactions between themselves, sharing learning opportunities and experience, and with other established sources of knowledge.

Such interactions range from internal transactions within multinational corporations, to arm's length transactions implemented by exchanges of devices such as long-term contracts, equity swaps, interlocking directorates, technological clubs and formal processes of technological co-operation, and purchase of technological assistance from Knowledge-Intensive Business Service firms which often have a global scope of action. Outsourcing of research activities and the procurement of Knowledge-Intensive Business Services play an increasing role in assessing the innovative capabilities of each firm.

The levels of technological procurement from other 'industrial' firms and the outsourcing of Knowledge-Intensive Business Services should be accounted for when assessing the actual amount of inputs invested in the process of research and learning. The procurement of technological knowledge from third parties, both at the international level and within countries, is an important component of the general process of accumulation of new knowledge.

Public policy interventions to sustain the acquisition of external technological knowledge and to implementing both domestic and international technology markets may be especially successful. Outsourcing of Knowledge-Intensive Services and the purchase of technological knowledge from other industrial firms could become an important recipient for policy interventions. The useful distinction between innovation policies and diffusion policies, introduced by Ergas (1987) may be extended to external disembodied knowledge and generally technological outsourcing.

REFERENCES

Antonelli, C. (ed.). *New Information Technology and Industrial Change*, Norwell, Massachussets: Kluwer Academic Publishers, 1988.

Antonelli, C. *The Economics of Localized Technological Change and Industrial Dynamics*, Norwell, Massachusetts: Kluwer Academic Publishers, 1995.

Antonelli, C. T*he Microdynamics of Technological Change*, London: Routledge, 1999.

Archibugi, D. and Michie, J. (eds.). *Technology Globalisation and Economic Performance*, Cambridge: Cambridge Univeristy Press, 1997.

Archibugi, D. and Michie, J. (eds.). *Trade Growth and Technical Change*, Cambridge, Cambridge University Press, 1998.

Archibugi, D. and Pianta, M. *The Technological Specialization of Advanced Countries*, Boston: Kluwer Academic Publishers, 1992.

Arora, A. 'Licencing Tacit Knowledge: Intellectual Property Rights and the Market for Know-How', *Economics of Innovation and New Technology*, 1995; 4, pp.41-60.

Arora, A. and Gambardella, A. 'Internal Knowledge and External Linkages: Theoretical Issues and an Application to Biotechnology', *Journal of Industrial Economics*, 1990; XXXVIII, pp.361-379.

Arora, A. and Gambardella, A. 'The Changing Technology of Technological Change. General and Abstract Knowledge and the Division of Innovative Labour',*Research Policy*, 1994; 23, pp.523-532.

Arrow, K.J. 'Economic Welfare and the Allocation of Resources for Invention' in Nelson, R.R. (ed.), *The Rate and Direction of Inventive Activity: Economic and Social Factors*, Princeton: Princeton University Press for N.B.E.R., 1962.

Arrow, K.J. 'Classificatory Notes on the Production and Transmission of Technical Knowledge', *American Economic Review*, 1969; P&P 59, pp.29-35.

Atkinson, A.B. and Stiglitz, J.E. 'A New View of Technological Change',*Economic Journal*, 1969; 79, pp.573-578.

Carlsson, B. and Eliasson, G. 'The Nature and Importance of Economic Competence',*Industrial and Corporate Change*, 1994; 3, pp.687-712.

Coe, D.T. and Helpman, E. 'International R&D Spillover',*European Economic Review*, 1995; 39, pp.859-887.

Cohen, W.M. and Levinthal, D.A. 'Innovation and Learning: The Two Faces of R&D'*Economic Journal*, 1989; 99, pp.569-596.

Cohen, W.M. and Levinthal, D.A. 'Absorptive Capacity: A New Perspective on Learning and Innovation', *Administrative Science Quarterly*, 1990; 35, pp.128-152.

David, P.A. *Technical Choice Innovation and Economic Growth*, Cambridge: Cambridge University Press, 1975.

David, P.A. 'Knowledge Property and the System Dynamics of Technological Change', *Proceedings of the World Bank Annual Conference on Development Economics*, Washington: The World Bank, 1993.

Engelbrecht, H-J. 'International R&D Spillover Human Capital and Productivity in OECD Economies: An Empirical Investigation', *European Economic Review*, 1997; pp.147y9-1488.

Ergas, H. 'Does Technology Policy Matter?' in Guile, B.R. and Brooks, H. (eds.),*Technology and Global Industry: Companies and Nations in the World Economy*, Washington: National Academy Press, 1987.

Frantzen, D. 'The Influence of Domestic and Foreign R&D on Countries' Total Factor Productivity', *Economie Appliquée*, 1998; 51, pp.131-155.

Geroski, P. 'Markets for Technology: Knowledge Innovation and Appropriability' in Stoneman, P. (ed.), *Handbook of the Economics of Innovation and Technological Change*, Oxford: Basil Blackwell, 1995.

Gibbons, M. et al. *The New Production of Knowledge*, London: Sage Publications, 1994.

Griliches, Z. 'Issues in Assessing the Contribution of Research and Development to Productivity Growth', *Bell Journal of Economics*, 1979; 10, pp.92-116.

Griliches, Z. 'The Search for R&D Spillover', *Scandinavian Journal of Economics*, 1992; 94, Supplement, pp.29-47.

Nelson, R.R. 'The Role of Knowledge in R&D Efficiency',*Quarterly Journal of Economics*, 1987; 97, pp.453-470.

Nelson, R.R. (ed.). *National Systems of Innovation*, Oxford: Oxford University Press, 1993.

Park, W. 'International R&D Spillover and OECD Economic Growth',*Economic Inquiry*, 1995; 33, pp.571-591.

Verspagen, B. 'Estimating International Technology Spillovers Using Technology Flow Matrices', Weltwirschaftliches Archiv, 1997; p.133.

Williamson, O.E. *The Economic Institutions of Capitalism*, New York: The Free Press, 1985.

---

[1]The assistance of Martin Marchesi in data collection and statistical analysis is gratefully acknowledged.

[2]Along similar lines however other indicators could complement trade: telecommunications traffic, linguistic homogeneity, education of skilled manpower can measure the proximity, hence exposure of each country to the technological environment of each other country and thus the likelihood that international knowledge spillovers can actually take place without the active participation of vendors.

[3]The substitution of equation (1) into standard specification of the technological production function:

$$(2) \quad Y = (K, L, T)$$

where T stands for the stock of inputs used in the generation of new technological knowledge leads, in fact, to equation (3). T is traditionally measured in international growth accounting exercises by means of the domestic R&D expenses, that is BERD, PERD or GERD.

[4]All relevant data are found in OECD STAN data bank.

[5]The analysis of the magnitude of TP with respect to BERD and GERD suggests that TP should be considered an essential component, together with BERD and PERD (Public Expenses in R&D; for PERD=GERD-BERD) of the total amount of resources invested by each country in the generation of new technological knowledge. The ratio of TP to BERD and GERD in this respect seems far more interesting than the traditional focus upon the so-called 'coverage rate' i.e. the ratio of payments to receipts of the technological balance of payments.

[6]Following standard procedures, the stock for physical capital has been built out of investments with a yearly 20% depreciation rate. The capital stock for both domestic and international technological variables has been built using respectively GERD, BERD and TP over a three year time period with a yearly 33% rate of depreciation.
[7]The countries considered include Germany, France, United States, United Kingdom, Canada, Austria, Belgium, Italy and Spain and have been sorted out from the OECD STAN data bank according to the length of the time series available.
[8]It seems clear that similar analyses should be implemented at the domestic level in order to appreciate the amount of resources invested by firms to acquire external knowledge from other domestic firms and their economic effects.

PART II:
MEASURING SERVICE INNOVATION

Chapter 5
Innovation, Measurement and Services: The New Problematique[1]

Rod Coombs and Ian Miles

5.1 INTRODUCTION

The received concepts of innovation, and the measurement techniques based upon those concepts, are firmly rooted in the study of manufacturing innovation. Are they well suited to analysing innovation in economies in which service sectors and service functions play such a dominant role?

Following this introduction, we outline the familiar facts concerning the scale and scope of the transformation of advanced economies created by the rise of 'services'. We consider some alternative definitions and taxonomic approaches that have been used to capture the 'peculiarities' of services and of service innovations.

This leads us to a stark choice in terms of how research should proceed on defining and measuring innovation in a 'service-rich world'. There are three choices:

1. **Assimilation**. The basic idea here is that services innovation is fundamentally similar to manufacturing innovation, and can be studied according to the methods and concepts developed for the latter. This is where most current measurement work is located. At most, minor modifications to conventional survey and other instruments are required; the more substantial change is that services are included in the population sampled!

2. **Demarcation.** This approach, in contrast, argues that services innovation is highly distinctive, following dynamics and displaying features that require novel theories and instruments. This approach, although not fully developed, has been the underpinning of specialised

studies of innovation in services, and is now being tacitly applied by the use of 'dual approach' surveys which adopt different questioning styles for manufacturing and service firms.

3. **Synthesis.** This approach, which also remains to be developed more extensively, suggests that services innovation brings to the fore neglected aspects of the innovation process which are widely distributed across the economy. Though these are displayed most obviously (at the present moment) in service firms and industries, at least some of them are liable to be increasingly significant across the board, and others are prevalent activities whose involvement in innovation processes remains largely unstudied.

We will explore these approaches in more detail, and then address the measurement challenges opened up by the three alternative approaches. We argue that the assimilation approach is too limited by far, although measurement based on this approach does at least confirm that services do innovate. But the approach does not give us sufficient leverage on contemporary innovation dynamics. Thus, researchers and policy makers are faced with a choice of perspectives. They can operate on the basis of assumptions that there are two parallel but separate worlds of manufacturing innovation and service innovation, with all their respective indicators. Or, there is the prospect of opening up a new world of innovation concepts and indicators that effectively transcend the distinction between 'manufacturing' and 'services' as sectors, at least *for the purpose of innovation studies.*

5.2 THE SERVICE CHALLENGE

Whether we focus on the outputs of the service sectors of industrial nations, or on the proportion of the employed population engaged in service functions, the growth of the service components of economic activity is one of the great trends of the latter half of the twentieth century. For a surprisingly long time, economists and sociologists largely neglected this growing significance of services. Much of the theory, and many of the statistical instruments, which we use to chart socio-economic change and to develop and assess policy measures, remain based on approaches developed to deal with a world in which manufacturing occupied a predominant role. Nowhere is this truer than in the field of innovation studies.

Despite a vast growth in the analysis of services since the beginning of the 1980s, it is still common to find services described as non-innovative, as at best late adopters of technologies generated by other sectors. In fact, services are hardly laggards in the use of new Information Technology (IT). On the contrary, some services are major users, accounting for the bulk of IT investment. Thus, about three quarters of all expenditure on IT hardware in the UK and USA stems from services. Such branches of the services sector as financial services are the major users of IT in almost every part of the world[2].

As well as being **users** of new technology, some services are highly innovative in their development of applications of IT, creating new system configurations and software, and often acting as the catalyst for developments elsewhere (for example, by selling on IT services such as software, databases, and telematics to others). Some services, too, play substantial roles in diffusion of new technologies (especially, but not only IT), and in fusing technological knowledge with knowledge of their clients' needs, via marketing, training, and consultancy[3].

A case can be made for the 'assimilation' approach being quite sufficient for capturing these roles as user and source of innovation. Established R&D and innovation surveys, in large part using the conventional instruments of the Frascati and Oslo manuals, **do** capture services innovative efforts. (The fact that no-one bothered to apply these tools to services until recently reflects the neglect of services mentioned at the outset of this chapter.)

Contrary to some claims about the existence of very specific innovation processes in services, some authors suggest that survey tools demonstrate little difference between manufacturing and service firms. They suggest for example that the distinction between product and process innovation is meaningful in the case of services, **contra** the claims of many service sector researchers (cf. Evangelista & Sirillli, 1998). These results are inherently interesting: but it is possible to overrate the similarities that are displayed and these data do not address all of the differences which the emerging literature on services innovation leads us to anticipate. For example, Sundbo's study of the organisation of innovation in services (Sundbo, 1988) makes a good case for there being many distinctive features of services organisation and trajectories[4]. Thus, services' organisation of innovation is much less centred on R&D departments and managers, and more oriented around project management, than has been found to be the case in manufacturing. Similar results are emerging from ongoing CRIC research, the Community Innovation Survey dataset, with strong evidence that services' innovation expenditures are also typically distributed in quite different ways across activities like R&D, training, and market development than are those of manufacturing firms.

Probably the most elaborate attempt to align services and manufacturing innovation – though still overtly very much a scouting exercise – is that made by Soete and Miozzo[5]. Pavitt had classified services as simply supplier-driven in his taxonomy of innovative firm and sector types[6]. Soete and Miozzo, in contrast, show that several contemporary service branches readily fit across the range of a slightly modified set of the categories. This suggests that the range of innovative practices in services is as wide as that in manufacturing. They distinguish three (or four) categories of service businesses:

I. **Supplier dominated sectors.** Major examples of these are public or collective services (education, health care, administration), and personal services (food & drink, repair businesses, hairdressers, etc.), together with the retail trade[7].

II. (a) **Production-intensive scale-intensive sectors.** These involve large-scale back-office administrative tasks. Such services are especially suited to the application of IT with the aim of (at least initially), reducing costs.

(b) **Network sectors.** Such sectors are dependent on physical networks (for example, transport and travel services, and wholesale trade and distribution); on elaborate information networks (for example, banks, insurance, broadcasting and telecommunication services). Public utilities, such as water and gas supply, may also be located in this group of firms, though these are not usually taken to be service sectors. Such services play a major role in defining and specifying innovations, influencing the suppliers of new technologies.

III. **Specialised technology suppliers and science-based sectors.** These include such services as software and specialised business services, laboratory and design services[8]. The main source of technology is the innovative activity of the services themselves. The innovations tend to be 'user dominated'.

This approach then, is both a demarcation and an assimilation of services versus manufacturing innovation. It demarcates types of service innovation, but *within* a framework of manufacturing innovation. Furthermore, it focuses on sectoral classes of innovating firm rather than on classes of actual innovation events.

The interest in service sector innovation, we suggest, reflects both the increasing quantitative dominance of services in our economies, and changes in the use of technology in services. As noted, IT is being widely employed in services: this technology is highly generic, being applicable to a vast range of information-processing activities. Such activities are characteristic of many services. Before discussing approaches to services innovation in detail, it will be helpful to examine the variety of activities that are classified as services, since the information-processing activities they feature, and the role of IT in their functioning, is extremely significant. This attempt to define services in a more **functional** way will serve as a useful orientation device when we come to consider the alternatives of radical demarcation of service and manufacturing innovation, and of radical assimilation of the two domains, in later sections of the chapter.

Varieties of Services

There are many ways of distinguishing among services, for example, in terms of standard statistical categories (transport and communications, education and research, etc.) and in terms of the markets they serve (for example, producer services, consumer services, state and community services). Soete and Miozzo demarcate among services in the first instance in terms of features relevant to innovation processes. But here we will focus on two dimensions that capture a great deal of the diversity among service activities. The first is the '**client-intensity**' or **interactivity** of the activity. A feature common to most service sectors is the greater reliance on interaction between clients and service suppliers than is the case for traditional mass production industries. This interaction involves exchange of information in a more or less elaborate form, and is clearly amenable to the application of IT. Nevertheless, service firms and sectors vary in their extent of client-intensity. Figure 5.1 below adopts a highly schematic approach to locating service activities in terms of client-intensity. In reality most service sectors include firms which specialise in more and in less client-intensive operations, though different branches are typically more or less standardised versus customised. (Figure 5.2 displays German data on this point; it should be noted that some quite standardised services can nevertheless be client-intensive, requiring high – if more passive – participation from the purchaser of the service.)

A second parameter that affects the technological innovation of services relates to their core processes. We here simply differentiate between services whose main functions are effecting physical transformations, those concerned with effecting changes in human beings, and those whose main activities involve effecting changes in the state or availability of information (including producing data, information and knowledge). Figure 5.1 presents an indicative guide to different services' location on this dimension, too. (This is necessarily indicative rather than accurate. We do not have data allowing us to make definitive allocations; the features are to some extent historically variable; and most branches of the service sector feature different niches, which vary in the extent to which services are customised to client requirements.)

A moment's examination of Figure 5.1 reveals that services may move over time up or down the columns, indicating a shift between the poles of commoditisation and specialisation. Indeed they may even exhibit market segmentation along this axis. Less often, there may be some blurring and movement between the columns, as in the use of telematics to deliver educational services.

The core transformative processes of services are clearly subject to different types of technology: different classes of knowledge may be brought to bear on effecting them. Thus, physical services - such as transport, repairs, domestic services - effect physical transformations, often involving movement through greater or lesser distances, the application of pressure or heat, and so on. Such services have long been able to make use of new technologies based upon the cheap provision of motor power through the internal combustion engine and electrical motors. The provision of the functionalities involved to final consumers may be a matter of purchasing services or one of self-servicing. Some of these services have been subject to serious competition from new consumer goods such as the motor car and household appliances[9].

Human services like health and education, in contrast, typically being produced together with and highly 'tailored' to individual clients, require different technologies. The core activity requires processing often highly specific data about individuals. This has required the active involvement of service workers. Even in health services, where significant strides have been made in pharmaceuticals, surgical equipment, and so on, there is typically much requirement for human interaction, and new technologies have only recently begun to support this (for example, patient records, medical smart cards).

Figure 5.1: A Classification and Impressionistic Mapping of Services

Level of client intensity/ customisation:	**Main Class of Transformation:** *Physical services*	*Human services*	*Information services*
Very high	Hotels	Surgery	Bespoke software
	Domestic service	Hairdressers	Management consultancy
		Counselling, Help and Advice services	
			Legal services
Moderate	Traditional restaurants	Mass education	Insurance
		Welfare services	
	Laundry		Accountancy
	Airlines		Real estate
	Retail & Wholesale Trade		Banking
			Telematic services
			Telephone services
	Fast food restaurants		Package software
			Broadcast radio/TV
Low	Mass public transport		
	Postal services		
	Freight transport		

Figure 5.2: Services and Standardisation: evidence from the German services survey

Sectoral variations among services in focus on specialised versus standardised services

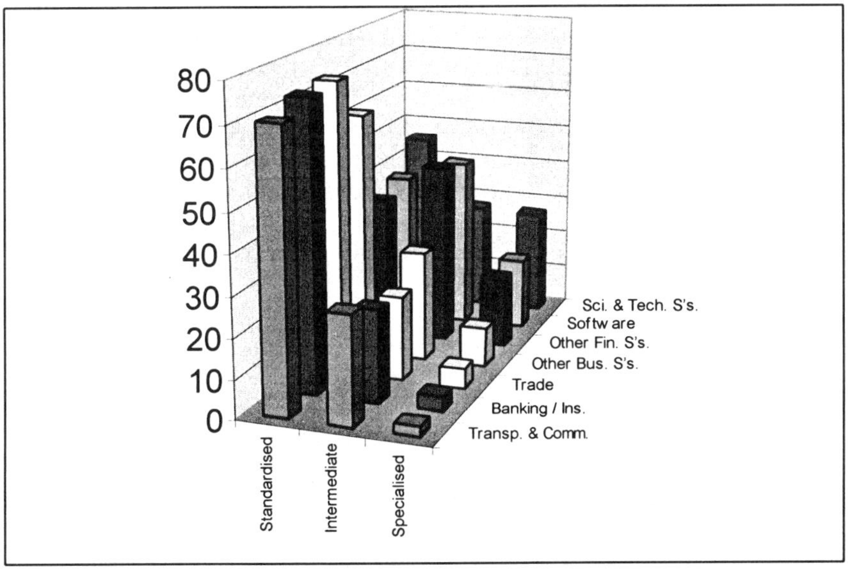

Proportion of firms mainly oriented toward different types of service output, reporting each class of innovation; German services innovation survey

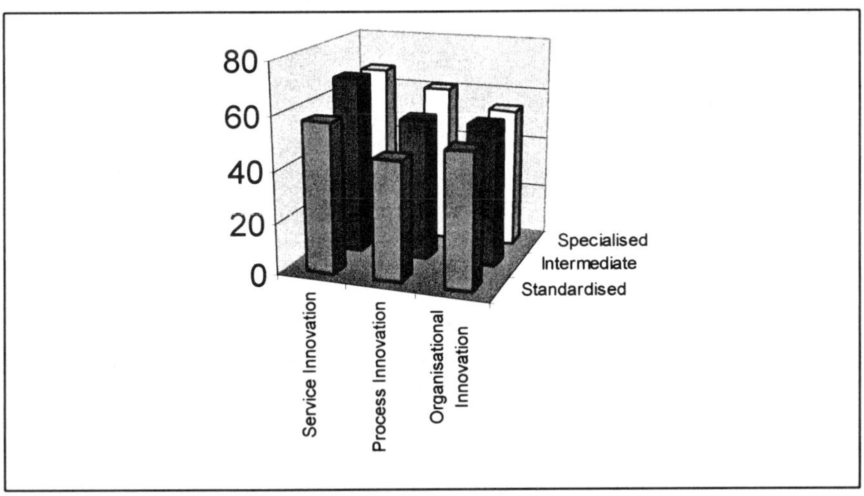

Source: Hipp, Tether and Miles (1998 forthcoming)

Information services have been much more amenable to electronics-based innovations, using electric power and electronic systems for communication (telephony and broadcasting), amplification, visual display, etc. These have been particularly important in entertainment, where again new consumer goods such as TV and hi-fi, supported by and supporting services such as broadcasting and the recording industries competes to medical informatics, from distance learning to interactive television. And all will see increasing use of IT in back office functions and (especially with the advent of mobile communications) in front-office and field work. Currently, some services are in the vanguard of IT use, but others are lagging. The continuing increases in familiarity, functionality and power of IT, together with the decreases in its price mean that its is liable to be diffused even more widely in the future, and is liable to be adopted by many of these 'laggards'.

5.3 DEMARCATING SERVICES

The previous section presented strong arguments for treating services as exhibiting patterns and modes of innovation that cannot be adequately captured with manufacturing categories. Here we push these differences still further, and consider the possibilities for a more radical demarcation of a separate domain of services innovation.

Section 5.2 also stressed that the services sector is extremely heterogeneous, in its use of technology as in other respects. Many commentators have nevertheless set out to describe common features of services. We would agree that there are features that apply to many services to a much greater extent that they typically do to primary or secondary sector industries. These features have presumably contributed to the specific innovation patterns of the sector. Admittedly, for every characteristic used to describe a 'typical' service, one or other prominent exception can be found. This is not just because of the range of activities indicated above. Much innovative effort is actually directed towards making services more like other sectors[10]. And manufacturing industry is also diverse, some of its activities being quite 'service-like'. However, it is still helpful to consider some of the specificities of services in order to think about issues of services innovation.

(1) **Technology and Plant.** Services have often been seen as having relatively heavy investment in buildings and light investment in technology. IT investment is changing this picture. And IT can be used to reduce the costs of buildings - for example, tele-services make it possible to reduce the number of local offices that are maintained. Service innovation therefore may be implicated in a distinctive trade-off pattern between service processes which are dependent on physical facilities and those which are not.

(2) **Labour.** The diversity of services is clear, in that some services have high proportions of professional staff, like doctors, lawyers, and teachers – while others other services are mainly based on unskilled, and often family, casual or part-time labour. In both cases, labour costs can be reduced by relocating essential operations to areas of cheap labour (using telecommunications to maintain co-ordination). Thus, we see the outsourcing of highly skilled software development and low-skilled data entry tasks to developing countries. The need for expensive and scarce skills may also be reduced through remote access to expertise and, possibly, by use of expert systems and other decision aids.

(3) **Organisation of Work.** Much service work involves craft-like production. There can be considerable discretion for the individual employee and limited management control of the pace and quality of work; and classically information has been relayed up and down the larger service organisation through an elaborate bureaucracy. IT Systems are being introduced to monitor activity, feed data from field and front-office workers directly into databases and Management Information Systems. Meanwhile while managers use telecommunications (and broadcast media in some large companies) to communicate with workers; these are elements of the 'flattening' of businesses (not only in the services sector). Service production is often non-continuous, with limited economies of scale. Often the products are 'one-offs'. But some services now assemble individually tailored products from components produced in a standardised way with consistent quality (for example, 'fast-food' chains, who have left the traditional craft of cookery behind and developed more assembly-line-like systems with standard components and high division of labour).

(4) **Immaterial/Intangible Product.** Many services are hard or impossible to store or transport (so consumption and production are liable to be coterminous, with the supplier moving to the user or vice versa). But IT (telematics) is being used for ordering, reservation, and - in software and information services - delivery of the service. Automated Teller Machines and similar systems allow for service delivery out of office hours.

(5) **Customisation of Product.** Many services are 'client-intensive', requiring inputs from consumers into consumption, delivery, design and production. Interchange of information about client requirements and product specifications is interactive, and often unpredictable and non-routine. IT applied to these interactions includes: Electronic Data Interchange as a standard system for remote input of orders and client details; software (data mining and more interactive methods) to analyse client requirements and match these to service products.

Another type of innovation involves self-service, wherein the client does some of the work of assembling the service product.

(6) **Marketing Problems.** The above features make it difficult to demonstrate many services to potential clients in advance of purchase (a particular problem for new suppliers and for new services in general). Efforts to overcome this problem include guarantees, quality standards, and demonstration packages.

(7) **Regulatory issues.** Professional and legal regulation is a common feature of some services - in part because of the problems of demonstrability; in part because traditionally some services were seen as nationally strategic and sheltered from international competition, and in part because some services like accountancy and environmental services are largely dictated by regulations. Current trends inspired by neoconservativism and globalisation have led to re-regulation and general flux in the constraints and opportunities faced by services. Regulatory institutions and service providers both often respond by demanding new performance indicators and diagnostic evidence.

Many of the features of services discussed above are bound to impact upon the innovation process. To the extent that these are sharply distinctive features of services, we might expect to see specificities in the choice of technology, and in the processes of innovation and development of new technical knowledge. Many earlier commentators saw services as supplier-driven sectors, absorbing (often belatedly) technologies developed in manufacturing sectors. Now however, the evidence for services' autonomous innovation is overwhelming. But it remains the case that these characteristics of services – most centrally those of intangibility and client-intensity – are important influences on innovation processes.

It is evident then that the organisation of innovation in services rarely takes the form that is typical in manufacturing industries. Innovation surveys and other approaches demonstrate that service firms rarely have R&D departments. (Indeed, a problem for measurement arises in that service managers frequently do not recognise their innovative activities as being classifiable as R&D.) Innovating services most often set up product or project development teams on an ad hoc basis. One consequence is that service sectors less often participate in networks of R&D managers and the like[11].

The two central (but not universal) features of services, intangibility and client-intensity, have influenced the major efforts to develop theories of service innovation which imply sharp demarcations between this and manufacturing innovation. The approach of 'the reverse product cycle' initiated by Richard Barras[12] takes the 'supplier-driven' approach to services on one step. It suggests that the absorption of new IT into services as a means of increasing service process efficiency provides a catalyst for services to undertake their own innovation trajectories. Thus from process innovations they move through a period of quality improvement to product innovations.

There is little suggestion here of novel measurement problems, it should be noted, in Barras' formulations. However, a re-reading of his account suggests that the pattern he describes is often equally well characterised as an evolution from *back-office* process innovation, through innovation in the *delivery* of services to clients, to *product* innovation. We have argued elsewhere (Miles, 1994) that delivery innovation is an exceptionally important feature of contemporary services innovation, especially but not solely through the application of IT. It is arguable that conventional innovation indicators based on the product-process distinction are overlooking, or misrepresenting, such historically specific innovation trajectories.

Relations between service suppliers and their clients include more than just the delivery of the service itself, as indicated above. Another approach to services innovation introduces the unappealing neologism servuction to describe the penumbra of such relations that surrounds process, product and indeed consumption[13]. In some applications of this approach, innovation in servuction is contrasted with innovation in production. As with the comments about delivery innovation above, it seems likely that many innovative activities involving interactions between supplier and client (and networks of clients) may be overlooked, underemphasised, or misrepresented by conventional measurement approaches. The sorts of activities that are involved here include, for example: marketing; transactions; input of information from clients for choice, design or customisation of the service; after-sales support; and so on. Customisation raises particular challenges for innovation measurement: it is usually explicitly sidelined in terms of R&D and innovation indicators, but is a critical process in many services, and much service innovation is intrinsically entangled with customisation of a continually evolving product. A related point is that with the high degree of co-design and co-production of service products, it may be difficult to locate the innovation within service supplier or client: it is not unusual, for instance, for service firms to site their staff within client organisations for periods of time. (Similar problems are raised in collaborative R&D more generally.)

A final issue in services innovation, connected with the intangibility of the service, product, is IPR. As Andersen and Howells (1998) note, the elements of IPR systems that were historically evolved to protect technological innovations have mainly been those connected with patents (though software is increasingly covered – uneasily – by copyright, and design rights are also of significance to services). Many commentators have argued that the weak IP regime in services has acted as a serious deterrent to innovation (for example, Sundbo most recently) though surveys and our own case studies lend little support to this. There is reason to think that the IP regimes do influence services' innovative activities in subtle ways however: a topic that has only recently begun to be explored (cf. our recent studies of IP and knowledge management in a range of knowledge-intensive business services[14]). We find distinctive strategies to be related to firm size and to the particular sorts of knowledge that is critical for the firms' competitiveness.

In Section 5.5 we return to the measurement challenges created by this radical demarcation approach. Before this, in Section 5.4, we consider the contrasting approach which starts with recognition of recent attention to the intertwining of services and manufacturing in the 'rainbow economy'. This approach does not try to 'fit' manufacturing and service innovation to each other, but to avoid drawing such a gross distinction. The manufacturing-services contrast is merely a monochrome appraisal of what is a much more diverse and variegated mixture of activities across the economy, and in innovation processes.

5.4 THE RAINBOW ECONOMY

We have seen that a strong case can be made for the continuing distinctiveness of service firms and industries, and for the likelihood that recognition of this implies modifications of standard perspectives on theorising and measuring innovation. However, these distinctive elements could be taken to suggest an alternative approach. This involves stressing the variety of activities encountered throughout the economy. The economic system is here seen as a web of intertwined functions, some of which for largely historical reasons are labelled services, some of which are labelled manufacturing (etc.). Any firm or sector is liable to feature a great many of these functions, though the proportions vary markedly – and it is the different mixtures that give sectoral distinctions their utility.

But now, rather than seeing services sectors as demanding new methods of innovation analysis, the argument will be that it is the service and service-like activities across all sectors of the economy that demand the new approaches. In other words, the focus on service sectors has served to throw light on neglected elements of the whole economy. (And in this context, we recall Giarini's emphasis that the concept of the service economy should not be reduced to the growth of service sectors, but seen as implying the rise of service concepts throughout the economy.) Firms and sectors are then seen to vary quantitatively in terms of the extent to which they engage in various service functions – marketing, design, after sales, transactions, delivery and so on. Each of these is a site of potential innovation alongside the conventional foci of the product-process distinction.

This approach suggests that there is little in the way of an ultimate rationale for strongly demarcating innovation analyses between manufacturing and service sectors. There may be a rationale for focusing on the (quantitative rather than qualitative) specificities of services in that this may be the most cost-effective way of identifying the elements of service innovations which have been neglected across the whole economy, simply because they are more frequently encountered in services. And the historical tendency of many services to be technology followers rather than leaders, together with the particular IP situation of many service innovations, may have also led to the emergence of distinctive forms of innovation management which need to be grappled with (though we suspect that in this respect that services' innovation is not all that different from innovation issues encountered in SMEs in all sectors). There may thus be virtue in pursuing detailed studies of services' innovation as a prelude to applying innovation measurement indicators on a more general basis.

There are good grounds for thinking that, just as service firms are 'industrialising' and acquiring characteristics more typical of manufacturing sectors, so many manufacturing firms' activities are becoming more service-like[15]. Thus it is particularly important to be aware of features of services innovation which may be increasingly characterising manufacturing sector innovations, too, and eluding capture by our conventional instruments.

Services have been an increasingly evident element in many other productive activities. In part this is because service functions, which were earlier 'hidden' inside other jobs, have been differentiated into specialised roles in the process of continuing division of labour. And in part it reflects the creation of new services functions associated with technological change and an increasingly complex economy. The value-chain of economic activities that are embodied in final and intermediate industrial products thus features a growing share of service activities, and increasing variety of service activities, supplied both as business services and as in-house inputs from service workers within companies. Thus there is growth both in the business services sold to industry, and in the share of white-collar workers in all sectors.

The greater part of most companies' costs (other than those for purchased materials) typically involves overheads. Even in manufacturing, more than two-thirds of all nonmaterial costs tend to be indirect or overhead expenses. Most overhead is actually a matter of services that the company is supplying internally[16]. This implies that in a large number of industrial sectors the production of physical goods constitutes only a small part of the overall process of value-adding. Within manufacturing, 75% to 85% of all value-added, and a similar percentage of costs, involves service activities.

This suggests that the major value-added to a product (as indicated by what purchasers are prepared to pay for it) is typically due less to its basic commodity value than to design features, perceived quality, and so on[17]. These features are added by services activities, inside or outside the manufacturing company. For instance, in the computer industry, the manufacture of the computer accounts for only 10% of the ultimate price; the bulk of the costs is taken up design, software development, distribution etc. Sports shoes have even lower 'material' cost components. Most companies, whether statistically defined as manufacturers or as services firms, are thus predominantly service providers!

The conclusion can be taken to a not illogical extreme – material products themselves are only physical embodiments of the services they deliver, or tools for the production of final services. (Quinn makes this point, and the argument was also developed by Gershuny and Miles, who differentiated between service products (such as a concert or railway journey) and service functions (for example, entertainment or mobility) which could be delivered either through services or though goods (for example, an audio system or a motor car).) Many managers we have interviewed over recent years have articulated such a view. Quinn proposes that value added is increasingly likely to come from technological improvements, styling features, product image, and other attributes that involve service functions and, often, specialised service firms and/or employees. Services' contribution to the process of value adding is thus growing. This trend may be expected to continue should the demand for differentiated and individual products increase, as is anticipated by many academic and industrial commentators (often under the rubrics of 'postFordism' and 'flexible specialisation').

In an important study focusing on service functions supplied by manufacturing companies to their customers, Mathé and Shapiro[18] (hence, M&S) amassed much evidence on this point. For instance, they begin by noting that:

> "In 1986, Japan's MITI White Book looked at international competition in new terms: the authors predicted that after-sales and delivery services would play an increasingly crucial role in the success of manufacturing firms...The French vehicle manufacturer Citroen declared 1988 the year of service; Renault shortly followed suited...executives from both Texas Instruments and Rank Xerox announced plans to offer an array of expanded services, citing a growing Asian industrial offensive... In today's competitive global environment, service is increasingly a critical element. This... is especially visible in markets for complex equipment, whose increasingly sophisticated customers are more and more apt to make comparisons between rival products based on cost-of-ownership criteria rather than simply on the basis of acquisition costs. Effective support services have begun to play an indispensable part of 'product performance' " (p.5).

M&S cite a number of studies indicating the important role of service elements in manufacturing, especially of high-tech products. For example:

- In a 1991 Arthur D. Little survey of 52 field service managers in high-tech equipment firms, service quality scored highest among customer concerns. (The next most important attribute was reputation, followed by product quality.)
- A survey of German mangers by Simon addressed was the competitive significance of service elements in manufacturing. These managers saw the role of service as growing substantially overt the next 10 years.
- M&S present case-study material, such as the example of how French flower-growers found themselves to be losing out to Dutch competitors, even in the Parisian flower shops. The flowers were nor superior. Rather the Dutch gained advantage by having analysed delivery constraints:

computerised order-taking, grouped orders, organised delivery itineraries, used shop keys to enter the shops and deposit the flowers before opening, offered new services (ready-to-sell bunching at the flower 'factory') and they guaranteed stable prices.

Consequently, in industrial as well as service sectors, competition (and innovation) is being intensified in those service elements surrounding the actual material product. These include, but go beyond, Teece's well-known 'complementary assets'[19]. The case can be made that in many sectors, competition is shifting away from how companies build their product to how well they serve customers before and after they produce and sell the products[20]. The sale transaction is only a point in a chain of events, many of which involve services. M&S distinguish between essential services (which must be provided for the firm to continue to exist) and amenities that are not necessary for the product to function but which add to its utility (and to the firm's differentiation and competitive edge). They also distinguish services which encourage or facilitate the sale of products – Public Relations, demonstrations - from those which increase the benefit or satisfaction from use of the product - after-sale repair and maintenance, pre-installation customisation, rush delivery, specialised training, product updates, helplines, etc. Likewise, service components are integral to product innovation. For instance, the process of upgrading products, i.e. creating increasingly differentiated, high quality production, aimed at the specific needs of market segments, depends upon these service functions. Manufacturers, then are increasingly deriving competitive advantage from elements in service processes associated with material production, and the associated human skills, logistic capabilities, knowledge bases, and other service strengths that competitors cannot easily reproduce, and that lead to greater demonstrable value for the customer[21].

The implication of all this is that service activities are an increasingly important feature of the whole economy. Innovation in service activities is thus liable to be of significance across the economy. And any challenges to measurement generated by the peculiar features of services innovation are thus liable to apply across the economy. Established measurement procedures may be missing out on important elements of innovation.

5.5 CHALLENGES TO MEASUREMENT

Simply recognising that services can be innovative has provoked changes in R&D and innovation surveys – revealing the important role of services in business R&D (over 25% in many countries) and innovation (disagreement over whether services are less or equally innovative). Even those approaches which assimilate services innovation to the manufacturing frameworks imply some modifications to the latter – Soete and Miozzo on network innovations, for example. These are unlikely to be adequately mapped by establishment-level surveys.

But the sharpest challenges are thrown up by the two approaches we have centred our discussion on. These are now contrasted in Figure 5.3.

5.6 CONCLUSIONS

The growth of the 'services sector', as traditionally defined, to a dominant position in the share of output of the industrial economies, means that we can no longer treat the innovation which goes on in this sector as negligible or as residual. Nor can we assume that it simply follows the patterns found in a manufacturing context, and can be explained by the concepts derived from this. Furthermore, although statistical constraints often force us to use received notions of sector, we have to recognise that the understanding of innovation processes and the consequences of services' activity as innovation intermediaries forces us to go beyond these notions of sector.

Many of the mostly widely remarked features of the evolution of technological innovation in the late 20th century – for example, mass customisation; the pervasive use of IT; - depend centrally on the *combination* of material and non-material innovation. We have to reconceptualise innovation and its consequences around this new paradigm. In essence, we are moving away from a model of innovation that puts all the emphasis on artefacts and *technological* innovation; and towards a model which sees innovation in terms of changes in *market relationships* but with major artefact and technological dimensions. Understanding 'service innovation' is therefore quite liable to lead to nothing less than a widening, or even a complete re-interpretation, of the concept of innovation itself.

Figure 5.3: Implications of the Two Approaches for Innovation Measurement

Measurement Challenges	**Sharp Demarcation**	**Rainbow Economy**
Concepts of R&D and Innovation	Problems specific to measurement in services. Conventional terminology inappropriate to and unrecognised by many services. Important role of organisational innovations, yet these remain poorly measured in received approaches.	Problems impact across all economy, though often particularly intense in service sectors. Organisation of firms in all sectors is such that innovative activities in marketing, distribution etc. are often not under purview of R&D managers, thus liable to be overlooked in surveys etc (and in firms' own strategic planning?)
Definition of R&D	Term not seen as appropriate even in many technology-intensive services (despite Frascati modification to include software). Role of customisation much more 410ambiguous than Frascati manual implies.	Need a new concept of 'investment in innovative activities' (defined as having the intention of altering the nature of the market offering of the enterprise, or of its underlying costs of doing business)
Definition of innovation	The term 'innovation' is not recognised at all by managers in some countries (but substitutes like 'something new' are too all-encompassing?). Product-process distinction liable to be misleading at best when it comes to delivery and other interactional innovations. Organisational innovation seems critical in many services, but is hard to quantify.	Need to distinguish among various (material and immaterial) changes which can occur in either 'sector'. Material changes may then largely be classified as standard into product, process (and delivery?) etc. Immaterial changes could productively be divided into those 'focused on client relationships' or 'focused on internal processes', with subcategories of each (e.g. transactions, product tracking, etc.)
Location of innovation	Innovations co-produced with clients may be attributed solely to the latter.	Innovative activity seen as intrinsically structured as 'participating in a process located in a network', usually not closely bound to one 'innovator'
Organisation of Innovation	Project rather than R&D management may require different respondents	Innovative activity diffused among functional units of firms
Distributed Innovation processes	Only capable of being demonstrated through complex input-output analysis. Makes things more complicated than they really are!	Networks vital: problems in identifying and measuring network level innovation (large technical systems and co-ordinated introduction of standards and peripherals, as well as collaborative innovative efforts in R&D etc.)
Economic effects of innovation	Difficult to trace because of inter-sectoral flows.	Sectors become less important than **chains and networks.** The **location of innovation also becomes the natural location for the identification of economic data**
Implicit underlying under-standing of innovation	New material artefacts which happen to consume immaterial inputs in their production	Entrepreneurial acts by firms which change existing market relationships. These may or may not involve material novelties.

REFERENCES

Andersen, B., Howells, J. and Miles, I. 'Intellectual Property Rights, Innovation and Services', CRIC and PREST, The University of Manchester, March 1998, prepared as an input to the *TSER SI4S Project "Services in Innovation, Innovation in Services"*.

Barras, R. 'Towards a Theory of Innovation in Services', *Research Policy*, 1986a; 15 (4), pp.161-173.

Barras, R. 'New Technologies and the New Services', *Futures*, 1986b; 18 (6), pp.748-772.

Barras, R. 'Interactive Innovation In Financial And Business Services: The Vanguard of the Service Revolution', *Research Policy*, 1990; 19, pp.215-237.

Belleflamme, C., Houard, J. and Michaux, B. *Innovation and Research and Development Process Analysis in Service Activities*, 1986, Brussels, EC, FAST Occasional Paper No. 116, September.

Business Monitor MO14, London: HMSO, April, 1995.

Chase, R.B. and Garvin, D.A. 'The Service Factory', *Harvard Business Review*, 1989; July, pp.61-69.

Eiglier, P. and Langeard, E. *Servuction*, Paris: McGraw-Hill, 1987.

Gershuny, J.I. and Miles, I.D. *The New Service Economy*, London: Frances Pinter, 1983.

Mathe, H. and Shapiro, R.D. *Integrating Service Strategy into the Manufacturing Company*, London: Chapman & Hall, 1993.

Miles, I. *Innovation in Services: Services in Innovation*, Manchester: Manchester Statistical Society, 1996.

Miles, I. 'Services, Systems and Innovation – Cinderella Comes to the Ball', presented at *Services and Systems Workshop*, CRIC, Manchester, March 18-19, 1998.

Miles, I. 'Services and Foresight', *Service Industries Journal*, 1999; April.

Miles, I et al. *Mapping and Measuring the Information Economy*, Boston Spa: British Library, 1990.

Miles, I., Kastrinos, N., Bilderbeek, E., den Hertog, P. with others at PREST and TNO. *Knowledge-Intensive Business Services: Their Role as Users, Agents and Sources of Innovation*, PREST, mimeo, report to the European Community DG12, European Innovation Monitoring Scheme, 1995.

Miles, I., Kastrinos, N., Flanagan, K., Bilderbeek, R., den Hertog, P., Huntink, W. and Bouman, M. *Knowledge-Intensive Business Services: Users, Carriers and Sources of Innovation*, forthcoming from the EC, Luxembourg, (DG13 SPRINT-EIMS).

OECD ICCP. *Usage Indicators - A New Foundation for Information Technology Policies*, Paris: Organisation for Economic Co-operation and Development, Information Computer Communications Policy, ICCP 31, 1993.

Pavitt, K. 'Sectoral Patterns of Technical Change: Towards a Taxonomy and a Theory', *Research Policy*, 1984; 13 (6) pp.343-373.

Quinn, J.B., Doorley, T.L. and Paquette, P.C. 'Beyond Products: Services-Based Strategy', *Harvard Business Review*, 1990a; March, pp.58-67.

Quinn, J.B., Doorley, T.L. and Paquette, P.C. 'Technology in Services: Rethinking Strategic Focus', *Sloan Management Review*, 1990b; 11 (2), pp.79-88.

Quinn, J.B. and Paquette, P.C. 'Technology in Services: Creating Organisational Revolutions', *Sloan Management Review*, 1990; 11 (2), pp.67-78.

Soete, Luc and Miozzo, Marcella. *Trade and Development in Services: A Technological Perspective*, Working Paper No. 89-031, Maastricht: MERIT, 1989.

Sundbo, J. *The Organisation of Innovation in Services*, Frederiksberg: Roskilde University Press, 1998.

Teece, D. 'Profiting from Technological Innovation: Implications for Integration, Collaboration, Licensing and Public Policy", *Research Policy*, 1986; 15 (6), pp.285-305.

Zemke, R. and Schaaf, D. *The Service Edge: 101 Companies that Profit from Customer Care*, New York: Plume (Penguin Books), 1990.

---

[1]This chapter is an output of the ESRC funded Centre for Research on Innovation and Competition.
[2]I. Miles et al (1990); OECD ICCP (1993).
[3]I. Miles, N. Kastrinos, K. Flanagan, R. Bilderbeek, P. den Hertog, W. Huntink and M. Bouman (forthcoming).
[4]Jon Sundbo (1998).
[5]Luc Soete and Marcella Miozzo (1989).
[6]Keith Pavitt (1984).
[7]We would note, however, that some large retailers are themselves significant influences on their suppliers. An obvious example is UK supermarket chains, which set quality and environmental standards, and identify new products, for their suppliers.
[8]The services 'R&D Consultancy, Technical Testing and Analysis' account for almost 10% of current UK R&D - cf. *Business Monitor MO14*, April 1995, London, HMSO.
[9]J.I. Gershuny and I.D. Miles (1983)
[10]I. Miles, N. Kastrinos, R. Bilderbeek, P. Den Hertog, with others at PREST and TNO (1995); I.Miles (1996).
[11]For evidence about the relatively poorer networking of services in innovation systems as displayed in the UK Foresight Programme, see I Miles (1999).
[12]R. Barras, (1986a, 1986b, 1990).
[13]C. Belleflamme, J. Houard and B. Michaux (1986); P. Eiglier and E. Langeard, (1987).
[14]I. Miles (1998); B. Andersen, J. Howells and I. Miles (1998)
[15]See references in footnote 9 above. The following paragraphs draw on work originally carried out with R Bilderbeek and P den Hertog of Dialogic.
[16]J.B. Quinn, T.L. Doorley and P.C. Paquette (1990a, 1990b); J.B. Quinn and P.C. Paquette (1990).
[17]There is no value judgement here as to whether these market decisions represent genuine additions to social well-being or the quality of life. A long debate has raged about the sovereignty of consumers as opposed to the creation of 'false' needs in the market. Advertising is among the service elements included in the cost of products, as is market research. But it is likely that there are more important forces shaping consumer opinion and choices, even though these too are socially constructed. This topic is the subject of ongoing CRIC work by our colleagues A. McMeekin and M. Tomlinson.
[18]H. Mathe and R.D. Shapiro (1993).
[19]D.Teece (1986).
[20]R.B. Chase and D.A. Garvin (1989).
[21]This is based on the formulation set out by Quinn et al – see footnote 16. For a lengthy exposition of 'The Service Edge', focusing on customer service in 101 US companies - including both manufacturing and service firms - see R. Zemke with D. Schaaf (1990).

Chapter 6
Rethinking Innovation Comparisons Between Manufacturing and Services: The Experience of the CBR SME Surveys in the UK[1]

Alan Hughes and Eric Wood

6.1 INTRODUCTION

There has in recent years been an increase in the use of large-scale surveys to measure the nature and extent of innovation activity. The most important development in a European context has been the development within the European Union of a harmonized Community Innovation Survey (CIS) (Archibugi et al, 1995). There have been two such surveys and a third is planned. They aim to cover the whole of the economy including both the manufacturing and service sectors. This commitment to an economy wide sectoral coverage has provided the occasion for a debate about the appropriate way to measure innovation activity in surveys covering both services and manufacturing (OECD, 1992; Evangelista and Sirilli, 1995, 1997; OECD-Eurostat, 1997; Kleinknecht, 1998). This debate is part of a wider literature on the extent to which service activity generally is misrepresented by measurement techniques and theoretical perspectives developed primarily with goods and specifically manufacturing activity in mind (Hill, 1977).

In the main, the literature on innovation measurement associated with the CIS survey, and the wider literature of which it is a part, has drawn a broad distinction between services and manufacturing as a whole. In the most recent, CIS separate survey instruments were in fact used for firms in these two broad sectors (see, for example, Foyn, 1998). We argue in this chapter that whilst a disaggreggated sectoral approach to the conceptualization and measurement of innovation activity is desirable, it is not necessarily most useful to base it around a broad division between manufacturing and services as conventionally defined. We argue, instead, that there is considerable variation in the content and nature of the innovation process within each of these sectors. Moreover, we argue that there are industrial groupings in manufacturing and in services respectively which have more in common with each other than they do with industrial groupings in their 'home sector'. This important feature of the innovation process is obscured if the debate focuses on distinctions between manufacturing and services broadly defined using conventional official statistics criteria. We illustrate our argument using a well known innovation typology due to Freeman (1979), which we apply to a large sample of small and medium firms in manufacturing and a subset of service industries. These firms were respondents to the biennial national University of Cambridge CBR survey of the SME sector in the UK (Cosh and Hughes, 1996, 1998). Since these surveys cover manufacturing as a whole but only business services our analysis is restricted to that part of the service sector. It is however the fastest growing and most dynamic component of service activity in the UK and other industrial economics (Vickery, 1998).

In the next section of this chapter we provide a brief overview of the debate on measurement problems associated with innovation activity in the service sector. In Section 6.3 we develop a framework for comparing innovation activity across groups of industries in the manufacturing and service sectors. In Sections 6.4-6.7 we compare innovation characteristics across these groups of industries using a variety of survey-based indicators of innovation activity. The final section contains our principal conclusions.

6.2 PROBLEMS IN MEASURING INNOVATION IN MANUFACTURING AND SERVICES

A number of arguments have been advanced to support the view that measures of innovation traditionally developed for the manufacturing sector will distort the measurement of service sector innovation activity (Miles, 1995; Miles et al, 1995; Barras, 1986; Gallouj and Weinstein, 1997; Evangelista and Sirilli, 1995). On the input side it is argued, for instance, that there is a bias towards human capital in service production processes. This may be linked to a greater scope in the service sector for 'process' innovation associated with changes in business organisation and work practices, rather than with changed plant and equipment. If this is so, then service innovation inputs will be inadequately reflected in indicators based on R&D expenditures, or investment in new capital equipment. More generally, it has been argued that there is a much greater simultaneity in service production and consumption than is the case with manufacturing. This then blurs the traditional manufacturing distinction between process and product innovation. Moreover, it leads to a situation in which 'product innovation' is intimately linked to changes in methods of delivery or distribution, because the product may often be defined by a particular delivery or distribution mechanism. This may, in turn, it is argued, pose problems in trying to distinguish product innovation from product differentiation. Taken together these arguments are often associated with a general concern that an emphasis in innovation measurement on the 'technological' aspects of product and process innovation will be ill-suited to the service sector. Moreover, it is argued that such an approach will systematically underestimate innovation activity in the service sector.

Many of these arguments are clearly matters of degree rather than substance. For instance, many service sector activities, such as energy, transport and retailing, clearly involve large-scale capital intensive production processes. It is also clear that service sector innovation may be intimately linked to investment in new plant and equipment and 'technologically' driven 'production' processes. This stretches all the way from the IT based revolution in supermarket check out and stock control activities, to the automated tellers in retail banking service provision, the delivery of Big Macs, and the provision of remote diagnostic services in health care (see, for example, Desai and Low, 1987; Barras, 1990; Gallouj and Weinstein, 1997). There is equally plenty of evidence for efficiency improvements in manufacturing associated with changes in work organization practices (Leibenstein, 1987; Frantz, 1987). These arguments are well acknowledged in the wider debate over the general nature of the distinction between goods and service activities. Their conceptual implications for the analysis of innovation are also the subject of a growing theoretical literature (Kleinknecht and Bain, 1993; Barras, 1986; Miles, 1995; Saviotti and Metcalfe, 1984). There have however been few disaggregated empirical attempts to assess the significance of the overlaps between manufacturing and services in the context of innovation activity. Our chapter is an attempt to begin to fill this gap in a way which spans the conventional manufacturing and services divide.

6.3 COMPARING INNOVATION ACTIVITY IN BUSINESS SERVICES AND MANUFACTURING

Our objective is to analyse innovation patterns in product groups both within and across the manufacturing and service sectors rather than simply drawing an aggregated comparison between the broad sectors taken as a whole. In order to create splits within the manufacturing and business service sectors, we ideally wanted a system of classification which met three criteria. Firstly, the system of classification should group sectors according to their innovation characteristics. Secondly, the system should be simple in the sense of creating a relatively small number of groupings. Thirdly, the system of sector grouping should be applicable in both the manufacturing and service sectors.

Pavitt's (1984) seminal paper identified four distinctive innovation groups within the manufacturing sector; supplier dominated; specialized equipment suppliers; scale-intensive and science-based. Pavitt et al (1989) revised the grouping slightly, splitting supplier dominated into producers of traditional consumer goods and suppliers of traditional intermediate goods. While these classification systems meet the first and second criteria, they are not straightforward to apply to services. While it may be possible to identify some matches in the service sector, for example for science-based and perhaps scale-intensive, it proves more difficult in cases such as traditional consumer goods and traditional intermediate goods.

Freeman (1978) has proposed another classification of innovator types which is more amenable for our purposes. Table 6.1 shows a summary of the Freeman (1978) system. Freeman's classification rests on a three-way split within manufacturing between capital goods and chemicals, basic materials and consumer goods. Freeman argues that within the manufacturing sector, the primary site for the development of product technology improvements are the traditional engineering sectors including capital goods and chemicals sectors. In Freeman's view, innovation in the basic materials sectors is primarily oriented toward factor saving rather than the development of product technology. In consumer goods, advertising and fashion-based design are typically more important than the development of product technology.

Table 6.1: Freeman's (1979) innovation groups for manufacturing industry

	Industry	Innovation characteristics
Group 1	**Capital goods and chemicals**	
	Aircraft and aero-engines	
	Electronics	Technical effort directed
	Other electrical	mainly to the design and
	Vehicles	development of new products
	Instruments	which have superior technical
	Chemicals	characteristics
	Machinery	
Group 2	**Basic materials**	
	Non-ferrous metals	
	Metal products	Technical effort is directed
	Stone, clay and glass	mainly to cost reduction
	Paper	through factor-saving
	Ferrous metal	innovation
Group 3	**Consumer goods**	
	Food	
	Textiles and apparel	Fashion-based design and
	Lumber and furniture	Advertising more important
	Other manufacturing	Than technical innovation

At one level, Freeman's taxonomy does not appear to meet our criteria of application in the service sector. Clearly, basic materials sectors do not have obvious correspondent's services. Nevertheless, Freeman's suggestion that the bulk of technological innovation within manufacturing is concentrated in a relatively small number of sectors may well be true for the service sector as well. Engineering consulting firms, R&D consulting firms, computer consulting firms, etc. constitute an "engineering sector" within services and may in the spirit of Freeman represent the primary site of innovation activity within services.

This approach permits a simple two-way split within both manufacturing and services. Table 6.2 shows how the 1995 CBR SME sample can be split in this way[2]. The sample of 576 firms is slightly weighted in favour of manufacturing firms which make up 56.1% of the total number. The proportion of firms in the "core engineering" sectors within manufacturing and services is similar. That is the proportion of capital goods and chemicals firms in manufacturing and the proportion of engineering and technical services firms in the business service sector are both approximately 38%. Thus, the bulk of the sample in both manufacturing and business services are not in the "core engineering sectors".

Table 6.2: Industry breakdown of SME respondents to 1995 CBR questionnaire

Industry Split	Number of firms	(percentage)
Manufacturing	**323**	**(56.1)**
• **Capital goods and chemicals**	**125**	**(21.7)**
Instrument engineering	10	(1.7)
Chemicals and chemical products	31	(5.4)
Electrical and electronic goods	27	(4.7)
Non-electrical machinery	57	(9.9)
• **Basic materials and consumer goods**	**198**	**(34.4)**
Rubber	3	(0.5)
Timber and timber products	11	(1.9)
Metals and non-metallic mineral products	12	(2.1)
Fabricated metal products	51	(8.9)
Pulp and paper	8	(1.4)
Publishing and printing	39	(6.8)
Textiles and clothing	36	(6.3)
Furniture	16	(2.8)
Food products and beverages	12	(2.1)
Manufacturing not elsewhere classified	10	(1.7)
Business Services	**253**	**(43.9)**
• **Engineering and Technical Services**	**101**	**(17.5)**
Computing services	46	(8.0)
R&D consulting and other technical services	55	(9.5)
• **Other Business Services**	**152**	**(26.4)**
Advertising	17	(3.0)
Financial and other business services	135	(23.4)
Total	**576**	**(100)**

In the analysis below, we make the following standard comparisons of innovation outputs, innovation inputs, and barriers to innovation. Firstly, we compare manufacturing as a whole with business services. Secondly, within both manufacturing and business services, comparisons are made between the "core engineering sectors" and the "other" sectors. Lastly, within both the "core engineering sectors" and "other" sectors, comparisons are drawn between those firms in manufacturing and those in business services. This combination of tests provides a useful picture of variations between manufacturing and business services as well as within manufacturing and business services. An alternative split between knowledge based sectors within manufacturing and services respectively is presented in Hughes and Moore (1998).

6.4 INNOVATION OUTPUT IN MANUFACTURING AND SERVICES

Figures 6.1-6.3 below provide different indicators of innovation output in terms of product and process innovations in manufacturing and business service sectors. For product and process innovations, the figures discriminate between imitation and novel innovations. Imitation and novel innovation were defined as follows. If a company indicated that it had introduced a product innovation, then it was defined as novel if they indicated that the product in question had not already been in use in the firm's own industry. Otherwise, if a product innovation was indicated to have already been in use in the firm's own industry then it was defined as an imitation innovation. (The actual innovation question asked is reproduced in the Appendix to this chapter.)

Figure 6.1 shows the proportion of firms which report imitative or novel product innovations in the period 1992-95. If one considers the business service and manufacturing sectors in aggregate, it can be seen that business service firms are marginally less likely than manufacturing firms to introduce imitation or novel innovations. In neither case are the differences significant.

Once the manufacturing and business service sectors are split in the manner described in the previous section, larger differences in the probability of a firm introducing a novel product innovation become apparent. The "core engineering sectors" within manufacturing and business services are more likely to have introduced novel product innovations in the period 1992-95. The probability of a firm in capital goods and chemicals having introduced a novel product innovation is 35.2% by comparison with 16.2% in basic materials and consumer goods firms. The corresponding figure for engineering and technical services is 26.7% and for other business services 15.8%. In both cases the differences are significant at the 5% level using the Mann-Whitney non-parametric test. A Mann-Whitney test of differences in the probability of a novel product innovation between the "core engineering sectors" in manufacturing and business services was rejected at the 10% level as was the test of differences between the "other" sectors in manufacturing and business services. The differences in the likelihood of a firm reporting an imitation product innovation are small and not significant.

The above results for manufacturing confirm Freeman's (1978) assessment that the capital goods and chemicals sectors are the primary site of development of new product technology. The results for business services imply that Freeman's model can equally be applied in the business services sector, as engineering and technical service firms are significantly more likely to have reported the introduction of novel product innovations by comparison with other business service firms.

Figure 6.1: Imitative and Novel Product Innovation in UK SMEs 1992-95

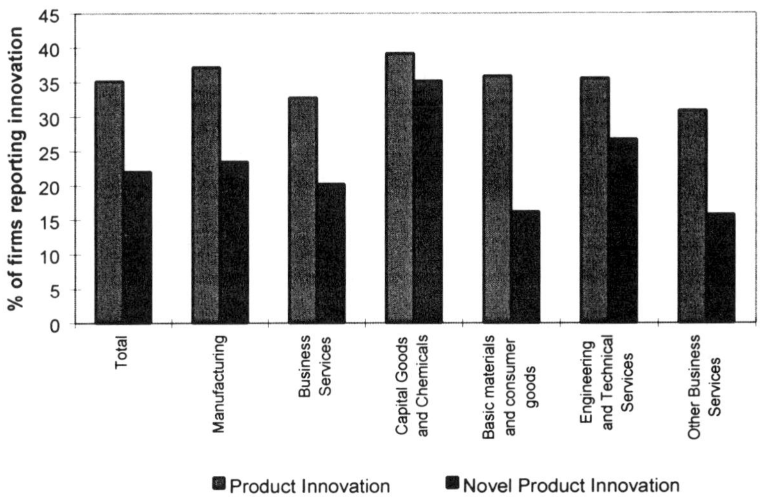

Figure 6.2 shows the comparisons for process innovation. In aggregate, business service firms are marginally more likely to introduce imitative process innovations and marginally less likely to introduce novel process innovations by comparison with manufacturing firms. Within manufacturing, capital goods and chemicals firms are less likely to introduce imitation process innovations. Likewise within business services, engineering and technical services are less likely to introduce imitative process innovations. These differences are not significant. Capital goods and chemicals firms are significantly more likely to introduce novel process innovations by comparison with the other sectors. This implies that taking the manufacturing and business service sectors together the "core engineering sectors" within manufacturing are the primary site for the development of novel process technology.

Figure 6.2: Imitative and Novel Process Innovation in UK SMEs 1992-95

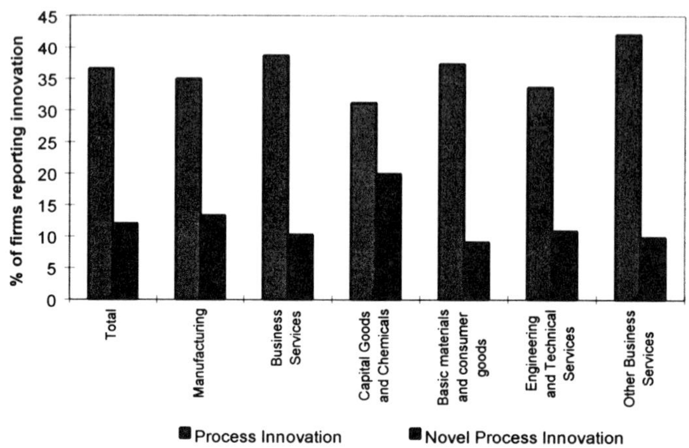

Figure 6.3 provides further comparisons on product innovation activity, showing the proportion of total sales made up respectively of products newly introduced in the last three years, products upgraded in the last three years and products unchanged in the last three years. Business service firms and manufacturing firms have similar proportions of their product ranges made up of newly-marketed products at 12.7% and 11.4% respectively. Business service firms have a significantly higher proportion of sales consisting of upgraded products by comparison with manufacturing firms and a correspondingly lower proportion of sales consisting of unchanged products.

Within both manufacturing and business services, firms in engineering have a significantly higher proportion of their sales consisting of newly-introduced products by comparison with other firms. In capital goods and chemicals, 13.9% of sales is made up of newly-introduced products by comparison with 9.9% in other manufacturing firms. The figure for engineering and technical services is 15.4% and that for other business services 10.9%. The differences in the proportion of newly marketed products between capital goods/chemicals and engineering and technical services and between basic materials/consumer goods and other business services are not significant. These results for manufacturing are once again consistent with the Freeman (1978) model, as are those for the business service sector. This serves to strengthen the above suggestion that, both in manufacturing and business service firms, it is the engineering firms which are the primary site for the development of new product technology.

Within manufacturing and business services, there are also significant differences in the proportion of upgraded products. In capital goods and chemicals, 25.1% of sales on average are made up of upgraded products by comparison with 14.9% in basic materials and consumer goods. This difference is significant at the 5% level. The figures for engineering and technical services and other business services are 31.4% and 25.9% respectively, which are significantly different at the 10% level. Amongst engineering firms, the difference in the proportion of sales made up of upgraded products between those in manufacturing and those in business services is not significant. For the remaining firms, however, other business service firms have a significantly higher proportion of sales made up of upgraded products by comparison with basic materials and consumer goods firms.

Figure 6.3: Proportion of Product Innovation, Product Upgrading and Unchanged Products in UK SMEs 1992-95

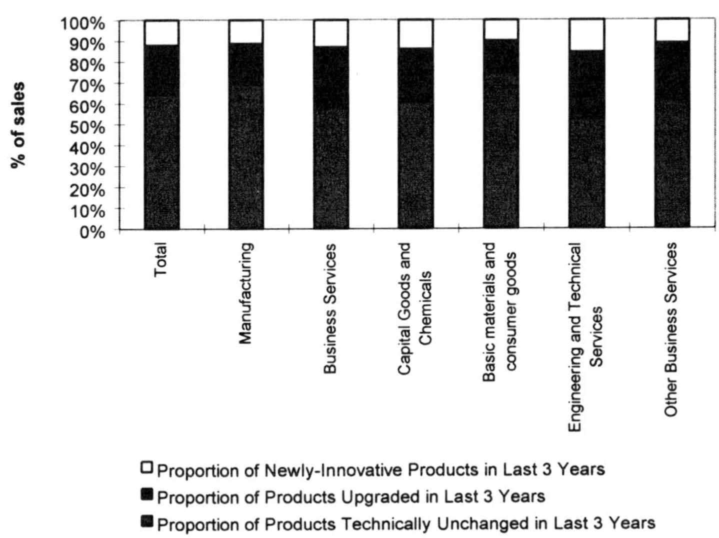

6.5 INNOVATION INPUTS IN MANUFACTURING AND SERVICES

Figure 6.4 provides two indicators of R&D activity. The first is the proportion of firms which report having any staff engaged in R&D either full or part time. For those firms which report having staff engaged in R&D, the figure shows the proportion of staff engaged either full or part time in R&D.

Figure 6.4: R&D Employment in UK SMEs 1992-95

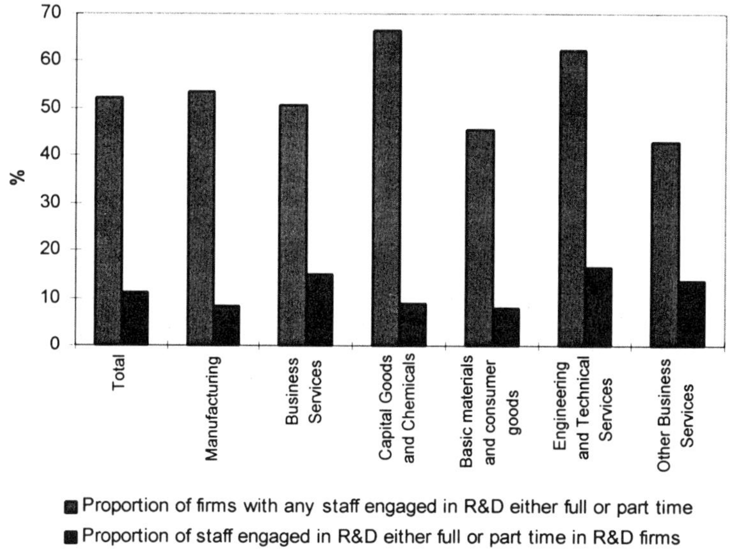

The proportion of firms in the business service and manufacturing samples which have staff engaged in R&D are similar at 50.6% and 53.6% respectively. Within manufacturing and business services, however, there is wide variation in the proportion of firms with staff engaged in R&D. In capital goods and chemicals firms the proportion is 66.4% by comparison with 45.5% in basic materials and consumer goods. Likewise, the proportion in engineering and technical services is 62.4% and in other business services 42.8%. In both cases, the differences are significant at the 5% level using the Mann-Whitney test. Amongst engineering firms, the difference in the proportion of firms with staff in R&D between manufacturing and services is insignificant. The same is true for the remaining firms outside of engineering, with the difference in proportion of firms with staff in R&D insignificant.

Amongst firms which do have staff in R&D, those in business services have a significantly higher proportion employed in R&D by comparison with manufacturing firms, with figures of 15.1% and 8.3% respectively. Within manufacturing and business services, the differences between the engineering and other firms in the proportion of staff in R&D is relatively small. Engineering and technical service firms have a significantly higher proportion of staff in R&D by comparison with capital goods and chemicals firms. Likewise, other business service firms have a significantly higher proportion of staff in R&D by comparison with basic materials and consumer goods firms.

Figure 6.5 shows the proportion of firms which employ technologists, scientists or higher professionals. For those firms which do employ such staff, the Figure also shows the proportion of their total staff complement who fall into those occupational categories. A significantly higher proportion of business service firms employ technologists, scientists or higher professionals by comparison with manufacturing firms. Amongst those firms with staff in that category, the proportion of such staff is significantly higher in business service firms by comparison with manufacturing firms, at 35.8% and 11.3% respectively.

Within manufacturing and business services, firms in the engineering sector have a higher proportion of staff who are technologists, scientists or higher professionals. These differences are significant at the 5% level using the Mann-Whitney test. The differences are much larger between capital goods/ chemicals and engineering/technical services, at 12.4% and 42.9% respectively. Likewise there is a large difference in the corresponding figures for basic materials/consumer goods and other business services at 9.7% and 29% respectively. These differences are not surprisingly also significant at the 5% level.

Figure 6.5: Technologists, Scientists and Higher Professionals as Proportion of Labour Force in UK SMEs 1995

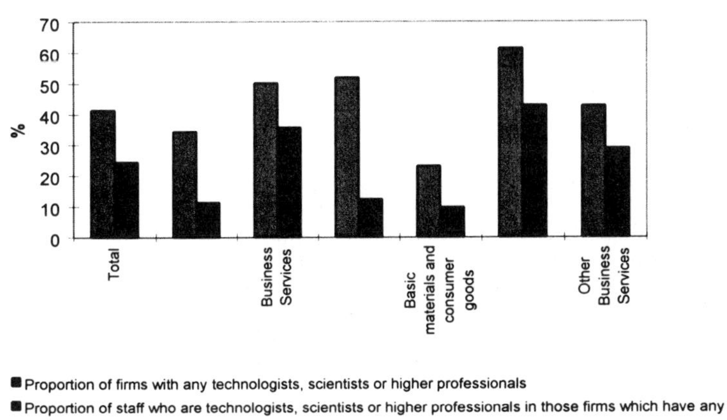

Table 6.3 shows an analysis of the importance of various sources of information for the innovation process. The sample of firms used in this analysis includes only those firms which reported either a product or process innovation in the period 1992-95, as non-innovating firms were not asked to respond to the questions on the sources of information for innovation.

Table 6.3: Sources of Innovation Information in UK SMEs 1992-95

Mean Scores

	Total	**Manufacturing**	**Business Services**	**Capital Goods & Chemicals**	**Basic materials & consumer goods**	**Engineering & technical services**	**Other Business Services**
Industry groups		**A**	**B**	**a**	**b**	**c**	**d**
Internal sources:							
Within the firm	3.71	3.69	3.73	3.95[b]	3.49	3.83	3.67
Within the group	1.30	1.32	1.28	1.50	1.17	1.24	1.31
External sources:							
Suppliers of materials and components	2.48	2.69[B]	2.22	2.73[c]	2.65[d]	2.16	2.26
clients or customers	3.49	3.53	3.44	3.66	3.43	3.65	3.30
Competitors in your line of business	2.71	2.70	2.73	2.77	2.64	2.92	2.60
Consultancy firms	1.52	1.46	1.61	1.50	1.42	1.51	1.69
Universities/higher education institutes	1.38	1.36	1.40	1.46	1.28	1.41	1.39
Technical institutes	1.35	1.37	1.33	1.45	1.31	1.32	1.33
Patent disclosures	1.24	1.25	1.22	1.39[b]	1.15	1.29	1.16
Professional conferences, meetings, professional journals	1.81	1.61	2.06[A]	1.58	1.64	2.15[a]	2.00[b]
Fairs/exhibitions	2.09	2.21[B]	1.93	2.31[c]	2.13	1.97	1.91
Trade associations	1.68	1.74	1.59	1.66	1.80	1.60	1.58
Chambers of commerce	1.23	1.27	1.17	1.30	1.24	1.15	1.20

Means are of ranked scores on a scale 0-5 (0 insignificant, 5 crucial)

B – Score is significantly greater than that in column B at the 5% level
A – Score is significantly greater than that in column A at the 5% level
a – Score is significantly greater than that in column a at the 5% level
b – Score is significantly greater than that in column b at the 5% level
c – Score is significantly greater than that in column c at the 5% level
d – Score is significantly greater than that in column d at the 5% level

Mann – Whitney U non parametric test used.

We can begin by considering manufacturing and business services in aggregate. It can be seen that for both manufacturing and business service firms, the most important external source if information for innovation is clients or customers. Information from suppliers of materials and components and from fairs/exhibitions is significantly more important for manufacturing firms by comparison with business service firms. Information from professional conferences, meetings, professional journals is significantly less important for manufacturing than for business service firms.

Information from suppliers of materials and components is significantly more important for capital goods and chemicals firms by comparison with engineering and technical services firms and for basic materials/consumer goods firms by comparison with other business service firms. Information from professional conferences, meetings, professional journals is significantly more important for engineering and technical service firms by comparison with capital goods and chemicals and for other business service firms by comparison with materials/ consumer goods firms.

Taken together these differences no doubt reflect the relatively greater physical nature of manufacturing production processes, and of inputs of parts and components in production, as well as the relatively greater human capital and disembodied knowledge content of business services. This echoes some of the more traditional bases for distinguishing in these general terms, between goods and services. For certain sources of information for innovation, similar levels of importance between manufacturing and business services mask differences within those sectors. Thus information from within the firm and from patent disclosures is significantly more important to capital goods firms by comparison with basic materials and consumer goods firms. The aggregate differences in the importance of these sources between manufacturing and business services are, however, insignificant.

6.6 BARRIERS TO INNOVATION IN MANUFACTURING AND SERVICES

Our survey data allows us to compare constraints on innovation activity, as well as innovation inputs and outputs. Our analysis of the data in Table 6.4 suggests that there is relatively little difference in aggregate between manufacturing and business service firms in the importance of different barriers to innovation. The only significant difference which is apparent in Table 6.4 concerns the lack of skilled personnel. It appears that innovation in business service firms is significantly less likely to be hampered by the lack of skilled personnel than is the case in manufacturing firms.

Table 6.4: Barriers to Innovation in UK SMEs 1992-95

Mean Scores

	Total	Manufacturing	Business Services	Capital Goods & Chemicals	Basic materials & consumer goods	Engineering & technical services	Other Business Services
Industry groups		A	B	A	b	C	d
Economic factors:							
excessive perceived risk	2.46	2.40	2.53	2.22	2.52[a]	2.73[a]	2.40
Lack of appropriate sources of finance	2.51	2.44	2.61	2.47	2.42	3.06[ad]	2.31
innovation costs too high	2.63	2.67	2.58	2.73	2.63	2.63	2.54
pay-off period of innovation too long	2.38	2.37	2.40	2.40	2.34	2.61[d]	2.26
Firm level factors:							
Firm's innovation potential (eg. R&D, design, etc. too small)	2.41	2.49	2.31	2.53	2.47	2.40	2.26
lack of skilled personnel	2.29	2.42[B]	2.13	2.49	2.37[d]	2.3	2.02
lack of information on technologies	1.85	1.89	1.79	1.91	1.88	1.77	1.8
lack of information on markets	1.98	2.01	1.94	2.00	2.01[d]	2.17[d]	1.79
innovation costs hard to control	2.08	2.06	2.09	2.19	1.98	2.27[d]	1.97
resistance to change in the firm	1.69	1.72	1.66	1.76	1.69	1.57	1.71
deficiencies in the availability of external technical services	1.62	1.65	1.57	1.75	1.59	1.63	1.53
lack of opportunity for co-operation with other firms and technological institutions	1.70	1.70	1.70	1.61	1.76[d]	1.92[ad]	1.56
Other reasons							
Lack of technological opportunities	1.56	1.60	1.52	1.56	1.62	1.56	1.49
no need to innovate due to earlier innovations	1.55	1.60	1.49	1.47	1.68	1.46	1.51
innovation too easy to copy	1.68	1.76	1.58	1.66	1.82	1.54	1.61
legislation, norms, regulations, standards, taxation	1.79	1.80	1.78	1.82	1.78	1.88	1.72
lack of customer responsiveness to new products and processes	1.85	1.87	1.83	1.82	1.90	1.90	1.78
uncertainty in timing of innovation	1.76	1.68	1.85	1.78	1.62	1.98	1.76

Means are of ranked scores on a scale 0-5 (0 insignificant, 5 crucial)

B – Score is significantly greater than that in column B at the 5% level
A – Score is significantly greater than that in column A at the 5% level
a – Score is significantly greater than that in column a at the 5% level
b – Score is significantly greater than that in column b at the 5% level
c – Score is significantly greater than that in column c at the 5% level
d – Score is significantly greater than that in column d at the 5% level
Mann – Whitney U non parametric test used.

More differences come to light in Table 6.4 when we consider the split between the manufacturing and service sectors. Within manufacturing, basic materials and consumer goods firms report a significantly higher constraint on innovation due to excessive perceived risk by comparison with capital goods and chemicals firms. Within business services, engineering and technical service firms report significantly greater constraints on innovation due to lack of appropriate sources if finance, pay-off period of innovation too long, lack of information on markets, innovation costs hard to control, and lack of opportunity for co-operation with other firms and technological institutions.

Differences also exist between the respective sectors within manufacturing and business services. Engineering and technical service firms report significantly greater constraints on innovation due to excessive perceived risk, lack of appropriate sources of finance and lack of opportunity for co-operation with other firms and technological institutions by comparison with capital goods and chemicals firms. Basic materials and consumer goods firms report significantly greater constraint on innovation due to the lack of skilled personnel and lack of information on markets by comparison with other business service firms.

This comparative emphasis on financial constraints by our engineering and technical services firms in consistent with evidence from other analyses (Cosh and Hughes, 1998; Hughes and Moore, 1998; Bank of England, 1998).

6.7 CONCLUSION

In terms of the probability of introducing novel product innovations, the proportion of sales made up of newly innovated products, the likelihood of a firm having staff engaged in R&D, and the likelihood of a firm employing technologists, scientists or higher professionals, it appears that the differences between the manufacturing and business service sectors are smaller than within them. In particular, engineering firms in both manufacturing and business services are significantly more innovation intensive by comparison with non-engineering firms in the same sector. And engineering firms in manufacturing and those in the business service sector show strong similarities in terms of their innovation intensity. More surprising perhaps is that manufacturing firms in basic materials/consumer goods and other business service firms show strong similarities in terms of their innovation intensity.

Despite the strong similarities between the corresponding groupings in manufacturing and business services, significant differences do exist. Amongst firms which have staff engaged in R&D, those in engineering and technical services employ a significantly higher proportion by comparison with those in capital goods/chemicals and those in other business services employ a significantly higher proportion by comparison with those in basic materials/ consumer goods. The same pattern applies for the proportion of staff who are technologists, scientists or higher professionals.

Engineering and technical service firms report significantly higher barriers to innovation due to excessive perceived risk, lack of appropriate sources of finance and lack of opportunity for co-operation with other firms and technological institutions. Basic materials and consumer goods firms report significantly higher barriers to innovation due to lack of skilled personnel, lack of information on markets and lack of opportunity for cooperation with other firms and technological institutions.

These results would seem to suggest that there may in fact be far more in common between the manufacturing and business service sectors in terms of their innovation activity than has previously been supposed. Recent suggestions that innovation activity in manufacturing and services is so fundamentally distinctive that entirely different sets of indicators are required to measure such activity in these sectors appear to be misplaced. This is not to suggest that substantial differences do not exist. On the contrary, applying the same set of measurement indicators across the two sectors usefully captures important differences while also highlighting the strong similarities. More significantly our analysis also reveals that there are important similarities between industry groupings drawn from across the traditional manufacturing and services divide.

Appendix

1995 CBR Survey Instrument

Section B: Innovative Activity

*In this section we would like you to tell us about your innovative activity. We are interested in innovation in products and processes which are **new to your firm**.*

*In answering the questions in this section, please count innovation as occurring when a new or changed product is introduced to the market (product innovation) or when a new or significantly improved production method is used commercially (process innovation), and when **changes** in knowledge or skills, routines, competence, equipment, or engineering practices are required to make the new product or to introduce the new process.*

*Please do **not** count as product innovation, changes which are purely aesthetic (such as changes in colour or decoration), or which simply involve product differentiation (that is minor design or presentation changes which differentiate the product while leaving it technically unchanged in construction or performance).*

B1. Has your firm introduced any innovations in products (goods or services) or processes during the last three years which were new to your firm? *(Please tick only **one** box in **each** row)*

	Yes	No
Products		
Processes		

If you ticked No for both products and processes please skip B2.

B2. If you introduced a product innovation, was it, to the best of your knowledge, already in use in other firms either in (a) your industry or (b) other industries? If you made more than one product innovation please answer with respect to your most important product innovation. *(Please tick only **one** box in **each** row)*

Product Innovation	Yes	No	Don't Know
(a) in use in your industry			
(b) in use in other industries			

B3. If you introduced a process innovation was it, to the best of your knowledge, already in use in other firms either in (a) your industry or (b) other industries? If you made more than one process innovation please answer with respect to your most important process innovation. *(Please tick only **one** box in **each** row)*

Process Innovation	Yes	No	Don't Know
(a) in use in your industry			
(b) in use in other industries			

References

Archibugi, D., Cohendet, P., Kristensen, A. and Schäffer, K.A. *Evaluation of the Community Innovation Survey*, Aalborg: IKE Group, SPRINT/EIMS Report, 1995.

Bank of England. *The Financing of Technology-Based Small Firms*, London, October, 1996.

Barras, R. 'Towards a Theory of Innovation in Services',*Research Policy*, 1986; Vol. 15, pp.161-173.

Barras, R. 'Interactive Innovation in Financial and Business Services: The Vanguard of the Service Revolution', *Research Policy*, 1990; Vol. 19, pp.215-237.

Cosh, A., Hughes, A., and Wood, E. 'Innovation: Scale, Objectives and Constraints' in Cosh, A., and Hughes, A. (eds.), *The Changing State of British Enterprise*, ESRC Centre for Business Research, University of Cambridge, 1996.

Cosh, A., and Hughes, A. (eds.). *The Changing State of British Enterprise*, ESRC Centre for Business Research, University of Cambridge, 1996.

Cosh, A., and Wood, E. 'Innovation Scale, Objectives and Constraints' in Cosh, A., and Hughes, A. (eds.), *Enterprise Britain: Growth, Innovation and Public Policy in the Small and Medium Sized Enterprise Sector 1994-97*, ESRC Centre for Business Research, University of Cambridge, 1998, pp.38-48.

Cosh, A. and Hughes, A. (eds.). *Enterprise Britain: Growth, Innovation and Public Policy in the Small and Medium Sized Enterprise Sector 1994-97*, ESRC Centre for Business Research, University of Cambridge, 1998, pp.38-48.

Cosh, A.D., Hughes, A., and Wood, E. 'Innovation in UK SME's: Causes and Consequences for Firm Failure and Acquisition' in Acs, Z.J., Carlsson, B., and Karlsson, C. (eds.),*Entreprenuership, Small and Medium Sized Firms and the Macroeconomy*, Cambridge: Cambridge University Press, 1999.

Desai, M. and Low, W. 'Measuring the Opportunity for Product Innovation' in de Cecco, M. (ed.), *Changing Money: Financial Innovation in Developed Countries*, Oxford: Basil Blackwell, 1987.

Evangelista, R. and Sirilli, G. 'Measuring Innovation in Services',*Research Evaluation*, 1995; Vol. 5, No. 3.

Evangelista, R. and Sirilli, G. 'Innovation in Services and Manufacturing: Results from the Italian Surveys', *ESRC Centre for Business Research Working Paper*, No. 73, December 1997.

Foyw, F. 'Community Innovation Survey 1997-1998', *Statistics in Focus*, Research and Development Theme 9-2/1999, pp.1-7, Eurostat, Luxemboug, 1998.

Frantz, R. *X-Efficiency: Theory, Evidence and Applications*, New York: Kluwer-Nijhoff, 1987.

Freeman, C. 'Technical Innovation and British Trade Performance' in Blackaby, F. (ed.),*De-industrialisation*, NIESR, London: Heineman, 1979.

Gallouj, F. and Weinstein, O. 'Innovation in Services',*Research Policy*, 1997; Vol. 26, pp.537-556.

Hill, T.P. 'On Goods and Services', *Review of Income and Wealth*, 1977; 4, pp.315-338.

Hughes, A. 'Finance for SME's: A UK Perspective',*Small Business Economics*, 1997; Vol. 9, No. 2, April, pp.151-166.

Hughes, A. and Moore, B.C. 'High-Tech Firms: Market Position, Innovative Performance and Access to Finance' in Cosh, A.D. and Hughes, A. (eds.), *Enterprise Britain: Growth, Innovation and Public Policy in the Small and Medium Sized Enterprise Sector 1994-1997*, ESRC Centre for Business Research, University of Cambridge, 1998, pp.86-98.

Kleinknecht, A. 'Measuring and Analysing Innovation in Services and Manufacturing', Paper presented at the workshop entitled "Conceptualising and Measuring Service Innovation" at the Centre for Research in Innovation and Competition (CRIC) at Manchester Business School in May 1998.

Kleinknecht, A. and Bain, D. (eds.). *New Concepts in Innovation Output Measurement*, London: Macmillan and New York: St. Martin's Press, 1993.

Leibenstein, H. *Inside the Firm: The Inefficiencies of Hierarchy*, Cambridge, Mass.: Harvard University Press, 1987.

Miles I. *Services Innovation, Statistical and Conceptual Issues*, Working Group on Innovation and Technology Policy, OECD (Doc. DSTI/EAS/STP/NESTI/ (95)12), Paris, 28 March 1995.

Miles, I., Kastrinos, N., Flanagan, K., Bilderbeek, R., Den Hertog, P., Huntik, W. and Bouman, M., 'Knowledge-Intensive Business Services: Users, Carriers and Sources of Innovation', Rapport pour DG13 SPRINT-EIMS, March 1995.

OECD. *Proposed Guidelines for Collecting and Interpreting Technological Innovation Data* (Oslo Manual), OECD, Paris, 1992.

OECD-EUROSTAT. *Proposed Guidelines for Collecting and Interpreting Technological Innovation Data* (Oslo Manual*)*, OECD, Paris, 1997.

Pavitt, K. 'Sectoral Patterns of Technical Change: Towards a Taxonomy and a Theory',*Research Policy*, 1984; Vol. 13, pp.343-373.

Pavitt, K., Robson, M. and Townsend, J. 'The Size Distribution of Innovating Firms in the UK: 1945-1983', *The Journal of Industrial Economics*, 1987; Vol. 35, No. 3, pp.297-316.

Pavitt, K., Robson, M. and Townsend, J. 'Technological Accumulation, Diversification and Organisation in UK Xompanies, 1945-1983', *Management Science*, 1989; Vol. 35, pp.81-99.

Saviotti, P.P. and Metcalfe, J.S. 'A Theoretical Approach to the Construction of Technological Output Indicators', *Research Policy*, 1984; Vol. 13, pp.141-151.

Sundbo, J. 'Modularization of Service Production and a Thesis of Convergence Between Service and Manufacturing Organizations', *Scandinavian Journal of Management*, 1994; Vol. 10 (3), pp.245-266.

Vickery, G. 'An International Study of Selected Business Services' in*Service Sector Statistics*, Proceedings of the Statistics User's Council Annual Conference November, Esher, IMAC Research, 1998.

---

[1] A previous version of this chapter was presented at a workshop entitled "Conceptualising and Measuring Service Innovation" at the ESRC Centre for Research in Innovation and Competition (CRIC) at the University of Manchester in May 1998. The authors acknowledge the financial support of the ESRC for the research underlying this work.

[2] The CBR sample and survey methodology are discussed in detail in Cosh and Hughes (1996, 1998) and Bullock, Duncan and Wood (1996). Innovation activity in the sample is discussed in Cosh, Hughes and Wood (1996), Cosh and Wood (1998) and Cosh, Hughes and Wood (1999).

Chapter 7
Service Innovation: What Makes it Different? Empirical Evidence from Germany

Brigitte Preissl

7.1 INTRODUCTION

The increasing importance of services, both in terms of employment and in terms of contribution to Gross Domestic Product (GDP), has generated an interest in the innovation potential of the sector. Since the overall dynamics of growth of an economy is more and more determined by the development of service activities, innovation in services becomes a crucial factor for an economy's innovativeness and competitiveness. It has long been accepted that services are not generally adverse to innovation. Furthermore, traditional concepts of services as low-tech, low-productivity industries with little impact on a country's economic performance had to be given up. The importance of services for national innovation systems goes beyond their own innovative activities. Services play a major role in the economy as catalysts, carriers and facilitators of innovation in other sectors and in other industries of the sector itself[1].

The rise of services to general respectability went along with an assimilation of service and manufacturing sectors[2]. Manufacturing processes can be disaggregated into series of services with only small components remaining for the actual transformation of materials. Thus, the characteristics of manufacturing activities and, thus, their management differ less and less from those of service activities. On the other hand, services develop features of 'industrialisation', such as the growing use of technology (mainly information technology), modularization, and standardisation.

The question that follows from obvious convergence tendencies is how far differences are actually wiped out and how significant specific characteristics of services still are. In the context of this chapter, the peculiarities of service production will be analysed and it will be asked whether they lead to dynamics of innovation that differ from those observed for manufacturing industries. However, it should be borne in mind that the service sector consists of a range of heterogeneous services. For some services automation processes made possible by the use of information technology (IT) are revolutionising processes of service provision and render them equal to industrial production processes, while others are hardly affected by the diffusion of IT. Differences between service industries can thus be more influential for innovation dynamics than those between two industries belonging to the manufacturing or to the service sector.

The purpose of this chapter is to point out aspects of the innovation process in services that are specific to this sector. It will thus concentrate on phenomena that distinguish service innovation from innovation in general. Besides these peculiarities there are many elements of innovation, its creation, dissemination and implementation, that show characteristics which are similar to those of manufacturing innovation. Many aspects of innovation theory that have dominated recent discussions, such as processes of appropriation of knowledge, learning organisations and changing qualification needs, do not seem to present significant differences between the two sectors. Often the size of companies is a more significant variable than the manufacturing-service dichotomy. Hence, some phenomena are more common in services than elsewhere, because they are characteristic for small companies and there is a relatively large number of small firms in the service sector. The following text will focus on those aspects that are specific to services because they reflect differences in productive activity between the two sectors. The arguments presented here are supported by empirical evidence gained from research in Germany. However, service innovation studies conducted in other EU countries suggest that there are no major differences, at least in a European context[3].

The chapter has three parts. In Section 7.2 characteristics of innovation processes, which have been generally accepted as basic concepts for an analysis of innovation in manufacturing[4], will be described. Section 7.3 will ask whether these concepts provide a valid framework for the analysis of service innovation by confronting them with the results of empirical analysis of innovation in German service industries[5]. Finally, conclusions for service innovation theory and research will be presented.

7.2 DETERMINANTS AND CHARACTERISTICS OF INNOVATION PROCESSES

Research on innovation has focused on manufacturing industries[6]. This has had an important impact on the central topics of discussion and on the main research findings as well as on the indicators and variables in innovation analysis. Features that have been identified as relevant for the understanding of innovations are:

- Impulses for innovation are typically expected from *technical developments, and inventions of new machines or materials*.

- Innovation results from research and development (R&D) activities. These activities aim at the generation of production techniques or new marketable goods. R&D does not necessarily have to be allocated within the innovating firm, but can also be bought in from other sources in national or transnational innovation systems.
- The focus on technology as the main motor of innovations has lead to innovation policies that concentrate almost entirely on supporting technical projects and technology related R&D activities[7].
- It is generally expected that innovations are about methods of production or goods that are *new to the market*. 'Markets' in this context are usually conceived as national markets or national economies.
- Being the first supplier of a new product or the first user of a new technology allows the initiator or generator of an innovation to gain competitive advantages and to realise extra profits for some time. Thus, the driving force behind innovation is competition and the prospect to leave competitors behind.
- In order to allow an innovator to exploit a new product or a new production method economically, and thus to stimulate innovation activities, innovations can be protected by patents.
- For purposes of analytical precision innovations are usually categorised as product, process or organisational innovations. These innovation categories differ with respect to driving forces, processes of development and implementation as well as with respect to impact on employment and productivity. They characterise different stages in the evolution of manufacturing industries and in the cyclical movement of economic activity levels. The other innovation types that have been suggested by Schumpeter in his classical contribution to the development of market economies[8], market and delivery innovation have been of minor importance in the innovation literature[9].
- Technology orientation in innovation studies means that innovation based on new delivery concepts, organisational changes and entry into new markets has been neglected.
- In accordance with the emphasis on technology and technical R&D as main sources of innovation the number of people qualified for technical and engineering jobs are taken as an indicator for the innovation potential of an economy.
- Process innovation ultimately leads to productivity growth and thus to a rise in profits[10]. Product innovation allows the realisation of extra profits.
- The strategic importance of innovations led to secrecy in R&D programmes and a general reluctance of companies to share knowledge related to innovation and to co-operate with outsiders. Innovations were therefore mainly resulting from internal learning processes or from patents bought from an external supplier.

Innovation indicators derived from these statements reflect the bias towards manufacturing industries. On the input side, R&D expenditure, personnel with science end engineering degrees and the acquisition of patents have been the most important indicators. On the output side, innovation intensity has been measured by productivity growth, turnover from new products and patents gained. Empirical research on service innovation shows that these indicators might give an insufficient and misleading picture of the innovation achievements and potentials of the sector. The validity of these generally acknowledged characteristics of innovation for the service sector shall therefore be challenged in the course of this chapter.

7.3 SERVICE INNOVATION IN GERMANY: THE EMPIRICAL EVIDENCE

7.3.1 Sources of information

The following analysis of service innovation is based on research that has been carried out in the context of four projects:

- Deutsches Institut für Wirtschaftsforschung, Department of Industry and Technology, Berlin (DIW/SI4S): Case studies in 25 service firms conducted for the SI4S project.
- Deutsches Institut für Wirtschaftsforschung, Department of Regional Studies, Berlin (DIW/Berlin): A regional survey on innovation in Berlin which contains a sample of 775 service firms.
- Zentrum für Europäische Wirtschaftsforschung, Mannheim (ZEW and ISI): A representative postal survey on innovation in services which provided information for almost 2 900 service firms.
- Institut für angewandte Innovationsforschung, Bochum (iAi): A study on growth potential and innovation deficits in services based on expert interviews.

The DIW/SI4S team had access to original data from the ZEW/ISI survey and from the DIW/Berlin survey. Additional information has been collected in eight expert interviews with academics, industry representatives and government officials. (Table 7.1 shows the main characteristics of these sets of information.)

Both ZEW/ISI and DIW surveys started with concepts of innovation and innovation measurement that were derived from the Oslo and Frascati manuals but were then modified according to pretest results because the originally adopted concepts did not lead to a satisfactory quality of data. The different regional scope (Germany and Berlin) makes the two surveys incomparable in a statistical sense. In addition, the DIW/Berlin study allows for a direct confrontation of manufacturing and service industries[11].

Table 7.1: Sources of information on service innovation in Germany

	year of data collection	size of sample	service industries covered	scope of research
ZEW	1995/1996	2900	wholesale/retailing, transport, financial services, related to financial s., software, technology consulting, R&D, other business services	innovation in services. representative postal survey, Germany, old and new Länder
DIW/ Berlin	1996	775	hard-and software, management consulting, architecture, technical planning, R&D, marketing, other business services	innovation activities in Berlin, all sectors, representative survey
iAi	1996	29 experts	Waste disposal, health, trade, security, others	innovation potentials and deficits from a management point-of-view
DIW/ SI4S	1997	25 + 8 experts	Retailing, publishing, R&D, transport, IT, management consulting, technical planning, marketing, public administration	case studies, selected service industries

7.3.2 Distinctive characteristics of service innovation

From the vast material collected in the context of the projects described above, those results will be presented in this chapter that show the specific characteristics of *service* innovation. The following topics have been chosen for this purpose:

- types of innovation
- definition of innovation
- product innovation
- process and organisational innovation
- technology and innovation
- R&D and innovation
- co-operation in innovation and in R&D
- barriers to innovation
- qualification

7.3.2.1 *Types of innovation*

Three cases can be distinguished (see Figure 7.1). Service innovation can be initiated by service firms and directed towards an anonymous market. It can also be stimulated by clients, who ask for a particular service which is not yet available in the firm's existing range of services (case 2). The client identifies a problem (possibly with the help of a service provider) and grants a contract to a service firm. In order to be able to offer a solution to the client's problem, the service firm has to create a new service. The impulse for the innovation thus came from the customer, whereas the innovation itself is carried out by the service provider. In the third case, service firms can function as carriers of innovation: here an innovation is implemented in the client's business with considerable input from the service provider, but the service itself does not have to be innovative. Most analyses deal with the first case, because it comes closest to manufacturing innovation where goods are being developed without any involvement from the demand side. However, the other types seem to become increasingly relevant in services, because they reflect the characteristics of services of being generated in the course of their provision and of involving customer participation[12]. Furthermore, the role of services as agents for the realisation of innovation in other sectors is being acknowledged in the literature as an important part of national innovation infrastructures[13].

Figure 7.1

Service firm				*Client firm*
(1) innovation	→→→	**service provision**	→→→	**use of service**
(2) service offer	←←←	**service contract**	←←←	**product identification**
innovation	→→→	**problem solution**	→→→	**use of new service**
(3) service offer	→→→	**service provision**	→→→	**innovation**

Definition of innovation

Pretests for postal surveys as well as the case studies conducted for the SI4S project have shown that managers in service firms are often not familiar with the term 'innovation'. Questionnaires try to avoid this problem by offering explanations and/or definitions or by using different, more general terms, like 'changes in the firm' or 'new services'[14]. These deficits in communication between researchers and service providers can more easily be overcome in case studies. Here, a quite common experience was that the interviewed person either stated that no innovations had been introduced in the company *or* that any case of service provision was an innovation, because the service was 'generated' anew for each client all cases were different and never completely repeatable. In the first case, often further inquiries revealed that the company had indeed been innovating without being aware of it. In the second case, in the course of the interview a distinction between actual innovations and processes of gaining new work experience with each service provision could be established as a basis for further discussion.

Most innovations that were identified in the SI4S case studies were either new to the company or new to the regional market, only exceptionally they were new to the national or international market. However, for most service companies, the relevant market is the local or regional, not the national economy. Therefore, activities which had not existed before in the area where the company meets its main competitors, are to be categorised as 'innovation', despite the fact that they are not new to a larger (national or international) market. In terms of the impact on productivity and competitiveness of the firm this makes sense, because both will be positively influenced by the new activity. In terms of the export potential of those service firms and thus the international performance of the German service sector, the contribution of this type of innovation is rather small. However, if local service companies are able to offer a service which is international 'state-of-the-art', although it is not innovative with respect to the German market, this makes it easier for these firms to compete with foreign competitors that enter the market. Furthermore, learning processes involved in providing new services can be used to have a better chance to get access to export markets.

As many others, German studies on innovation in services use the traditional distinction between process and product innovation and add organisational innovation. At first sight, the two large surveys ZEW/ISI and DIW/Berlin show contrasting findings about the relative importance of the two main types of innovation. The ZEW/ISI study concludes that process innovation is the dominant form of innovation in services, and that product innovation is much less important[15]. The Berlin innovation survey suggests the contrary[16]. Here many firms could not relate process innovation to their activities, and therefore the authors concluded that this innovation type was not really relevant for services - at least compared with manufacturing. However, Berlin service firms were quite active with respect to organisational innovations. Between 30% and 47% of the firms (depending on the industry) in the sample had engaged in this type of innovation in the last three years[17] whereas in the ZEW/ISI sample organisational innovation was by far the least important innovation type. Process innovations were so insignificant in the service sample in Berlin that no useful results could be derived from the respective data, and this innovation type was not analysed any further. On the contrary, the SEW/IS report concludes that between 33% (in services auxiliary to finance in East Germany) and 60% (in transport services in the West) of service innovations were related to *processes* of service provision.

It is quite obvious that a large part of the discrepancies go back to different definitions of process and organisational innovation. The Berlin study categorises phenomena, which for ZEW/ISI are process innovation, as 'organisational' and thus, not surprisingly, comes up with a much lower process innovation figure. If both types of innovation are taken together, the differences between the two surveys almost disappear[18].

However, this example might also lead to the conclusion that in service innovation the categories product, process and organisational innovation are not adequate to analyse innovation activities and need further specification. Contrary to what one might expect due to the usual characterisation of services as economic activities in which product and process are inseparable, in the DIW case studies[19], it did not seem too difficult to separate process from product innovation. The problem with the distinction between process and organisational innovation seems to be related to the specific characteristics of services. In manufacturing 'processes of production' refer to the technical side of what happens in the factory along the assembly line, whereas 'organisation' refers to administration, departmental structures, marketing and logistics. The two spheres can be analytically separated. In services, however, 'processes of production' are often nothing else but the 'organisational' settings in a service firm. It therefore does not seem appropriate to distinguish between 'process' and 'organisation' in service innovation.

Delivery innovations are an important type of service innovation because they affect the core of service provision: the customer-supplier-interface. Automation of services as a typical delivery innovation can generate a new service and usually constitutes a major change in the process of service provision. An increasingly important example is *self-service information systems*. If information is provided via an electronic data base instead of being delivered by a person, the service rendered is 'the information', and, in principle, the content of the service does not change whether the information is given by a person or by a computer system. However, the computer might give access to additional services, and thus provide a slightly different service from the one that existed in a personalised form before. In this case, delivery innovation goes along with product innovation.

Product innovation

The analysis of product innovation in service firms led to a set of results which point at the special features of service innovation as compared to manufacturing innovation:

- *Services are not as well defined as material products, therefore the transition from one service to another can be difficult to identify.*

The volatility of service provision enhances this effect. New services can emerge in the process of provision due to special requirements that were not visible before. Often service innovations of this type are not perceived as an 'innovation', since they are not the result of a planned and deliberate action. The 'novelty' might then be forgotten, because it is not codified, and no physical 'sample' or blue print exists. However, the new service might also be repeated in a similar situation so that the service provider becomes aware of an innovation opportunity. Only then the new service will be codified, integrated in firm strategies and identified as an innovation.

- *Many product innovations consist of minor changes of or additions to existing services.*

However, these innovations may well be based on extensive learning processes and thus be characterised by a high degree of appropriability. This appropriability is rather derived from lead time, learning-curve effects as well as costs and time to reproduce the service innovation, than on patents or secrecy[20]. Two examples shall illustrate this. The first relates to an innovation in IT services that consists of the customisation of existing software. Software consultancy firms have discovered that often complex new software programmes cannot be easily used by potential adopters and offer the implementation and integration of software in existing IT systems as an IT consultancy service. Obviously, due to the required expertise in IT and software handling, the service cannot be easily and quickly imitated by competitors. In the second example, the addition of vacuum packaging in food stores as a new customer service, imitation is not difficult, but additional sales efforts[21] as an accompanying measure might increase appropriability. In both examples, service innovators make intensive use of technology as an input, but the service innovation itself is non-technological (a consultancy service in the first, and a retail service in the second example).

- *Only a few service innovations are new to the market or "pioneer services".*

Due to this fact, service innovations do not show the distinct characteristics of a patentable novelty. The area of reference for many services is the local or regional market, despite tendencies of internationalisation that become increasingly relevant for others. Therefore, the impact of innovations that are only 'new' to a local market, on company strategy and performance as well as on competitive dynamics in the local markets, does not differ much from that of 'true' innovations. The crucial factor for the innovation decision is still the intensity of competition in the market for a new service. The heterogeneity of the service sector does not allow for a generalised statement here, since competitive conditions as well as the geographical extension of relevant markets vary greatly between industries. The thesis of high industry specificity[22] of the impact of competition on innovation is particularly valid in this respect. The industries which did produce service innovations that were new to the national or even a wider market, during the period covered by the research (1994 to 1997), were mainly software and environmental services.

- *Service innovations can be attributed to the service provider or to its client*

This problem is not only one of proper classifications. The attribution of the innovation to one party pre-determines the possibilities to benefit from a new service and, thus, has an important impact on incentives to innovate. A prominent example is technical planning, where service innovation mainly consists of the *delivery of blueprints* for innovative solutions to manufacturing firms or clients in other service firms. Here the innovation might involve an innovative service but it does not have to. As a rule, the innovation should be attributed to that party which holds the right to exploit the results. However, this rule becomes unsatisfactory, if the exploitation by one of the parties is subject to conditions; for example, rights might be limited in time or can be exercised only in certain circumstances or in certain markets. Furthermore, due to the intangible nature of knowledge and information, a service firm can hardly be prevented from using knowledge gained in the process of generating an innovation for one client, in other similar cases working for another client. The fact that a particular solution has been produced once, and hence the knowledge is with the service provider, can lead to the same kind of competitive advantage that a fully codified and appropriated innovation would have for the service firm. Thus, even if technical consultants have agreed to transfer rights concerning an innovation to their client, part of the effect of the innovation stays with the service provider.

- *By far the largest group of service innovations are a combination of services which integrate various functions that were previously offered by a whole variety of firms.*

The new type of service that emerges in this context results from an increasing demand for *innovative problem solutions.* This phenomenon reflects a trend in the economy to move away from the purchase of isolated goods and services which are combined into company specific product/service packages according to individual customer needs by the buyer's management team, towards the contracting out of 'problems' to be resolved by one service firm. This service firm then combines the necessary inputs from different service and non-service industries and provides turn-key solutions in a one-stop-shopping scenario.

- *A considerable share of service innovations are directly related to IT.*

New services emerge in three different fields, where services function as generators, carriers and facilitators of technology diffusion and technological innovation:

(a) new software developments,
(b) IT integration services (provision of IT *systems*) and
(c) assistance in handling and adopting IT products.

These new services are often linked with the distribution of IT products. They respond to a need to accompany the implementation and exploitation of IT with expert knowledge (in technological as well as in management areas) which is generally not available in the user firm. The dynamics of product innovation in IT related services are strongly dependent on the path of technological development in IT manufacturing and diffusion patterns of IT.

Process and organisational innovation

Since service innovation surveys did not provide meaningful results for the empirical distinction of process and organisational innovation, both types are treated together here. However, in the SI4S case studies, different categories of process innovation could be identified:

A first group is centred around the *re-organisation of procedures* of service provision. Especially, in larger service firms, like banks or insurance companies, which have a larger range of options for organisational arrangements, innovation affects *centralisation or decentralisation* of decision-making competencies and responsibilities.

Another type of re-organisation is the *outsourcing of support functions*, like accounting, cleaning or printing services. Outsourcing strategies belong to a more general set of *changes in the division of labour* between departments and individuals, and between companies with different specialisations. An example for an internal re-arrangement of competencies is the attribution of personal consultants to bank clients which has substituted functional competence structures in all major German banks. Clients thus refer to one employee only for any service they require, and do not have to go from one counter to the next for different service needs. In most cases these innovations can only be realised with the support of IT, though they are not directly 'caused' by the technology, but motivated by much wider strategic considerations[23].

Re-arrangement of service organisation is a necessary step for the *introduction of modularised services*. The division of services into modules prepares the standardisation and 'industrialisation' of service provision. On the one hand, modularization requires to define a service (by splitting it up in well-specified components), and to introduce factory-like procedures in service provision. On the other hand, modularised services can be consumed in individual packages without having to re-generate the whole service concept for each individual customer. Clients can choose between full or basic services, simple or sophisticated versions. Thus, processes of differentiation and customisation are paradoxically combined with codification and standardisation.

The second group of process innovations comprises those innovations that are based on *automation*. Here, personal services which were previously provided by human beings are delivered by machines, like ticket vending machines or computer-operated train timetables. Other innovations in this group aim at increasing productivity in the provision of services through automation, for example, computer assisted editing or the use of standard software packages to fill in tax declaration forms.

Automation processes are new to most service industries. Innovation in this field requires careful planning of the balance between human and machine supported services. In manufacturing, it might not make a considerable difference, whether an object is hand-made or comes from a fully automated conveyor belt, because essentially the product will serve the same functions. In services, however, automation alters the quality of service, and this may have a negative effect on customer satisfaction. It can cause a sense of neglect and insecurity in the client. In this case, accompanying measures will be needed, like emergency phone numbers (with real people at the other end of the line) and continuous monitoring of the acceptance of the system.

IT-related re-organisation of business procedures is another form of process innovation which is present in almost any service firm, except the very small ones or the very new ones. With respect to the motivation for and the impact of the adoption of IT, services do not seem to differ much from manufacturing.

Undoubtedly, IT adoption moves service provision closer to industrial processes of production, it opens up opportunities for rationalisation and productivity gains. Especially smaller service firms in traditional services, like retailing or consultancy, however, consider the use of IT rather as a 'must' that cannot be avoided because of the risk to loose out in competition, than as an opportunity for innovation.

7.3.2.2 *Interaction between service innovation and technology*

The following scheme shows the different ways in which technology adoption and innovation interact in services (see Table 7.2)[24].

Table 7.2: Interaction between technology and services in innovations

channels of interaction	**innovation examples**
substitution of service provision by technology	self-service information terminals electronic entry control
definition of service through technology	support of Internet adoption SAP/R3 customisation
determination of service through technology	development of new software for process control computer based consultancy IT training
diffusion of technology through services	designing Internet presentation for client SAP/R3 support assistance in desktop-publishing
production of new technologies through services	blue prints for recycling development of biogas systems

In the case of *substitution of services by technology*, it is left to the customer to handle the technology. The acceptance of this new active part is crucial for the success of an innovation. The service provider implements a more or less user-friendly system and provides instructions for the consumer on how to use it[25]. Labour saving strategies are the main motif for this kind of change.

Definition of services by technology mainly refers to technologies which require complementary services to result in a marketable product. Internet technology, for example, has no use-value without services to make it accessible and services to be delivered via its channels. Here technology is a necessary infrastructure which sets the framework for mushrooming service innovations; on the other hand, the technology remains meaningless without the service[26].

The technology provides options. In order to use them, it has to be adapted to the user's needs in a process of customisation. Here services have an active role in the diffusion of technology. Thus, a new division of labour between technology providers, service firms and their customers in innovation processes emerges. Apart from the role of services in technology diffusion in other firms and industries, the adoption of technology is one of the central drivers of process and product innovation in service firms themselves[27].

Empirical analyses have shown that all forms of interaction between technology and service innovation are relevant, and there is no dominant pattern. Obviously, the central technology in service innovation is IT, however, in environmental services, biotechnology and recycling techniques become increasingly important. Nonetheless, all technologies adopted and developed in service industries are supported by or operated with IT components. This emphasises the strong absorption capacity of services for IT as well as the high information intensity of most services[28]. IT is the most powerful driver of change in services. Partly it simply enhances the industrialisation of services and makes the gap with manufacturing disappear. However, often more complex changes are induced. IT revolutionises the handling of and the access channels to information. This leads to a great potential for the automation of services that deal with the processing, transmission and diffusion of information. At the same time, the demand for knowledge which enables people to handle rapidly changing technologies and infinite flows of unstructured information, select the relevant bits and turn them into meaningful production inputs, rises dramatically. In addition, there seems to be scope for a 'human touch' to be added to services which lack a personal component after they have been taken over by computers[29]. This involves a need for service providers with high qualification levels who are able to provide comprehensive, context-based interpretation and customisation of random information. Thus, IT related innovation in services result in processes of standardisation and individualisation.

7.3.2.3 *R&D and service innovation*

R&D activities provide the most commonly used indicators for innovation in manufacturing. The concept of R&D comprises basic research for the development of new technologies and new production materials and their preparation for use in manufacturing processes. From negligibly low R&D figures for services it seemed to follow that services were innovation laggards[30]. Because of the low technology intensity in services, it has also been assumed that R&D was not relevant for these industries. Recent developments throw considerable doubt on the validity of these assumptions: (1) the low technology content hypothesis does not hold any longer in the light of intensive use of IT in services, and (2) empirical research suggests that R&D, measured as size and expenditures of R&D departments, might not be the best indicator for service innovation. The negative message coming from official R&D statistics is difficult to interpret, however, because it is not clear whether the obvious lack of R&D in services is due to actually lower R&D intensity or just the result of bad coverage of the service sector in R&D documentation[31].

In the Frascati manual, R&D is a precisely defined activity; the concept is well established in manufacturing contexts. However, respondents in postal surveys for *service innovation* often are not familiar with such R&D concepts[32]. The Frascati manual considers only activities performed in a separate department or by distinct groups of researchers as R&D. This leads to a severe negligence of R&D activities in services, where these functions are in most cases dispersed in the company and performed along with the process of service preparation and service provision[33]. Some surveys therefore operate with wider definitions[34].

The ZEW/ISI report presents an indicator which gives the share of R&D expenditure in total expenditure for an innovation. This indicator shows that on the average only 13% of service firms had spent a substantial share of expenditures for innovations on R&D[35]. Other components, like the conception of new services, tests and market introduction, improvement of procedures, follow-up investment and qualification have led to substantial expenditures much more often.

In the SI4S case studies, often managers in knowledge intensive business service firms claimed to do a substantial amount of R&D, but it turned out that this was done on behalf of a customer who got full possession of the results and not as an activity that eventually led to new products or new production procedures within the R&D performing firm. Statistically, this R&D therefore should be attributed to the client industry.

Resources for R&D were a significant limiting factor. Often R&D was considered to be too expensive to be affordable for rather small firms. In the current recession R&D projects are being postponed until the liquidity position allows money to be put aside for these purposes.

With the exception of industries that are specialised in R&D and technical planning, R&D in service firms concentrates on the conception of new services, new organisational and procedural solutions for service provision, and on IT implementation. It is less concerned with technical basic research. In all service firms that were included in SI4S case studies, R&D is *informally* organised. In the larger service firms, R&D activities are either allocated in operations research, strategic planning or IT departments. In smaller firms, they are performed in ad-hoc project teams. Neither large nor medium-sized or small German firms have R&D departments. There is a clear concentration of R&D on IT applications and software. In some cases, co-operation with other service providers enables firms to conduct expensive research and to use knowledge which is not available in-house. There is no evidence that service firms tend to contract out research, which however, may be increasingly true for manufacturing firms, especially for research intensive industries. As could be expected, patents do not play a significant role in service R&D. One reason for this is that it is difficult to define a patentable service innovation, and imitation can hardly be controlled; another reason is the short duration of innovation cycles which do not allow for lengthy patenting procedures.

7.3.2.4 *Co-operation in innovation planning and realisation*

Co-operation and networking is not a trend that emerges independently from other trends in service markets. Accordingly, networking per se is not a strategic concept, unless it follows from other market requirements or strategic aims. However, there are market developments that require a higher degree of co-operation than service firms were used to engage in. For example, increasing demand for problem solutions can only be met if service providers with varying competencies combine their resources and capabilities. As a result, a networked group of service providers will have a competitive advantage over isolated service firms that still have to build up contacts with firms that are capable to participate in service provision.

A specific form of co-operation in services is the participation of clients in the conceptualisation and realisation of an innovation[36]. This is a phenomenon which gains increasing importance along with the customisation and individualisation of services. The fact that services are often being 'created' while they are provided, facilitates the involvement of customers in the process. The intensity of interaction depends on the endowment of both parties with the relevant knowledge and the willingness of the client to see the service as a co-operative product. Thus, often the success of co-operation is considered to rely on good personal relationships between the service providers and their counterparts in the client firm. Accordingly, a lack of participation of the customer can make a project fail. Since German manufacturers often have little experience in external service provision, they expect a turn-key problem solution and do not see why they should have an active part and put their own resources in a service that they have already paid for. This rather new form of dealing with inputs requires learning processes in the client firm which can slow down the dynamics of service innovation.

Some smaller technology-oriented service firms are afraid that they might loose out in co-operation with larger partners, and their research results might be appropriated by potential competitors. Others think that benefits from commonly generated innovations might not be shared fairly.

7.3.2.5 *Barriers for service innovation*

Barriers to innovation in service firms can be grouped in four categories[37]:

1. Market risks
2. Financial restrictions
3. Legal and bureaucratic barriers
4. Restrictions within the company.

Market risks comprise uncertainty about the realisation of an innovation, the development of demand and competition as well as risks in cost calculation. Financial restrictions can be divided into a lack of internal funds on the one hand and difficult access to bank loans and grants on the other. Internally, innovations can be hindered by a lack of personnel and physical resources and by resistance from management and employees.

Barriers to innovation differ between small and large firms[38]. In small service firms a lack of capital is mentioned by almost half of the firms as an important reason for reluctance to innovate, whereas only 16% of large service providers seem to have this problem. Cost of innovations was a barrier for far more than half of the small firms, and still for 44% of the larger ones. For both groups internal resistance and lack of know-how were the least important categories. Legal and bureaucratic barriers seemed to concern larger firms more than the smaller ones[39].

In the case studies, the most frequent spontaneous answer to the question what hindered innovation was a lack of venture capital. Only after the presentation of a list of possible barriers by the interviewer, other factors emerged. However, shortage of finances was still seen as the outstanding problem. The attitude of financial institutions towards service firms applying for innovation capital is determined by typical manufacturing innovations. The same seems to hold for public support programmes: Investment in human resources, in blue prints and the development of innovative service concepts is not an asset banks are prepared to lend money against, because there is no inherent material security (like buildings, machines or patents). More guarantees have therefore to be provided by the lender even for relatively small sums of money. Interestingly, in the service firms' perception, publicly supported venture capital did not seem to offer any more advantageous or less risk-adverse conditions. Obviously, financial markets and public grant administrators do not consider service innovation as a profitable investment.

Public support programmes are too slow to ease the problem. In markets with quick innovation cycles, long application procedures and extremely long decision making processes in public entities prolong times to market and thus can have a negative effect on the success of an innovation. Often money is effectively paid only after the project has been completed. Especially in projects with long development times, large sums have to be advanced by the innovating firm, and liquidity throughout the development of new services is a crucial factor. Schemes that reimburse expenses at the end of a project are therefore not attractive for many service firms. As a consequence, they often prefer to rely on their own resources or not to innovate at all.

Usually innovation support programmes do not provide grants for market research and market introduction. Yet 24% of the firms in the ZEW/ISI sample claim that expenses for this part of an innovation are of eminent importance. In the case studies, costs of exploring markets and of preparing the introduction of a new service, were among the factors that made firms hesitate to engage in innovations.

Software development requires the engagement of highly qualified personnel over a long period of time. Since software cannot be patented (only algorithms), securities for bank loans are an important problem. Once the new programme has been developed, it can be multiplied with almost no additional cost; but often markets are too insecure to make this a valid guarantee for the provision of grants in the initial phase. Software markets have extremely short product cycles. Therefore, high investment in development must be accompanied by quick sales of large numbers of copies in a short time. This increases the risk of investment in software and makes moneylenders reluctant to support software companies.

Other barriers to innovation that were described in the case studies related to political factors, technical problems, absorptive capacity of markets, patenting and qualifications (see Figure 7.2). Some of the factors mentioned in the case studies obviously derive from the personal experience of the interviewed and are not 'objective' barriers. However, the (more or less subjective) judgement of managers about the prospects for innovative activities is often the major determinant of innovation decisions.

Figure 7.2: Barriers to innovation - Statements of service providers

Political factors	*Technical problems*
- broadcasting regulation requires multimedia service providers to deal with 16 Länder regulators - publishers are not allowed by retailing laws to market books directly through electronic channels - data protection laws limit the use of customer data for marketing purposes - procedural regulation in public services prevents the introduction of customer-friendly innovations - pending legislation in health care and in environmental services makes strategic innovation planning of innovations	- high risk in the development of blueprints for new technologies, because there are no 'tests' - lack of machines for the purposes of small companies - lack of experience with 'young' technologies innovation is more complex than expected, and plans are abandoned - ministries that decide about R&D support lack technological knowledge
Absorptive capacities of markets	*Qualification*
- managers in manufacturing are conservative and not open for innovative service concepts - potential customers in manufacturing lack a 'service mentality', they are reluctant to outsource services and they do not co-operate easily with service providers - clients are not prepared to pay an adequate price for high quality services - lack of knowledge about the customer's market: can he gain a competitive advantage, if he has access to an innovative service?	- qualification is not a sufficient condition; experience on the job is equally, if not more, important - employees lack management thinking; training in management skills would be helpful, but is too expensive - traditional vocational training does not enable young people to react flexibly to new requirements - problem of keeping up the necessary continuous dynamics in the acquisition of knowledge - qualification levels will be a problem in the future, given the dramatic cuts in public investment in education.
ZEW/ISI 1996, Eickelpasch/Pfeiffer 1997, Preissl 1997.	

7.3.2.6 *Personnel and qualifications*

As expected services are rather labour intensive. Personnel costs made up for between 20% and 80% of total costs in the case study sample (43% on the average). In knowledge intensive service industries the main assets are people with particular qualifications. These qualifications must be flexible in the process of innovation. Qualifications in services are likely to vary with knowledge intensity. Table 7.3 gives an overview of qualification levels in German service firms. The figures emphasise the high importance of vocational education and training and of non-specified qualifications[40]. Only in software as well as in technical consultation and R&D does higher education prevail. Even in financial services and auxiliary activities more than half of the employees went through vocational training and only relatively few (16 resp. 18%) hold university degrees.

Table 7.3: Configuration of qualifications in German service industries (averages in p.c.)

	university degrees		college for further education	vocational education + training	other
	sciences	business			
Wholesale	6	4	13	57	20
Retailing	3	4	11	67	14
Transport/communi-cation	2	3	9	54	33
Financial services	4	12	13	57	14
Auxiliary to financial services	6	12	15	56	11
Software	33	9	20	31	7
Technical consulting/ R&D	40	7	12	32	9
Other business services	10	11	13	44	22

Source: ZEW/ISI 1996, p.59, Table 7.4.

The DIW/SI4S case studies have shown, however, that these formal degrees only provide a set of basic skills which are necessary, but not sufficient conditions for an adequately qualified workforce. In the context of innovation, flexible adjustments as well as mechanisms and techniques to acquire new skills and qualifications are the relevant variables.

Three types of qualification schemes are to be distinguished:

- programmes to enhance general skills to prepare employees for new challenges, for example, motivational courses, TQM, customer service courses, management skills;
- learning about an innovation and about how to realise it, for example, information seminars on SAP, conferences on new management concepts, like Efficient Customer Response - EPR, or on corporate design in the Internet;
- learning to work with new processes and new services, for example, introduction into computer supported customer care systems, or in desktop publishing.

These three re-qualification schemes reflect the different components of qualification needs in innovation processes. The first set usually involves external courses (or internal courses held by external consultants or trainers). The second set can be used as a tool to gain impulses and information about innovation and to provide first ideas on how to handle the implementation of an innovation. In the third case, the decision about an innovation has been made and the company is in the process of putting it into effect. Training in this phase should enable employees to master new processes and to provide a new type of service. Types two and three are often organised, encouraged or supported by suppliers of IT hardware and software. Type three can also consist of courses in rather basic skills, like communication or presentation techniques which enhance innovation by providing new concepts and informing about new ways of doing business.

Knowledge about innovation can also come from experience in the work process. Especially consultants and R&D firms emphasised the fact that their work routines involve continuous learning. In these firms, personnel are specifically qualified to develop information into new knowledge. The development of skills 'that cannot be taught' is an important element in companies' innovation processes. Practical experience in the same firm or at least in a similar job in another firm is a scarce and highly appreciated qualification. Only very slowly qualification measures include standardised knowledge, like courses on ISO 9000 quality standards.

In all smaller firms within the case study sample the dominant form of re-qualification is learning-by-doing. Mutual support and co-operation among employees take care of the spread of knowledge and skills within the company. In contrast, larger firms train people in formalised external and internal courses (in all three types). It can be assumed that the reasons for these differences between SMEs and large companies are (1) a lack of finances for expensive formal re-qualification in small firms, and (2) the need to provide standardised well organised formal qualification programmes in large firms in order to guarantee a harmonised and co-ordinated implementation of innovations in large organisations.

7.4 CONCLUSIONS

In the following, conclusions will be drawn with respect to service peculiarities that support the claim for a separate 'theory of innovation in services'.

- Service innovation shows conceptual characteristics that differ from the generally accepted innovation criteria. Services are rarely new in the context of a national or international market. Service industries differ among each other in terms of their relevant markets; and a criterion that requires that an innovation is new with respect to at least a national market does to cover the majority of innovative activities in services.

- In services it is not unlikely that innovations are initiated and stimulated by a service provider's client. In many cases active participation of customers in the conception and realisation of an innovation is a necessary condition for success.

- Services have an important role as catalysts, carriers and facilitators of innovation. Thus service innovations have a complementary role in national innovation systems which is due to their characteristic form of knowledge processing.

- Process innovation can be analytically separated from product innovation, however, it does not seem to make much sense to differentiate between organisational and process innovation.

- Innovation indicators have to be revised. On the input side, it has been shown that the common understanding of R&D does not take into account the specific informal nature of R&D in services. Service sector R&D is not directed towards new machines and materials, but towards new procedural patterns and new service concepts. On the average, service innovation involves less expenditure on R&D as innovation in manufacturing. As a consequence, other innovation inputs, like investment in market tests or in training and qualification, become more important. The codification of non-technological knowledge as an essential service innovation input has only just started.

  On the output side, the most striking difference to the manufacturing world is the irrelevance of patents for the protection of innovations. Obviously, in services, other mechanisms take over this protective function. Difficulties to measure service output and service quality contribute to the difficulty to measure the impact of innovation in a service firm or a service market.

- The eminent importance of knowledge as an input in service innovation requires specific appropriability concepts. Knowledge is an intangible asset, and, in the same way as information, it does not disappear in the seller's mind when it is sold.

- Understanding service innovation at the current stage of convergence between processes of production in services and in manufacturing involves an understanding of the dynamics of differentiation and customisation on the one hand and standardisation and codification on the other, which characterises recent developments in service processes. Modularization seems to be an important trend with specific implications for the direction of innovation efforts.

Most of the characteristics listed above are due to the basic nature of services; they derive from the fact that in many services production and consumption fall together, but also from the high knowledge intensity which creates special features of service innovation. Where technology challenges the traditional service characteristics, and process and product fall apart geographically or in time, it still cannot be taken for granted that the resulting activity will not differ substantially from manufacturing activities. Volatile procedures, complex codification of processes and results, as well as specific rules in the use of technology and in processes of industrialisation deserve particular attention. It might turn out, however, that new boundaries have to be drawn across service and manufacturing sectors to categorise industries according to characteristics in innovation dynamics, since some services may be more similar to certain manufacturing industries than to other services.

REFERENCES

AD-Employ. *Employment Trends Related to the Use of Advanced Communications*, Synopsis Report, PREST, Manchester, April, 1995.

Bundesministerium für Bildung, Wissenschaft, Forschung und Technologie (BMBF+T) (1997): Bundesforschungsbericht 1996 (Federal Report on Research in Germany).

Coombs, R. 'Technological Opportunities and Industrial Organization' in Dosi et al (eds.),*Technical Change and Economic Theory*, London: Pinter, 1988, pp.295-308.

denHertog, P. and Bilderbeek, R. *Recent Innovation Patterns in Services in the Netherlands*, TNO Report STB 97-05, April, 1997.

denHertog, P. and Bilderbeek, R. *Innovation In and Through Knowledge Intensive Business Services in the Netherlands*, TNO Report, STB 98-03, January, 1998.

Dosi, G., Freeman, C., Nelson, R., Silverberg, G. and Soete, L. (eds.). *Technical Change and Economic Theory*, London: Pinter, 1988.

Dosi, G. 'The Nature of the Innovative Process' in Dosi et al (eds.),*Technical Change and Economic Theory*, London: Pinter, 1988. pp.221-238.

Eickelpasch, A. and Pfeiffer, I. *Zukunftssicherung durch Innovation - Profil, Potential und Strategien der Unternehmen in Berlin*, Berlin: DIW, 1997.

Gallouj, F. and Weinstein, O. 'Innovation in Services',*Research Policy*, 1997; 26, pp.537-556.

Hales, M. *Producer Services and Manufacturing Production*, A Report for the OECD Science, Technology and Industry Group, July, 1997.

Hauknes, J. *Innovation in the Service Economy*, STEP Report No. 7/96, Oslo, December, 1996.

Hauknes, J. *SI4S (Services in Innovation - Innovation in Services): Services in European Innovation Systems*, Final Report, Oslo, 1998.

Landauer, T. *The Trouble with Computers, Usefulness, Usability and Productivity*, Cambridge, Mass.: MIT Press, 1995.

Miles, I., Kastrinos, N., Flanagan, K., Bilderbeek, R., denHertog, P., Huntink, W. and Bouman, M. *Knowledge-Intensive Business Services: Their Roles as Users, Carriers and Sources of Innovation*, PREST, December, 1994.

Miles, I.and Boden, M. 'Are Services Special? - Services and the Knowledge Based Economy' in Miles, I. (ed.), *Services, Innovation and the Knowledge-based Economy*, (forthcoming 1998)

Preissl, B. *Service Innovation in Germany*, Report to the SI4S Project, Berlin: DIW, August, 1997.

Preissl, B. 'New Landscapes in the Service Sector - Information and Communication Technologies and Structural Change' in Madden, G. and Macdonald, S. (eds.) *Information Technology and Development*, Elsevier, 1998a.

Preissl, B. 'Service Innovation in Germany - Information Technology as a Driving Force', *Vierteljahrsheft zur Wirtschaftsforschung des DIW*, 2/98, 1998b.

Schumpeter, J. *The Theory of Economic Development*, Cambridge, Mass.: Harvard University Press, 1912 (Reprint 1934).

Strambach, S. *Knowledge-Intensive Services and Innovation in Germany*, University of Stuttgart, Institute of Geography, July, 1997 (mimeo).

Zentrum für Europäische Wirtschaftsforschung and Fraunhofer Gesellschaft - Institut für Systemforschung und Innovation, ZEW/ISI, 1996, Innovationstätigkeit im Dienstleistungssektor, Mannheim Karlsruhe.

---

[1] See Miles et al (1994) and denHertog and Bilderbeek (1997).
[2] See Miles and Boden (1998); Gallouj and Weinstein (1997).
[3] See Hauknes (1998).
[4] See also Gallouj and Weinstein (1997), p.537.
[5] This part of the chapter is based on research for the project "Service in Innovation - Innovation in Services (SI4S)" conducted on behalf of DG XII of the European Commission in 1997.
[6] Hauknes (1996), p.106.
[7] See, for example, BMB+F (1997).
[8] Schumpeter (1912).
[9] See, for example, Gallouj and Weinstein (1997).
[10] This thesis has been challenged, especially with respect to the effects of innovation based on information technology, by the observation of a so-called productivity paradox. See, for example, Landauer (1995).
[11] It has to be kept in mind, however, that Berlin's economy is a special case because of the long lasting isolation of West Berlin and the direct confrontation of a transition economy in the East and highly protected economic structures in the Western parts of the city. In addition, the disastrous financial situation of the Berlin Senate resulted in a severe intensification of the current recession.
[12] See Gallouj and Weinstein (1997), p.541.
[13] See, for example, denHertog and Bilderbeek (1998), Strambach (1997) and Miles et al (1995).
[14] This approach has been adopted in national surveys conducted by SI4S teams, whereas the ZEW/ISI solution was to give a simple definition in the questionnaire in order to help respondents to categorize their activities correctly.
[15] ZEW/ISI (1996), p.8. All further figures referring to this survey are quoted or adapted from ZEW/ISI (1996).
[16] Eickelpasch and Pfeiffer (1997), p.33 and p.37. All further figures referring to this survey are quoted or adapted from Eickelpasch and Pfeiffer (1997).
[17] Eickelpasch and Pfeiffer (1997), p.120.
[18] To a certain extent, differences in the results might be due to sample designs and regional diversity, and not caused by definitional inconsistencies.
[19] See Preissl (1997).
[20] On appropriability conditions see Dosi (1988), p.230-231 (citing Levin et al, 1984).
[21] See Dosi (1988), p.230.
[22] See Coombs (1988), p.296.
[23] It goes beyond the scope of this chapter to discuss whether process innovations in services follow an (IT-based) technology-push or a (market-related) demand-pull pattern. Case studies suggest, however, that the driving forces vary between industries, and the answer depends on the impact of IT in individual services. See Coombs (1988), p.295, and Preissl (1998a, 1998b).
[24] The scheme is an elaboration of Hauknes' "channels of interaction" (Hauknes, 1996, pp.109-110).
[25] Hauknes points out that substitution can also occur in the internal organization of a company, for example in the computerization of library services in a consultancy firm (Hauknes, 1996, p.109).
[26] Hauknes gives broadcasting as an example: here physical networks and content providers are perfect complements and each of them gains its meaning and its economic value from the other (Hauknes, 1996, pp.109-110).
[27] See Preissl (1998b).
[28] See Preissl (1998a).
[29] AD-Employ (1995), p.4.
[30] See, for example, Hauknes (1996), p.51 and p.56.
[31] See Hauknes (1996), p.57, and Hales (1997), p.28.

[32] This problem occurred in the ZEW/ISI survey (ZEW/ISI, 1996, p.24, note 17). See also Eickelpasch and Pfeiffer (1997), p.42.
[33] See Hauknes (1996), p.57 and Preissl (1997), p.54.
[34] This holds, for example, for the two German surveys (ZEW/ISI and DIW/Berlin), but also for two Dutch surveys conducted in 1988 and in 1993 (see Hauknes, 1996, p.60). ZEW/ISI try to compensate for the inappropriate Frascati definition by constructing a *broad* and a *narrow* notion of R&D. In the broad concept all firms that claim to engage in R&D, are counted, in the narrow concept only firms that also have some sort of an R&D department. With this distinction, it was found that 16% of all firms in the broader sense and 7% in the narrow sense had engaged in R&D in the reference period. See ZEW/ISI (1996), p.25.
[35] The lowest share was reported for transport services (5%), the highest for software firms (43%.). See ZEW/ISI (1996), p.25 and Preissl (1997), p.51.
[36] See also Gallouj and Weinstein (1997), p.541.
[37] ZEW/ISI (1996), p.69.
[38] All figures from ZEW/ISI (1996) data.
[39] See ZEW/ISI (1996), p.71, fig.8.2.
[40] In contrast with the ZEW/ISI interpretation it might be argued that the category ‘others’ does not necessarily contain only people with lower qualification levels than vocational degrees. They might also include qualifications for which the other categories are not adequate, for example, degrees gained abroad, or non-formal training.

Chapter 8
Information Flows and Knowledge Creation in Knowledge-Intensive Business Services: Scheme for a Conceptualization

Christiane Hipp

8.1 INTRODUCTION

8.1.1 New Modes of Knowledge Creation

In recent years the contribution of knowledge and information to the competitiveness of nations was analysed from different perspectives and on different levels[1]. Hertog and Bilderbeek (1997) argue that knowledge is a key resource in the economy today for two reasons. Firstly, the competitiveness of most industries depends on the innovative knowledge creation and transforming process, and secondly, knowledge-intensive industries contribute to the economic growth and competitiveness of countries. Guinet (1997, pp.173-174) summarises first results of a project on "Knowledge Flows in National Innovation Systems" currently undertaken by the OECD (selection):

(1) There is a clear trend towards higher knowledge intensity in all economic sectors, and there are indicators that higher knowledge intensity leads to better performance at the firm, sectoral and aggregate level.

(2) Flows of tacit knowledge have an important positive effect on innovation performance, and particularly on the ability of firms to detect, adapt, and use new knowledge and technologies.

(3) In most countries 'clusters' exist, in which firms interact through closely linked knowledge networks and which show above-average performance in terms of international competitiveness, growth and employment. These clusters are not necessarily high-technology or R&D-intensive, but they are - without exception - knowledge-intensive, when one takes into account the intensity of knowledge interactions among sectors.

Gibbons et al (1994) emphasise the need for a new understanding of the innovation process which goes far beyond either the linear or the chain linked model normally used in innovation theory. This new approach should be able to illustrate the increasing importance of group-entrepreneurship and network dependencies (Hanusch and Canter, 1993; Jacobs, 1997).

We can conclude that structural change - from a technology-based industrial economy to a service economy where the knowledge creation is one of the most important resources - is also reflected in the shift from a scientifically and technologically based innovation process to new forms of knowledge creation which are to a less degree dominated by technology. Instead, they are more transdisciplinary; including intellectual and social aspects.

8.1.2 Role of Knowledge-intensive Business Services (KIBS)

Different studies[2] show that knowledge-intensive business services play an important role in the changing knowledge infrastructure of the national system of innovation and "*become the prime source of sustained high value-added (...). In each case the producer services sector uses specialist knowledge to provide solutions which give products, even mass produced ones, their specific market edge*" (Gibbons et al, 1994, p.121)[3].

For the economic performance and the competitiveness of the national innovation system it is important to increase the information stock and to make the existing stock more socially useful. David and Foray (1995) emphasise that the efficient distribution and utilisation of knowledge cannot be expected to rise automatically. It is necessary to concentrate more on the knowledge distribution process than on the creation process of new information. The ability to transform information into knowledge as well as to receive and use knowledge for solving economic problems is at least as important for change and growth as the process of learning and discovery[4].

Recently, enterprises in manufacturing have started to divide themselves into smaller, more flexible units, to become members of firm networks and to learn "*from co-operation with specialised companies in the service sector*" (Hammerer, 1996, p.4). KIBS are important agents in the development of new knowledge (for example, R&D). These companies also assist in the widening of knowledge as their interaction with clients leads to greater client understanding. One reason for the use of business services is to avoid the costs of acquiring and maintaining such up-to-date knowledge in-house (Miles et al, 1994).

Due to their 'boundary spanning' role and their ability to create new and combine existing knowledge, KIBS play a significant role as knowledge-broker in the national innovation system. Some of the most dynamic industries for employment growth in Germany are the technical consultants and the providers of other corporate business services (Licht et al, 1997, p.12). This indicates the growing importance of these services as well as the increasing intertwining of manufacturing and business services.

8.1.3 Research Questions

Unfortunately, up to now, there are only rudimentary theories about the knowledge creating process in the service sector. Traditional approaches try to describe and analyse the innovation process in manufacturing. Barras (1986) was the first to develop an innovation theory for the service sector. The innovation process in the industrial goods sector usually at first represents a product innovation before improving the product quality and eventually leading to process innovations[5]. Based on this, Barras suggests a reverse product life cycle for the third sector, starting with the adaptation of new technologies to improve service processes before improving the quality of existing products, and finally developing new service products with the support of the new technology.

Barras helps us to understand the interactions and interdependencies between process and product innovation in the service sector and their influence on the internal learning process. He also emphasises the role of technology[6] in the service innovation process, but he does not shed light on the knowledge creation process in detail and different forms of information and knowledge flows, learning processes, and innovation networks in the economy. So a more detailed analysis of the innovation process in the knowledge-based economy that takes into account the dynamics of structural change and the heterogeneity of the emerging and existing services and manufacturing industries is still needed.

Based on the theoretical literature, I will offer a new, industry-independent definition of KIBS which represents the role of KIBS in the national innovation system to a greater extent. This approach also takes into account the fact that the official industry classification is not perfect and in more than 50% of all cases does not represent the real activity of the companies. The data and indicators from two German surveys in the service sector carried out in 1995 and 1997 enable me to analyse the characteristics of KIBS in terms of their internal knowledge creation process, the use of technology, external sources of knowledge and the innovation output. The hypothesis is: 1) that knowledge-intensive business services are highly integrated in an innovation network of clients, customers, and competitors; 2) that they play the role of a knowledge-broker, i.e. that they make existing knowledge useable for other companies; and, 3) that they have their own, internal knowledge-creating processes. This is a first step towards a new description of the innovation process for KIBS, as well as a first empirical step towards a new conceptualisation of the service sector.

8.2 DEFINITION OF KNOWLEDGE-INTENSIVE BUSINESS SERVICES

8.2.1 Existing Definitions of Knowledge-intensive Business Services

It is difficult to find a clear definition of KIBS because these service companies combine service activities for other companies (not private households) with a knowledge-intensity which make them hard to grasp. The literature is rather vague: different authors emphasise different aspects. The variety of terms reflect a variety of points of emphasis - some authors focus more on the performance, some on professional functions, the formal qualification, or the differences in wage structures (NIW, 1995).

Like many others, Machlup (1962) describes services in terms of their knowledge activities. All companies which sell knowledge belong to the professional services. This comprises legal services, engineering services, accounting and auditing services, and some medical services. Miles et al (1994, p.7) argue in the same direction. They define KIBS as business units which "*involve economic activities which are intended to result in the creation, accumulation or dissemination of knowledge*". They distinguish between KIBS I, which are traditional professional services (liable to be intensive users of new technology) and KIBS II, which are new technology-based KIBS.

A different approach is used by Alvesson (1993, p.1004). He assumes that knowledge-intensive business services offer "*products and services to other organisations so that these conform to the institutionalised expectations of their environments*"[7]. He distinguishes - like Hertog and Bilderbeek (1997) - between the products and services on the one hand and the KIBS as carriers of advanced knowledge on the other hand.

For Strambach (1997) knowledge-intensity means that companies do not - like cleaning enterprises - offer routine services. Her analysis differs explicitly from approaches where knowledge-intensive organisations are defined with the help of the number and the proportion of knowledge workers (defined by formal education level) inside the company. Strambach suggests that the 'education-approach' does not take into consideration other forms of knowledge, like tacit knowledge inside the firm, or the ability of companies to learn, and to adopt all kinds of knowledge from outside the organisation. The problem with her definition of KIBS is that she does not define clearly what is and what is not routine.

In 1984, Pavitt developed his famous taxonomy which characterises sectoral patterns of innovation and technical change (Pavitt, 1984). He distinguishes between three different types of companies: (1) science-based companies (chemical, electronics industry); (2) production-intensive companies: divided into specialised equipment producers (mechanical, instrument industry), and scale-intensive firms (metal manufacture and vehicles industry); and, (3) supplier-dominated firms (textile industry, service industry). Pavitt classified the whole service industry as supplier-dominated, without R&D, receiving their innovations from outside the sector.

Because Pavitt's approach does not reflect the variety and heterogeneity of the service sector in terms of its innovation activities, Soete and Miozzo (1989) developed a new adapted taxonomy (see Figure 8.1). In the new supplier-dominated cluster one can find public, personal and distributive services. The second cluster is divided into production- or scale-intensive services and network services. These firms depend on large-scale back-office administrative tasks (banking) as well as on physical and information networks (transport and telecommunication). The last cluster contains specialised technology suppliers and science-based services. In this category, Soete and Miozzo (1989) classify firms which do have own innovation activities, using and developing new technologies[8].

Figure 8.1: Changing Views of Innovation in Services

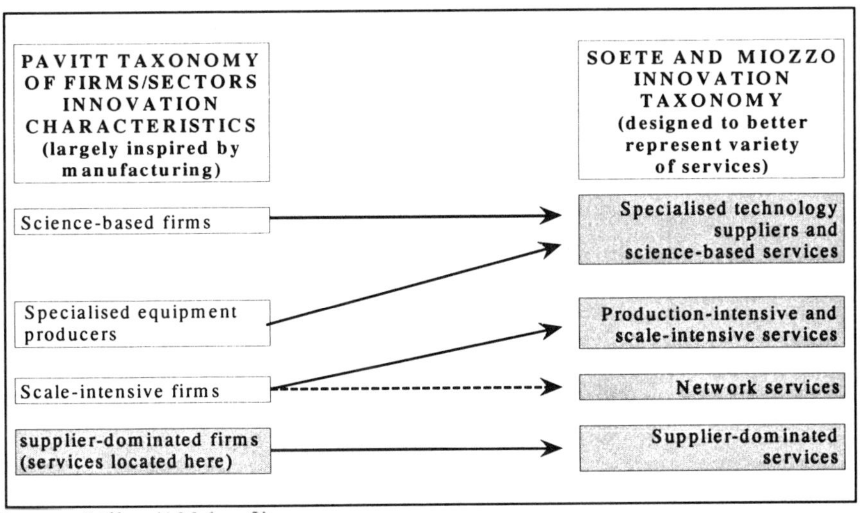

Source: Miles (1996, p.9)

Soete and Miozzo never tested their approach empirically so their analysis is a good starting point for a new definition and empirical test of KIBS which are located in the category of specialised technology suppliers and science-based services.

There is one main problem with the existing theoretical and empirical approaches. Normally all authors relate their definitions to special service industries, which might be characterised by the suggested distinctions. But the measurement of such knowledge-intensive business services by official sources represents a universal dilemma (Wood, 1996). Often these categories (for example, NACE classification) include consumer (and not business) services, even on the four digit level[9]. The classifications on national and international level also vary. This is especially true for business services. Furthermore, Grupp (1997) and Scherer (1982) show that firms group more around new products than in terms of industries. In the new knowledge-intensive economy this effect will be enlarged. In addition, especially in the service sector, a lot of new services appear which do not fit into the traditional classification scheme. Thus, taking also into account the weakness of the classification itself, it is not very useful to analyse KIBS empirically with the help of the official industry classification. A new definition of KIBS is needed.

8.2.2 A New Definition of Knowledge-intensive Business services

Although there are conflicting statements, it is possible to highlight some peculiarities of service products, commonly used in the literature: 1) heterogeneity of different service industries; 2) close interaction between service provider and customer ('integration of the external factor'); and, 3) highly intangible content of service products and processes (information, knowledge) and therefore the need for knowledge-/information-creating and -transforming processes. These peculiarities mean for the KIBS: 1) that their innovation activities are not comparable to other service industries or activities; 2) that they require more supplier-user interactions than many other information and communication services; and, 3) and that they are highly integrated in the knowledge-creation and -diffusion network.

Because of the problems mentioned above and because we talk about the role of KIBS in the knowledge infrastructure as well as the new mode of knowledge production, I will define KIBS in a new way. We can summarise that KIBS are characterised by the ability to receive information from outside the company and to transform this information together with firm-specific knowledge into useful services for their customers. Von Hippel (1988) was one of the first to draw attention to the role of users in the innovation process. The new mode of knowledge production as well as the service literature underline the importance of the integration of the customer into the production and innovation process (for example, Franke, 1991; Miles et al, 1994; Strambach, 1994). So the first characteristic feature is the existence of a close link between the customer (manufacturing or services) and the innovation process of the service firm (see Figure 8.2).

The only remaining problem is to find an indicator for knowledge-intensity. Roelandt and den Hertog (1996) propose indicators which separate knowledge bases, knowledge flows and competitiveness. They emphasise the importance of measuring the transfer and absorptive capacity, types of knowledge and knowledge carriers, as well as the potential role of intermediaries. So I will define 'intense' as the capability to integrate different sources of information and knowledge in the intra-firm's innovation process. To consider the knowledge-intensity in terms of scientific or technological knowledge and to be in line with the typology offered by Soete and Miozzo

(1989), we assume that KIBS have a close link to the science base embodied in universities or other research institutes.

Figure 8.2

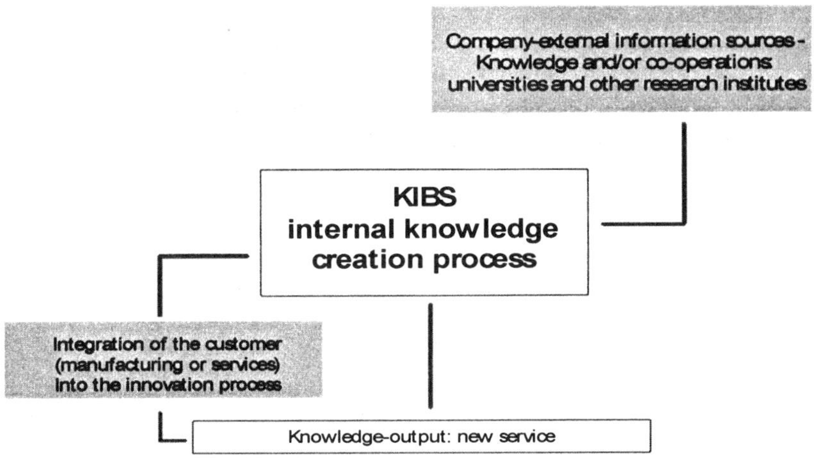

The advantages of this new, industry-independent definition are:

- The definition automatically takes into account that KIBS are intermediaries between knowledge-producers and knowledge-users.
- Indicators provided by the innovation survey allow for a clear distinction between KIBS and other companies. This distinction does not depend on the official industry classification.
- The definition takes into consideration the (disembodied) information and knowledge flow between different organisations and the ability to receive tacit and codified knowledge from outside (science base and customer). This overcomes the weakness of qualification indicators (for example, level of employees with university degree).

8.3 EMPIRICAL ANALYSIS OF KNOWLEDGE-INTENSIVE BUSINESS SERVICES

8.3.1 Description of the Service Innovation Survey in Germany: Methodology

To illustrate the theoretical work, I will use data taken from two German innovation surveys in services. The surveys are carried out on behalf of the German Ministry for Research and Technology (BMBF) and are jointly conducted by the Centre for European Economic Research (ZEW), the

Fraunhofer Institute for Systems and Innovation Research (FhG-ISI), and INFAS.

For the first written survey, the questionnaire was sent out to more than 11,000 German service companies, asking for data ranging from turnover, employment, human capital structure, innovation and R&D efforts, the supplier/customer interfaces, the relevance of certain technologies, obstacles to innovation, and certain other firm characteristics. As a sampling frame we used the database of CREDITREFORM, the largest credit-rating agency in Germany. This data set provides us with background information on all companies, for example, ownership structure, number of employees and sector affiliation (for a detailed description see Licht and Stahl, 1995). Based on this information we decided to apply a stratified sampling procedure; the strata are made up by region, size and industry. In 1995, 2,900 companies returned the questionnaire. From the non-respondent firms, a random sample of 1,100 enterprises was drawn, and these were contacted once again. The non-response study is designed to reveal any distortions in the implemented sample, and to correct them if necessary (Licht et al, 1997, pp.91-94).

In 1997, we started a second survey with a revised, CIS-standardised questionnaire, asking the same companies again. Since the service sector is very dynamic, we had to add new companies to the sample while others had to be removed because they had folded. For this chapter, I used data from companies who answered in the first as well as in the second survey. So I analyse companies which already existed in 1995, and which are still in business.

8.3.2 Descriptive Analysis: General Overview

Now, after two surveys, the data is available for 3,845 German service companies. Around 1,000 firms answered both questionnaires. 15% of them are defined as KIBS. Less than one third did not innovate during the last three years, while 73% carried out product, process or organisational innovations. In the material I analyse only innovators, that is, I compare innovative firms which are not KIBS with innovative KIBS[10]. In the end the sample consists 785 innovative companies, which is biased in the sense that there are no companies founded after 1994.

The distribution of KIBS by industry is not completely industry-independent (see Figure 8.3). 40% of the science and technology services and one third of the software companies are KIBS (but not all of them!)[11]. In banking/insurance and other financial services there are only 14% (but anyhow there are some!). This result promotes the underlying assumption that KIBS need to be defined as industry-independent if you want to take into consideration their specific knowledge characteristics.

Figure 8.3: Knowledge-intensive Business Services by Industry (unweighted)

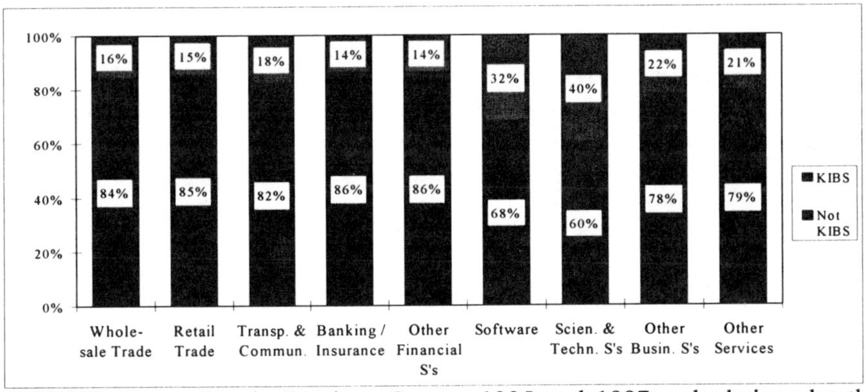

ZEW/ISI: Innovation in Services, Survey 1995 and 1997, calculations by the author

8.3.3 Analysis of the Knowledge-creation Process

There are different dimensions of knowledge. First, it can be classified into explicit and tacit knowledge[12]. Explicit knowledge is articulated in formal language while tacit knowledge can hardly be formalised. Second, knowledge can be embodied or disembodied. Embodied knowledge is linked to humans, technology or equipment. Disembodied knowledge is written or stored knowledge. Third, knowledge has different levels of proprietary (Grupp, 1997, p.320). The last dimension is important if one looks at knowledge as an output of the innovation process and at the incentive structure for innovation. In this chapter, the emphasis is put on the knowledge-creation process, where knowledge is regarded as input factor and the proprietary dimension is neglected.

Gibbons et al (1995, p.24) point out that the innovation process cannot be understood without opening the 'black box' of knowledge creation. Nonaka and Takeuchi (1995, p.71 and p.84) suggest a five-phase model of the organisational knowledge-creation process which consists of four major processes of knowledge conversion ('knowledge spiral'): (1) socialisation (from tacit to tacit); (2) externalisation (from tacit to explicit); (3) combination (from explicit to explicit); and (4) internalisation (from explicit to tacit). The survey offers some possibilities to measure different forms of knowledge which lead to different knowledge conversion. Table 8.1 gives an overview of a theoretical concept offered by Nonaka and Takeuchi and the assignment to relevant indicators.

Socialisation is a process of sharing experiences and thereby creating tacit knowledge. In this case, learning takes place not through language, but through observation, imitation, and practice. Through interactions with customers and other collaborating organisations, tacit knowledge is shared. The survey comprises questions that ask for external sources of knowledge, for the importance of market research, and for the importance of knowledge based on personal experience.

According to Nonaka and Takeuchi (1995, p.66), externalisation is called the process of concept creation that "*holds the key to knowledge creation, because it creates new, explicit concepts from tacit knowledge*". The survey specifically asks for the organisation of the internal process of research and development.

Combination means creating new concepts through networking codified information and knowledge. In this process, existing information is exchanged and reconfigured. An example of combination is formal education and training in schools. Indicators for the combination process are the qualification level and the training of the employees as well as explicit external sources of knowledge.

With respect to the measurement of the internalisation process, the internal codified knowledge is the only indicator provided by the survey. Documentation helps individuals to internalise what they experience, although documentation is not a necessary prerequisite for internalisation.

Table 8.1: Assignment of Knowledge Conversion to Survey Indicators

	Knowledge conversion	Description	Indicators from the service Survey
Socialis-ation	tacit ⇒ tacit	• Process of sharing experiences • Learning through observation • Interactions with customers before product development and after market introduction	• External sources of information For innovation: customer and collaborating organisations • Market tests • Knowledge based on personal Experience
External-isation	tacit ⇒ explicit	• Creates new, explicit concepts from tacit knowledge • Holds the key to knowledge creation	• Internal research and Development • Development of new concepts
Combin-ation	explicit ⇒ explicit	• Individuals exchange and combine knowledge through media such as documents, meetings, telephone conversations, or computerised communication networks • Reconfiguration of existing information through sorting, adding, combining, and categorising of explicit knowledge	• Qualification level of employees • Training of employees • External sources of information for innovation: patent disclosures, journals, conferences, computer-based information networks, published or other public knowledge
Internal-isation	explicit ⇒ tacit	• Closely related to learning by doing • Documentation helps individuals internalise what they experienced	• Internal codified knowledge

Source: Nonaka and Takeuchi (1995, 62-70)

The descriptive analysis of the suggested indicators is shown in Table 8.2. These results give a first idea how the innovation process of KIBS differs from other innovative service companies.

Table 8.2: Empirical Comparison of KIBS and Non-KIBS (unweighted)[13]

		KIBS	Non-KIBS
Socialisation	External sources of knowledge for innovation		
	• Suppliers**	42%	20%
	• Competitors**	62%	32%
	• Enterprises within the group*	46%	36%
	• Marketing and consultants**	41%	24%
	• Fairs and exhibitions*	33%	24%
	Collaborating organisations:		
	• Suppliers**	16%	2%
	• Competitors**	17%	6%
	• Enterprises within the group**	23%	6%
	• Marketing and consultants**	9%	3%
	Market research**	32%	21%
	Knowledge based on personal experience (only "very important")**	61%	49%
Externalisation	Internal research and development**	47%	17%
	R&D continuously**	35%	12%
	R&D occasionally**	12%	5%
	Organisation of R&D		
	• R&D departments*	23%	12%
	• R&D project groups*	51%	66%
	• Others	26%	23%
	Development of new service concepts*	49%	38%
Combination	Qualification level of employees: Mean (Std. dev.)		
	• university degree in natural or technical sciences**	22% (27)	11% (20)
	• university degree in economics or social sciences	8% (12)	8% (13)
	• technical college degree	12% (15)	11% (14)
	• completed apprenticeship	41% (27)	47% (28)
	• without degree or training	18% (26)	23 %(28)
	Training linked to innovation**	71%	37%
	External sources of knowledge for innovation:		
	• patent disclosures**	11%	2%
	• conferences and journals**	72%	32%
	• computer-based information networks**	46%	17%
	• published or other public knowledge**	52%	31%
Internalisation	Internal codified knowledge**	61%	42%

ZEW/ISI: Innovation in Services, Survey 1995 and 1997, calculations by the author

In KIBS, the socialisation process is based on different external sources like marketing, consultants, suppliers, competitors, and enterprises within the group. Additionally, they use market research more often than non-KIBS. Knowledge based on personal experience is very important for the creation of knowledge for KIBS. A major difference between KIBS and non-KIBS is reflected by the degree of internal research and development. KIBS conduct R&D more often than non-KIBS and they do so more continuously. Furthermore, KIBS are characterised by an intensive externalisation process. KIBS also use explicit external sources of knowledge more often, such as conferences and journals, computer-based information networks, etc. Apart from this, typical features of their combination process are the great importance given to training as well as the large share of employees with a university degree in natural or technical sciences. The conversion from explicit to tacit knowledge through internal codified knowledge is significantly more common for KIBS than it is for other innovators. So far we have focused on the internal knowledge-creation process. In the following we will shift our point of emphasis to the role of technology.

8.3.4 Descriptive Analysis: Role of Technology

The existing traditional literature on innovation neglects services because of the low technology-intensity and the inability to develop or to use technological innovations for products and processes (Pavitt, 1984). Recent studies, however, underline the interdependencies between the technical-economic paradigm in manufacturing and in particular services. For instance, Moulaert et al (1991) point out that technical consultants do have a significant influence on technological change. Also Miles (1996) assumes, as well as Hauknes and Miles (1996), that some new services play an important role in transferring and creating technological knowledge, and knowledge related to user's assimilation of new technologies. Gallouj and Gallouj (1997) identify five different relations between services and technologies: (1) Substitution relation (replacing human capital by technical capital); (2) identity relation (consubstantiality between tool and services, for example, electronic mailing); (3) determination relation (technological innovation determines the emergence of new service functions); (4) diffusion relation (services help to diffuse technological innovations); and, (5) production relation (services produce technological innovations).

The service survey is not able to measure these different relations in detail but Figure 8.4 shows the overall importance of technologies for innovative activities in the service sector. The picture indicates that all of the different technologies are more often important to KIBS than to other innovators. The differences are significant almost everywhere (exceptions: transport/logistic technologies, medical and bio-technologies). This is an additional indicator for the fact that KIBS act as an intermediary between the science base and the economy. Furthermore it shows that KIBS do have innovation processes which are technologically based, far from being supplier-dominated.

8.4 CHARACTERISTICS OF KNOWLEDGE-INTENSIVE BUSINESS SERVICES

The following probit-model (see Table 8.3) introduces other characteristics of KIBS - in addition to the knowledge-creation process and the use of technology. The dependent variable of this model is the distinction in KIBS and non-KIBS, while the independent variables show the peculiarities of KIBS. The model is controlled by industry and size.

Figure 8.4: KIBS and Technology (unweighted)[14]

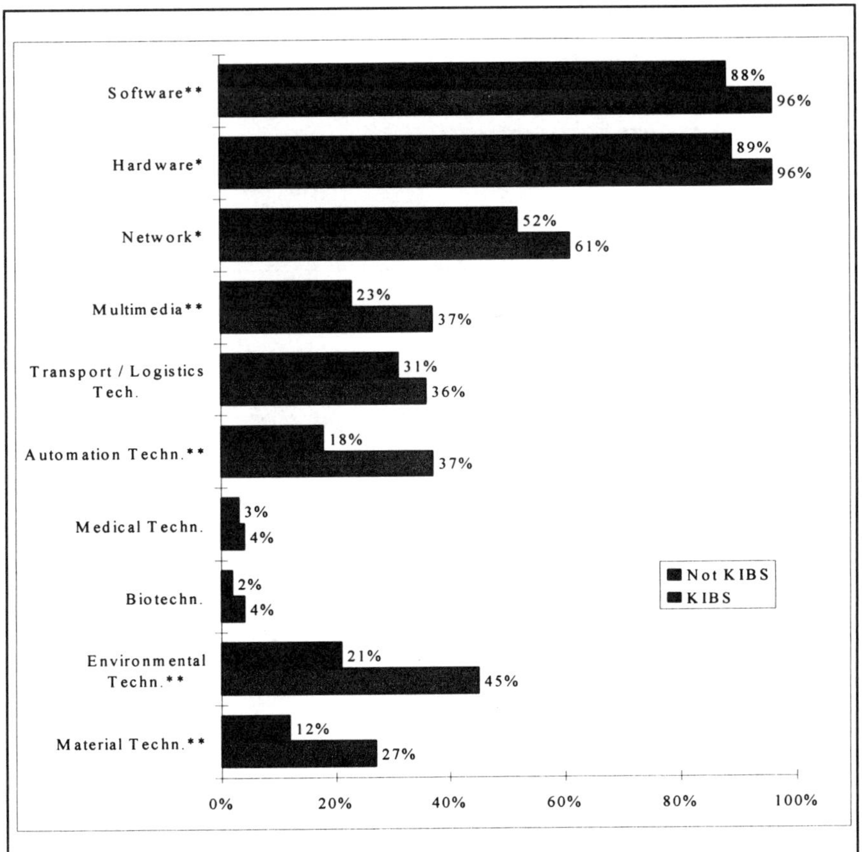

ZEW/ISI: Innovation in Services, Survey 1995 and 1997, calculations by the author

Table 8.3: Characteristics of KIBS (unweighted)

Characteristics of KIBS		**dF/dx**	**P>|z|**
INDUSTRY	Wholesale trade	-0,05	0,178
	Retail trade (Base)		
	Transportation/communication	-0,02	0,629
	Banking/insurance	-0,05	0,260
	Other financial services	-0,05	0,383
	Software	-0,03	0,531
	Science and technological services	0,17	0,070
	Other business services	-0,01	0,783
	Other services	-0,01	0,854
SIZE	1-19 employees	0,24	0,000
	20-49 employees	0,04	0,284
	50-249 employees (Base)		
	250 and more employees	0,13	0,012
KNOWLEDGE-CREATION PROCESS	Suppliers	0,12	0,001
	Marketing, consultant enterprises or private research institutes	0,10	0,003
	Co-operation with competitors	0,16	0,006
	Research and development	0,09	0,037
	Employees with university degree	0,05	0,087
	Highly increasing need for employees with university degree in natural or technical sciences	0,08	0,016
	Training directly linked to innovation	0,08	0,010
	Journals, patent disclosures	0,07	0,011
TECHNOLOGY	High performance computer communication networks	0,05	0,066
	Acquisition of machinery and equipment linked to innovations	-0,10	0,002
	Environmental technologies	0,06	0,081
	Multimedia	0,11	0,029
	Transport/logistic technologies	0,15	0,002
OTHER SUPPORT	Receiving government support for innovation activities	0,39	0,000
INNOVATION OUTPUT	Replacing services being phased out	0,08	0,015
	Fulfilling regulations, standards	0,15	0,000
	Reducing unit labour costs	-0,07	0,034
	Improving the productivity of the customer	-0,05	0,037
Factors HAMPERING innovation	Innovation costs are too high	0,07	0,009
dF/dx is for discrete change of dummy variable from 0 to 1 P>|z| are the test of the underlying coefficient being 0	Prob>chi2 = 0,0000; Pseudo R2 = 0,47 Number of obs = 440		

ZEW/ISI: Innovation in Services, Survey 1995 and 1997, calculations by the author

Most of the industries in the model are not significant. Only the science and technological services are more likely to be KIBS. This is due to the problems of the official statistics and of classification of industries in general[15].

Compared to firms with 50-249 employees, small and big companies have a higher tendency to be KIBS. This result is contrary to Soete und Miozzo (1989) who assume that the relative size of innovating science-based firms is small. Wood (1996) also regards certain large consultant organisations as KIBS, although KIBS are mostly small or medium-sized, including many self-employed individual consultants. König et al (1996) point out that bigger companies have a higher degree of diversification so that the propensity to co-operate with other external institutions is also higher. Furthermore, the authors show that small service companies use external sources of innovation more often – in relation to their size - than big firms. This phenomenon is explained by Teece (1988). Bigger firms have a wider range of internally available knowledge and therefore the necessity for external knowledge is lower. The empirical results are also in line with Gallouj and Gallouj (1997), who assume that besides small firms there are many large firms in the category of specialised and science-based services.

The knowledge-creation process is characterised by a more formal and institutionalised research and development process, integrating all kinds of tacit and explicit knowledge from vertical (suppliers) and horizontal (competitors), co-operation partners and other external knowledge sources. These different external sources underline the importance of KIBS in the innovation network, characterised by group-entrepreneurship and learning-through-networking.

The acquisition of machinery and equipment has a negative effect which underlines the assumption that KIBS do not mainly passively receive technological innovations from other sectors. In the underlying question companies had to compare different innovation expenditures in terms of their importance. Compared to other expenditures, like training or development of new service concepts, the acquisition of machinery and equipment is less important. That means that, for KIBS, buying technologies from outside is less important than their own technological development process. Especially information and communication technologies play a dominant role in their activities. Communication networks are very important to collect data from different sources and help to intensify service producer and customer interactions. Multimedia is probably the new technology with the highest potential for market growth and the emergence of new services; environmental problems represent one of the important challenges of economies and societies. "*Where our societies are confronted by major technological challenges, as in the case of IT and "clean" technology, new services are playing an important role in transferring and creating technological knowledge, and knowledge related to user's assimilation of new technologies*" (Miles, 1996, p.2).

In Germany the institutional conditions for technological services (for example, engineering services) are very good in terms of education (technical universities) and government programs (Strambach, 1997). This is illustrated by the significantly higher use of government programs by KIBS. Yet Strambach emphasises that other forms of service products (for example, management consulting), which are less technical-dominated, are also important knowledge providers for the innovation network. For them, universities and government research programs do not offer much support. It might be possible that the definition of KIBS given in this chapter does not cover low technology-intensive (but knowledge-intensive) business services to a certain extent.

The innovation output of KIBS is characterised by new service products which replace services that are phased out. On the one hand, this indicates that the relatively high proportion of small companies might not have the capacity to differentiate[16]. On the other hand, this might show that the product cycle for new services in this field is very short, reflecting the high speed of development of new IT products. Regulations and standards are correlated with environmental technologies. In Germany there are many (regulation-driven) movements to more 'green based' companies and KIBS support this tendency with new adopted services. The reduction of labour and production costs is not a goal of the innovation activities of KIBS. They do not follow a price strategy. This means that the users are more sensitive to the performance of the technologies and services than to their price. Another characteristic goes in the same direction. KIBS do not provide services which improve the productivity of their customers. KIBS rather offer service products which help the clients to develop or improve their products in terms of quality and performance.

8.5 CONCLUSION

This chapter shows that KIBS play an important role in the knowledge-based economy because they help to distribute information and create knowledge. KIBS combine and transform tacit and explicit knowledge to create new services. The information and knowledge is coming from different sources and co-operation partners. Internal knowledge is embodied in qualified employees or disembodied in internal codified knowledge.

The empirical analysis shows that the category of specialised technology suppliers and science-based services offered by Soete und Miozzo (1989) can be characterised by KIBS to a certain extent. The results permit the assumption that KIBS not only distribute information, but transform it into and create new knowledge, where technology plays a dominant role. So KIBS are integrated in the new mode of knowledge production.

Strambach (1997) points out that KIBS find themselves pulled in two directions: towards the customers and the science-based organisations. The network relations cause synergy effects and help to reduce uncertainty. But this leads on the other side to the establishment of structures and relations where the willingness to take risks for radical new products on new markets is shrinking (Leonard-Barton, 1995, p.187). Future research needs to concentrate on the dangers and challenges of the new mode of knowledge-creation as well as the contribution of KIBS for technical and economic change and the competitiveness of national innovation systems.

REFERENCES

Alvesson, M. 'Organizations as Rhetoric: Knowledge-Intensive Firms and the Struggle with Ambiguity', *Journal of Management Studies*, 1993; Vol. 30, No. 6, November, pp.997-1015.

Barras, R. 'Towards a Theory of Innovation in Services',*Research Policy*, 1986; Vol. 15, No. 4, pp.161-173.

David, P.A. and Foray, D. 'Accessing and Expanding the Science and Technology Knowledge Base', *STI Review*, 1995; No. 16, Paris: OECD, pp.13-68.

Franke, D-P. *Dienstleistungsinnovationen: eine prozessorientierte Analyse dienstleistungsbezogener Neuerungen auf der Grundlage des 'Integrationsansatzes'*, Bergisch Gladbach; Köln: Eul, 1991.

Gallouj and Gallouj, No titel,*SI4S Internal Working Paper*, IFRESI-CNRS, 1997.

Gibbons, M., Limoges, C., Nowotny, H., Schwartzman, S., Scott, P. and Trow, M. *New Production of Knowledge: Dynamics of Science and Research in Contemporary Societies*, London: Sage, 1994.

Grupp, H. 'External Effects as a Microeconomic Determinant of Innovation Efficiency', *International Journal of the Economics of Business*, 1997; Vol. 4, No. 2, July, pp.173-187.

Guinet, J. 'Knowledge Flows in National Innovation Systems',*Industrial Competitiveness in the Knowledge-Based Economy - The New Role of Governments*, Paris: OECD, 1997, pp.173-178.

Hammerer, G. 'Intangible Investments in Austria',*OECD Workshop on New S&T Indicators for a Knowledge-based Economy*, 19-21 June 1996, Paris, DSTI/STP/NESTI/GSS/TIP(96)4.

Hanusch H. and Canter, U. 'Neuere Ansätze in der Innovationstheorie und der Theorie des technischen Wandels - Konsequenzen für eine Industrie- und Technologiepolitik' in Meyer-Krahmer (Hrsg.), *Innovationsökonomie und Technologiepolitik. Forschungsansätze und politische Konsequenzen*, Heidelberg: Physica-Verlag, 1993, pp.11-46.

Hauknes, J. and Miles, I. 'Services in European Innovation Systems - A Review of Issues',*Step report*, Oslo: Step Group, 1996.

Hertog, P. den and Bilderbeek, R. 'The New Knowledge Infrastructure: The Role of Knowledge-Intensive Business in National Innovation Systems", *TNO-report*, STB/97/21, Apeldoorn: TNO, 1997.

Hippel, E. von. *The Sources of Innovation*, New York and Oxford: Oxford University Press, 1988.

Jacobs, D. 'Wissensintensive Innovation: Das Potential des Cluster-Ansatzes',*The IPTS Report*, 1997, No 16, July, pp.24-31.

König, H., Kukuk, M. and Licht, G. 'Kooperationsverhalten von Unternehmen des Dienstleistungssektors' in Helmstädter, E., Poser, G. and Ramser H.J. (eds.), *Beiträge zur angewandten Wirtschaftsforschung*, Berlin: Duncker & Humblot, 1996, pp.217-243.

Leonard-Barton, D. *Wellsprings of Knowledge. Building and Sustaining the Sources of Innovation*, Boston, Massachusetts: Harvard Business School Press, 1995.

Licht, G., Hipp, C., Kukuk, M. and Münt, G. *Results of the German Service-Sector Innovation Survey 1995*, Mannheim and Karlsruhe: ZEW, ISI, 1997.

Licht, G. and Stahl, H. 'Entreprise Panels Based on Credit Rating Data' in Eurostat (ed.), *Techniques and Uses of Entreprise Panels,* Proceedings of the First Eurostat International Workshop on Techniques of Enterprise Pancls, Luxembourg, 1995, pp.163-177.

Machlup, F. *The Production and Distribution of Knowledge in the United States*, Princeton: Princeton University Press, 1962.

Miles, I. et al. *Knowledge-Intensive Business Services: Their Roles as Users, Carriers and Sources of Innovation*, Manchester: PREST, 1994.

Miles, I. *Innovation in Services: Services in Innovation*, Working Paper: University of Manchester, 1996.

Nonaka, I. and Takeuchi, H. *The Knowledge-Creating Company*, Oxford, New York: Oxford University Press, 1995.

Moulaert, F., Martinelli, F. and Djellal, F. 'The Functional and Spatial Division of Labour in Information Technology Consultancy Firms in Western Europe" in Daniels, P.W. and Moulaert, F., *The Changing Geography of Advanced Producer Services: Theoretical and Empirical Perspectives*, London, 1991.

Pavitt, K. 'Sectoral Patterns of Technical Change: Towards a Taxonomy and a Theory',*Research Policy*, 1984; Vol. 13, No. 6, pp.343-373.

Polanyi, M. *The Tacit Dimension*, Gloucester (Mass.): Peter Smith, 1966, reprinted 1983.

Roelandt, T.J.A. and Hertog, P. den. 'Assessing the Knowledge Distribution Power of National Innovation Systems', *OECD workshop on New S&T Indicators for a Knowledge-based Economy*, 19-21 June 1996, Paris, DSTI/STP/NESTI/GSS/TIP(96)4.

Scherer, F.M. 'Inter-Industry Technology Flows in the United States',*Research Policy*, 1982; Vol. 13, No. 4, pp.343-373.

Smith, K. 'Interactions in Knowledge Systems: Foundations, Policy Implications and Empirical Methods' in OECD, *STI Review*, 1995; No. 16, pp.69-102.

Soete, L. and Miozzo, M. 'Trade and Development in Services: A Technological Perspective', *MERIT*, 1989, No. 89-031, Maastricht: MERIT.

Strambach, S. 'Knowledge-Intensive Business Services in the Rhine-Neckar Area',*Tijdschrift voor Economische en Sociale Geografie*, 1994; Vol. 85, No. 4, pp.354-365.

Strambach, S. 'Wissensintensive unternehmensorientierte Dienstleistungen - ihre Bedeutung für die Innovations- und Wettbewerbsfähigkeit Deutschlands',*Vierteljahreshefte zur Wirtschaftsforschung*, 1997; DIW, 66. Jg., Heft 2, pp.230-242.

Teece, D.J. 'Technological Change and the Nature of the Firm' in Dosi, G. et al (eds.),*Technical Change and Economic Theory*, London: Pinter Publishers, 1988.

Utterback, J.M. and Abernathy, W.J. 'A Dynamic Model of Product and Process Innovation', *Omega*, 1975; Vol. 3, No. 6, pp.639-656.

Utterback, J.M. *Mastering the Dynamics of Innovation*, Boston: Harvard Business School Press, 1994, paperback, 1996.

Wood, P. 'The Strategic Role of Knowledge-Intensive Services for the Transmission and Application of Technical Change and Management Innovation',*Fourth Framework Programme*, Targeted Socio-Economic Research, Area 1, INTERIM Report 1, Department of Geography, University College London, 1996.

ZEW. *ZEW news,* 1998; February.

---

[1]To define 'knowledge' I will use the definition by Nonaka and Takeuchi (1995, preface, pp.8-9). They classify knowledge into two different kinds: 1)*Explicit knowledge* which can be articulated in formal language including grammatical statements, mathematical expressions, specifications, and manuals. This kind of knowledge can be transmitted across individuals formally and easily; and, 2) *Tacit knowledge* is harder to articulate with formal language. It is often personal knowledge embedded in individual experience and involves intangible factors like personal beliefs, perspectives, and the value system. The difference between knowledge and information lays in the value for the user. Information can flow freely but is in many cases of no value for most organisations while knowledge is usable to the receiver. The interaction between tacit and explicit knowledge as well as the transformation of information into knowledge are the key dynamics of knowledge creation.

[2]Hauknes and Miles (1996), Miles (1996), Smith (1995), Strambach (1997), Wood (1996), ZEW 1998.

[3]Machlup (1962, S., p.19) identifies some services as highly productive for the economy because of their role as knowledge transfer agencies but he neglects the ability of services to create own knowledge. Machlup argues that "*some of the mass of knowledge, general-systematic or particular-concrete, which is of only transitory relevance has nevertheless great economic value. Certain services of specialists in particular kinds of transitory knowledge have a market value, not because it takes especially scarce qualifications to acquire this knowledge, but because the "division of knowledge" is a great time-saver and thus a highly productive arrangement in the economy*".

[4]Gibbons et al (1994), Hammerer (1996) as well as Roelandt and den Hertog (1996) argue similarly.

[5]Utterback and Abernathy (1975), Utterback (1994).

[6]There is a difference between technology and knowledge. Technology combines technical equipment and machinery as an artefact as well as technical knowledge while knowledge includes also non-technical (like organisational and social) aspects.

[7]Alvesson (1993, p.1003) argues that formal organisations are obliged to have the 'right' structures like personnel departments, management development programmes and modern technologies. Otherwise legitimacy problems arise. The concepts of organisational work are institutionalised, i.e. are taken-for-granted as legitimate, in society. In his opinion there is a decoupling of efficiency and response to institutional issues in order to attain legitimacy.

[8]See also Gallouj and Gallouj (1997).

[9]However, for the service sector statistical data is normally still not available on a disaggregate level.

[10]Question 22 in the survey from 1997 and question 26 in the survey from 1995 deal with the sources of knowledge for innovation. KIBS are companies which answered that customers either coming from manufacturing or services *and* universities or other research institutes are important or very important knowledge sources. Since we asked these two questions *only* for innovative firms, we do not have answers from non-innovators.

[11]The distribution of KIBS by industry (cross tabulation) has a significant Pearson Chi-square of 99%. The adjusted residual is 2.1 for KIBS and software and 4.2 for KIBS and science/ technology services (n=785).

[12]See Polanyi (1966), Gibbons et al. (1994).

[13]The qualification level of the employees is given in average figures (standard deviation) while all other answers are based on a Likert scale ranging from 1 (not important) to 5 (very important). The numbers in the table show which percentage of the companies regards the respective feature as important or very important. Differences between KIBS and Non-KIBS: ** means significance level of 99%; * means significance level of 95% (adjusted residual (t-value) >= 2).

[14]Differences between KIBS and non-KIBS: ** means significance level of 99%; * means significance level of 95% (adjusted residual >= 2).

[15]As mentioned above in section 8.2.

[16]However, Soete and Miozzo (1989) assume that science-based companies are characterised by product differentiation.

Chapter 9
Indicators of Manufacturing and Service Innovation: Their Strengths and Weaknesses[1]

Alfred Kleinknecht

9.1 INTRODUCTION

Public policy is increasingly concerned about promoting innovation in order to stimulate economic growth, employment and ecological sustainability. Clearly, there is an increasing need to measure and assess innovation and technological change and to increase our knowledge about driving forces behind (and socio-economic consequences of) innovation. In the past, publicly available, internationally comparable and reliable data on innovation and technological change have been extremely scarce. Only in very exceptional cases have researchers had access to firm-level innovation data. As a consequence, many interesting theoretical hypotheses have been poorly examined and political decision making has often been guided by intuition rather than by knowledge.

However, since the beginning of the 1990s, notably since the 1992 pilot round of the *Community Innovation Survey* (*CIS*), some progress has been made in collecting micro-data on innovation. The *CIS* undertook a more elaborate measurement of innovation inputs (total innovation costs, including non-R&D costs), as well as an attempt to measure newly developed indicators of the output side of the innovation process. As compared to R&D and patents, the new output indicators have the advantage that they measure innovation directly, i.e. they measure products and services introduced in markets.

Drawing from the experience in the Netherlands, we discuss and compare the following indicators: R&D and patents, total expenditure on product innovation, announcements of new products in trade journals and sales of imitative and innovative products. Section 9.2 addresses the strengths and weaknesses of each of these indicators. Section 9.3 gives some descriptive information about them. In Section 9.4 we summarise a multivariate analysis of determinants of innovation, using two conceptually different innovation output indicators. In Section 9.5 we round up with conclusions.

9.2 TRADITIONAL AND NEW INNOVATION INDICATORS: THEIR STRENGTHS AND WEAKNESSES

9.2.1 R&D efforts

R&D efforts can be measured by expenditures on R&D (as a percentage of a firm's total sales) or by the number of persons carrying out R&D (as a percentage of total employment in a firm). Being available since the 1950s, R&D figures still are the most popular innovation indicator.

Strengths

The R&D indicator has several strong points. First, data on R&D have been collected at regular intervals since the (late) 1950s. R&D data are compiled by the secretariat of the OECD that undertook numerous efforts towards international harmonisation of data collection. Besides data by sectors and size classes, time series are available and the data are frequently used by policy analysts for inter-country, inter-industry and inter-firm comparisons. While it is often complained that R&D data can not be split by technical field, it is, in recent years, increasingly possible to subdivide R&D by product versus process efforts. This subdivision is important for empirical analyses of the impact of innovation on firm performance since product (other than process) innovation efforts seem to be crucial for firm growth, employment and profits (see, for example, Brouwer, Kleinknecht and Reijnen, 1993; Geroski et al, 1993; or Bleichinger et al, 1997). Finally, analysis of intersectoral technology flows is, in principle, possible (although it has some drawbacks).

Weaknesses

Although R&D data measure an important input to new knowledge production, they have a number of disadvantages. First, R&D is only an input of the innovation process. Since inputs can be used more or less efficiently, it would be desirable to have indicators of the output side of the innovation process. In addition, R&D is only one out of several inputs. Other (non-R&D) inputs include product design, trial production, market analysis, training ofemployees, or investment in fixed assets related to product innovation. An illustration of the relative importance of non-R&D innovation expenditures in German manufacturing is given by Felder et al (1996) and for Italian manufacturing by Sterlacchini (1996). Brouwer and Kleinknecht (1997) report an estimate of product innovation expenditures for manufacturing *and* service industries in

The Netherlands. They show that product- (and service-) related R&D is only about one quarter of the total product innovation expenditure in Dutch manufacturing and service industries in 1992, the share of R&D in total innovation expenditure being higher in manufacturing than in services. This confirms, at a national scale, what has been expected from many individual case studies: R&D data tend to underestimate innovation in services. Clearly, the definition of R&D according to the Frascati-Manual of the OECD is somehow biased towards 'hard' (manufacturing) technology and is weak in capturing 'softer' service innovation efforts. This is a plea in favour of developing innovation indicators that are better suited to capturing not only the *development* of *new* (hard) technology but also the creative *application* of already *available* technology.

Another problem with R&D data relates to measurement. There is evidence that standard R&D surveys tend to severely underestimate the small-scale and often informal R&D activities in smaller firms. While, using an identical Frascati-definition of R&D, innovation surveys that include simplified questions about R&D usually find many more small and medium-sized firms reporting some R&D than is found in the official surveys in which a more complicated questionnaire is used (Kleinknecht, 1987; Sterlacchini, 1990). In The Netherlands, it also turned out that the numbers of firms that receive R&D subsidies tremendously exceed the numbers of R&D performing firms according to the official R&D survey. However, the figures from the subsidy records come close to the estimates produced from the innovation survey (Kleinknecht and Reijnen, 1991). The undercounting of small firm R&D has obvious consequences for our judgement about the relationship between firm size and innovation, and it may still be disturbing for comparisons across sectors, regions or even countries as far as shares of small firms differ between the units to be compared.

The striking difference between numbers of firms that report some R&D in official R&D surveys and numbers of firms that receive R&D subsidies[2] is still another source of measurement bias. During the 1980s and 1990s, several countries introduced subsidy schemes that allocate substantial funds to promotion of innovation in small and medium-sized firms. As a consequence, when being contacted for R&D surveys, SMEs may feel increasingly pressed to report R&D that they did not report in the past. This may lead to striking increases in numbers of R&D performing firms, and we do not know to what extent the reported increases are real.

Additional measurement problems relate to the interpretation of the definition of R&D. When testing a draft version of the harmonised *Community Innovation Survey* (*CIS*) in five different countries, we discovered that, in some countries, firms tend not to read the lengthy and complex Frascati definition of R&D. For example, in Norway, a number of respondents reported intuitively more R&D than they wished to report once they had read the Frascati definition (Kleinknecht, 1993, p.158).

Another problem with R&D data relates to the need for secrecy that often hampers research. Notably in small countries like The Netherlands which have a few very big firms, R&D data will often have to be published at relatively high levels of sectoral aggregation in order to prevent inferences about the very large firms. This poses a great problem, because many topics in innovation research can only be addressed meaningfully if at least fairly disaggregated sector data (but preferably: micro data) are available.

Finally, splitting of R&D data by regions tends to be hard to do. In some cases, all of a company's R&D may be reported by the holding company, whereas the R&D efforts are actually scattered all over the country. In multi-plant companies, it may happen that some data reporting units do little or no R&D themselves but they do take advantage of R&D done by other firms in the conglomerate. This can have severe consequences when firm level R&D is related to firm performance. A similar problem can even arise at the country level, also known as the 'Singapore effect': Singapore itself has a fairly modest R&D potential, but there is nonetheless a high rate of new product introduction by affiliates of multinationals which take advantage of R&D done by their mother or sister companies elsewhere in the world.

9.2.2 Patents and patent applications

Patents are often used as an (intermediate) output measure of innovation. Compared to R&D, the use of patent data has always been a second best solution, in spite of the use of superior patenting databases that have become available in the course of time (see the survey by Griliches, 1990).

Strengths

Very long historical time series are available, and these series show only minor disturbances by occasional changes of patent laws or by major law court decisions. Patent databases are publicly available, and contain classifications by technical field. Patent records offer a detailed overview of technical knowledge at a certain point in time. It is even possible to assess the relative importance of patents by means of citation analyses. Regional disaggregation of patent data is also possible, although this meets difficulties in some cases (for example, very large firms).

Weaknesses

It is obvious that the patent indicator misses many non-patented inventions and innovations, and some types of technology are not patentable. On the other hand, what is the share of patents that is never translated into commercially viable products or processes? And can this share be assumed to be constant across branches or firm size classes? Moreover in some cases, patent figures can be obscured by strategic behaviour: A firm may not commercialise the patent but only hold in order to prevent a competitor to use it.

Two other (minor) shortcomings should be mentioned: (1) patents are not always easily classified by economically relevant industry or product groupings; and, (2) some patents can reflect minor improvements of little economic value, while others prove extremely valuable and the question is whether such differences are adequately captured by citation analyses. Moreover, it has often been argued that the propensity to patent may differ across industries depending on the relative costs of innovation versus imitation. If imitation costs are relatively low (as, for example, in pharmaceuticals) firms will have a strong incentive to seek patent protection. The opposite holds if imitation costs are comparatively high.

While such arguments are well-known, there are some recent new insights with respect to the propensity to patent a product innovation. First, findings by Levin et al (1977) for the US and by Brouwer and Kleinknecht (1999) for The Netherlands suggest that, on average, firms do not consider patent protection as the most important means to appropriate benefits from innovation. Factors such as 'a time lead on competitors', 'secrecy' or 'keeping qualified people in the firm' tend to rank much higher.

Second, Arundel and Kabla (1998) have recently demonstrated that there are indeed considerable differences in the propensity to patent across sectors. Their findings from a database of Europe's largest firms have been independently confirmed by Brouwer and Kleinknecht (1999), using the *CIS*-I (1992) database that is representative of firms with 10 and more workers in The Netherlands. In a more detailed multivariate analysis, the latter confronted patent applications of firms to a measure of their innovative output, finding significant differences in the propensity to patent in three dimensions, i.e.:

- Smaller firms have a *lower probability* to apply for at least one patent. However, if they patent, they apply for *higher numbers* of patents. Apparently, small firms have a threshold problem, the first patent being the most expensive in terms of information costs.
- Firms that collaborate on R&D are patenting more intensively than is done by non-collaborators. Seemingly firms wish to protect the most precious parts of their knowledge before engaging into collaboration with a partner.
- Firms in high technological opportunity sectors tend to have a higher propensity to patent than firms in low technological opportunity sectors.

The above implies that, when using patents as an innovation indicator, we will underestimate innovation in low technological opportunity sectors and among innovators that do not collaborate on R&D. Moreover, we underestimate the rate of small innovators, while overestimating the innovation intensity of those that innovate.

Summarising the above and reminding notably the severe measurement problems with respect to R&D and the recent findings on the propensity to patent, it can be concluded that these two indicators are probably worse than is their image. It is therefore a good message that some alternatives became recently available. These will be discussed in the following.

9.2.3 Total innovation expenditures

Other than R&D expenditures, figures on total innovation expenditures cover a larger variety of inputs into the innovation process. A first attempt to measure non-R&D innovation expenditure has been undertaken, at a European scale, during the first round of the *Community Innovation Survey* (*CIS*). Estimates for Italy by Sterlacchini (1996) and for The Netherlands by Brouwer and Kleinknecht (1997) suggest that R&D budgets are just about one quarter of total product innovation expenditures, about half of these expenditures being taken by investments in fixed assets related to product innovation. Students of the service sector should note that fixed investment related to product innovation takes a higher share in services than in manufacturing, i.e. the R&D share in total innovation expenditure is systematically lower in services. Clearly, the new indicator is much richer than the classical R&D figures. However, it needs to be emphasised that questions about non-R&D inputs are difficult to answer accurately since many firms do not keep such records. As a consequence, item non-response rates were high and many firms indicated, in a separate question, that their answers were 'rough estimates' rather than 'fairly exact figures' (Brouwer and Kleinknecht, 1997). Inclusion of this indicator in a questionnaire is likely to have a negative impact on the overall response rate.

9.2.4 Sales of imitative and innovative products

This indicator is based on a firm's assessment in a postal survey of its new product introductions. Firms are asked to subdivide their present product range into products that, during the last three years,

- remained essentially unchanged,
- underwent incremental change,
- were subject to radical change or were introduced entirely new.

Subsequently, they are asked to estimate the share of these three categories of product in their last year's total sales. There are two other interesting dimensions in these data. First, one can ask whether the new products were 'new to the firm' (i.e. already known in the firm's sector) or whether they were 'new to the sector' (i.e. not previously introduced by a competitor). The former category can be interpreted as imitations, the latter as 'true' innovations. Secondly, we obtain figures about incremental innovations versus 'full' innovations.

Strengths

This is a direct measure of successful innovation; measuring innovations that were introduced into the market and that resulted in a positive cash flow. The distinction between 'true' innovations ('new to the sector') and imitative innovations ('new to the firm') as well as the distinction between incremental and 'full' innovations offers research possibilities that did not exist as long as our horizon was confined to R&D and patent data. We can, for example, estimate multivariate models that relate a firm's R&D input to its innovative output. This can say something about the more or less efficient use of R&D inputs and about factors that influence that relationship (see the summary below). Another advantage is that, as far as large firms respond to innovation surveys at a more disaggregated level, regional disaggregation of output indicators can be done more easily than in the case of R&D figures. Finally, although the measurement of innovation output focused on manufacturing in the recent round of the *CIS*, extension to services sectors is possible with only minor modifications as has been demonstrated in The Netherlands (see Brouwer and Kleinknecht, 1995).

Weaknesses

Indicators from postal innovation surveys sometimes suffer from low response rates (and response can be selective), which can make it difficult to produce figures that can be interpreted as national totals. In this context, it is another weakness that many firms can give only 'rough estimates' of the share in sales of innovative products. Moreover, figures on shares in sales of innovative products may be sensitive to the business cycle (which may be misleading in some circumstances), and intersectoral technology flows are hard to assess with this indicator. Inter-firm and inter-country comparisons are possible but notably intersectoral comparisons may be problematic since the length of life cycles differs between branches. Branches which typically have products with shorter life cycles will tend to have (ceteris paribus) higher rates of new product introduction, and vice versa. This implies that, in innovation surveys, one should always include a question about the average length of the life cycle of the firm's most important products. This information is an important control variable in multivariate estimates. Ideally, one should correct tables about sectoral differences in new product introduction by the average length of life cycles by sector.

9.2.5 New product announcements

Besides asking for shares in sales of newly introduced products, researchers have also tried to measure innovative output by systematically screening new product announcements in trade and technical journals. Such a data collection has been undertaken by The Futures Group for the US Small Business Administration in 1982 and has lead to a number of interesting new studies of the relationship of firm size, market structure and innovation (see, for a survey, Acs and Audretsch, 1993). Early in the 1990s, several attempts to collect similar data in Europe have been undertaken (see several chapters in Kleinknecht and Bain, 1993).

Strengths

As the previous indicator, this is a direct measure of the market introduction of new products or services. The data are relatively cheap to collect, and firms are not bothered with time consuming questionnaires, i.e. data collection does, in principle, not suffer from non-response problems. Since the data are taken from published sources, the subsequent use of the data is not hampered by privacy problems. It is possible to split the data by type of innovation (for example, new products, improvements of existing products, product differentiation, etc.), by degree of complexity or other dimensions.

It is an important advantage of these data that it allows covering innovation in (very) small firms at reasonable costs (firms below 10 or 20 employees tend to be neglected in postal surveys, in order to keep sample sizes and survey costs within reasonable limits). Data collection results in addresses of innovators which can be valuable for various purposes (for example, for indepth case studies). It is even possible to make comparisons over time, because the database can, in principle, be extended to the past (as far as the same journals - with the same formula - are available). A broad coverage of sectors (including services) is possible, while many other indicators tend to be confined to manufacturing. Intersectoral technology flows (from innovation 'producers' to innovation users) can be identified relatively easily. Finally, reliable regional disaggregation of data (which is so difficult in the case of R&D) is possible. New products announced in trade journals are, in principle, assigned to the unit which brings the new product on the market. Although this unit need not always be identical to the unit which developed the new products, this indicator may give a more realistic picture of the regional spread of innovation activity than is given by R&D figures.

Weaknesses

New product counts depend on adequate journal selection and the number of innovations will depend on the number of journals covered. Therefore, inter-country comparisons have to be limited to comparison of ratios (for example, the share of small firms, of certain sectors or regions in the total number of innovations). This implies that statistical properties of the database appear dubious since standard statistical procedures (clearly defined population, sample and raising factors) are not applicable. Moreover, publication policies of journals and the public relations policy of firms may influence data collection. Only *published* products and service innovations are measured[8]. We assume that firms have an incentive to publicise their product innovations and use the possibility to have them reported in a journal. There is no such incentive for (internal) process innovations (on the contrary!). This means that process innovations are not adequately covered. One can wonder, however, whether this latter shortcoming is crucial. As to the impact of *process* innovations we have least some conventional productivity figures. The real gap in statistical data collection (and in our economic knowledge) relates to *product* and *service* innovation!

9.3 DESCRIPTIVE COMPARISON OF ALTERNATIVE INNOVATION INDICATORS

The R&D and patent data in Table 9.1 are consistent with the above argument about a technological (hardware) bias of these two indicators. In all size classes, percentages of innovative firms in services are *lower* than in manufacturing. The same holds for percentages of firms that announced a new product in a trade or technical journal. However, percentages of firms that realised some sales with imitative or innovative products do not appear to differ systematically between manufacturing and services (Table 9.2), and the same holds for shares in total sales of imitative and innovative products (Table 9.3). The tables suggest that there is a remarkable discrepancy between the old and the new innovation indicators.

As a qualification, one should note that the definition of the output indicators differs slightly between manufacturing and services. While in manufacturing it was asked whether the firm was busy with the 'development and/or introduction of *technologically* new products or services', the service sector questionnaire used the same phrase, but leaving out the word 'technologically'. This was intended to encourage the reporting of 'soft' innovations that may be based on the original use of *existing* technology (rather than developing new technologies). Allowing for the latter, it turns out that manufacturing and service firms hardly differ in innovation performance. This is rather different from the impression gained from R&D and patent data.

With respect to service sector innovations, it is interesting to note from our collection of new product announcements that the largest part of innovations developed by service sector firms are to be classified as product (rather than service) innovations. While 31.3% of all innovations found in trade journals stem from service sector firms (and 68.7% from manufacturing firms), only 4.8% of all innovations were actually classified as 'service innovations' (Kleinknecht et al, 1993: p.64 and p.71). This discrepancy is mainly due to engineering firms and research laboratories (being classified as services) that develop innovations for manufacturing clients.

The descriptive tables give interesting hints about the coverage of the new indicators compared to the older indicators. Because of the already mentioned 'hard technology bias' of the patent and R&D indicators, it is no surprise that these two indicators suggest that innovation intensity in services is considerably lower than in manufacturing. However, judging from the share-in-sales indicators, services show an almost equal degree of innovativeness. On the other hand, there is also an interesting discrepancy between the two sorts of output indicator: (1) new product announcements in trade journals; and, (2) shares in total sales of innovative products from the Community Innovation Survey (CIS). While only 2.5% of the manufacturing firms (and 0.7% of the service firms) announced a new product, we find that much higher shares of manufacturing and service firms have 'true' innovations ('new to the sector') and the shares of firms having imitative innovations are even higher.

Table 9.1: Firms which performed R&D and firms which applied for patents in manufacturing and services (national totals for The Netherlands, 1988)

size classes	% of firms having some R&D		% of firms applying for patents	
(employees):	manufacturing	Services	manufacturing	services
10-19	21.2	12.4	2.4	0.8
20-49	34.6	14.0	4.4	1.7
50-99	62.6	26.8	8.1	2.4
100-199	76.9	46.5	17.1	5.7
200-499	86.3	48.9	23.7	5.3
500 and more	89.4	64.3	34.3	9.3
totals	42.4	17.9	7.2	1.7
Source: Brouwer & Kleinknecht 1996: 102				

Table 9.2: Percentages of firms which announced a new product or service (in 1989), or which realised sales due to imitative or innovative products (in 1992) in manufacturing and services (national totals for The Netherlands, 1989 or 1992)

	percentages of firms that announced new products		percentages of firms that realised some sales of : imitative products*		innovative products**	
size classes:	manufactg.	services	manufactg.	services	manufactg.	services
10-19	1.5	0.7	12.7	23.9	3.2	7.2
20-49	1.7	0.4	22.5	45.4	6.2	13.7
50-99	3.2	0.8	51.6	48.6	13.5	15.9
100-199	5.7	1.3	67.8	72.0	30.6	20.9
200-499	6.6	4.6	60.0	93.2	15.8	34.5
500 and more	13.4	4.1	99.0	99.0	52.9	43.1
totals	2.5	0.7	-	-	-	-

* new to the firm (already known in the sector)
** new to the sector (not earlier introduced by a competitor)
Source: Brouwer & Kleinknecht 1996: 104

Table 9.3: Shares in 1992 sales taken by products which were incrementally improved, radically changed or newly introduced during 1990-92 (national totals in The Netherlands; only innovating firms)

	shares in sales of imitative products*		shares in sales of innovative products**	
size classes:	**manufacturing**	**services**	**manufacturing**	**services**
10-19	42.7	46.5	25.5	20.8
20-49	39.3	38.9	14.1	11.7
50-99	39.0	37.3	12.8	8.2
100-199	39.4	35.5	10.5	6.8
200-499	38.5	39.0	8.6	12.5
500 and more	40.1	35.5	12.2	8.2
totals	39.8	40.5	14.7	12.5

* new to the firm (already known in the sector)
** new to the sector (not earlier introduced by a competitor)
Source: Brouwer & Kleinknecht 1996: 105

Does this reveal a fundamental conceptual or measurement problem? Not necessarily. First, it is intuitively plausible that percentages of firms that realise sales of *imitative* products are much higher than those of firms that realise 'true' innovations. The finding that imitative products take, on average, higher shares in total sales also sounds plausible. However, how to explain the big discrepancies between the share-in-sales indicators and the new product announcements? One should be aware that the latter have undergone a much tougher screening process. Firms must have sent out a press release, announcing a new product and a journal editor must have found this press release interesting enough for publication. Moreover, the percentage of innovators found will also depend on the intensity of our search process. In our data collection process, we decided to have each major branch of manufacturing and services covered by one journal. Using a higher number of journals would have increased the costs of the research project, and, of course, it would have increased the number of innovators found (see, for details, Kleinknecht and Bain, (eds.), 1993, pp.190-195).

Therefore, the differences found in the above tables do not necessarily imply that the three types of output indicators are inconsistent. One way of examining their consistency is to analyse which are the driving forces behind innovative behaviour by estimating the same 'innovation production model' for all three indicators. Should the factors promoting (or hampering) innovation be more or less the same for all three indicators, this would be a positive signal. It would imply that, for example for policy making, it does not really matter which of the indicators is used.

9.4 ANALYSING NEW INDICATORS

In the following we summarise multivariate analyses of three types of innovation output indicators which have been measured independently and which are, of course, conceptually rather different:

- imitative products and services (as a share in total sales)
- innovative products and services (as a share in total sales)
- new product and service announcements in journals

Other than the descriptive tables above, the multivariate analyses allows one to take into account a larger number of factors influencing innovation. Our model estimates answer a double question:

1. What factors influence the probability that a firm will innovate at all? (i.e. that the firm has *some* sales of innovative products or will announce *at least one* new product or service)

2. Given that a firm innovates: which factors influence the share in a firm's total sales taken by innovative products (or the number of new products)?

Before presenting a sketch of outcomes, we need to mention some qualifications. A first qualification relates to the interpretation of the share in sales taken by innovative products. This share could be taken as an indicator of innovation intensity, implying that a high share indicates a high rate of new product introduction during the last three years. However, a high share may also give an indication of the speed of diffusion. A firm may have introduced only a few innovations that experienced high sales figures (possibly quickly replacing the firm's old products).

The most important qualification is that the database does, in many cases, not allow the use of adequate time lags between explanatory variables and dependent variables. For example, we 'explain' innovation output in 1992 by R&D efforts or R&D collaboration in 1992, assuming that R&D efforts or patterns of R&D collaboration will not change much over time. Such shortcomings can only be overcome once we have panel data. When explaining new product announcements, the situation is a bit better. In this case, the data allow for a one year time lag between dependent and independent variables, since the 1989 new product announcements were linked (at the firm level) to a database from our 1988 innovation survey (see Brouwer and Kleinknecht, 1996, for details).

Moreover, the data on shares in sales are somewhat noisy. We found that most firms tend to report 'rough estimates' rather than 'fairly exact figures', and they tend to fill in 'round' figures (for example, 10%, 25%, etc.). While the estimates of the first part (probability to be an innovator) were quite easy to do, we had a number of problems in estimating the second part (innovation intensity). When including a larger number of explanatory variables, the model did not converge. Notably when explaining 'true' innovations, the low number of innovating firms caused problems (for details, see Brouwer and Kleinknecht, 1996). We therefore tentatively included a number of interesting variables and omitted them again when they turned out to have a poor explanatory power.

Table 9.4: Factors influencing announcements of innovative products (or services) in trade journals and sales of innovative and imitative products (or services). Stylised summary of tobit estimates

	***imitative* products**		***innovative* products**		**new product announcements**	
Explanatory variables:	**firm has some sales?**	**percentage in total sales?**	**firm has some sales?**	**percentage in total sales?**	**at least one new product?**	**how many ?**
R&D-intensity	+ +	+ +	+ +	o	+ +	O
Permanent R&D	+ +	+ +	+ +	o	+ +	o
small business presence	o	+ +	o	o	o	o
firm size/manufacturing	+ +	- -	+ +	- -	+ +	+ +
firm size/services	+ +	- -	+ +	- -	+ +	o
located in central region	o	o	+ +	o	+	+ +
sales growth (1990-92)	+ +	+	+	+ +	n.a.	n.a.
dependent on mother company	- -	+	-	+ +	o	+
concentration on IT	+ +	o	+ +	o	o	o
Concentration on biotechnology	o	o	+ +	o	o	o
R&D collaboration	o	o	o	o	o	o
Acquisition of technological knowledge	o	o	o	o	o	o
average length of life cycle	n.a.	- -	n.a.	-	n.a.	n.a.
dummy for high techn. opportunity sectors* in manufacturing[@]	+ +	+ +	+ +	o	+ +	+ +
dummy for service firms[@]	-	- -	- -	o	o	+

* we take Pavitt's (1984) 'science-based', 'scale intensive' and 'specialised suppliers' industries as high technological opportunity industries and his 'supplier-dominated' industries as 'low technological opportunity sectors'.
[@] reference group: low technological opportunity sectors in manufacturing

Source: Summary of estimates by Brouwer and Kleinknecht (1996:110-115).

Table 9.4 (above) summarises the signs of the estimated coefficients (see for a full documentation Brouwer and Kleinknecht, 1996). The results can be summarised as follows:

- as expected, a firm's R&D intensity tends to have a positive influence on its innovation output, although the results are less clear-cut when explaining shares in sales of 'true' innovations and numbers of innovations; here the above-mentioned caveat about the low number of 'true' innovators should be reminded, before drawing strong conclusions;

- the same holds for the nature of a firm's R&D efforts: firms having R&D as a *permanent* (rather than occasional) activity tend to be more successful innovators, underlining arguments in the literature about the necessity of a historically continuous accumulation of a firms technological knowledge (for example, Dosi, 1988). The variable on permanent versus occasional R&D efforts may also repair somewhat for our use of R&D intensity (of only one year), rather than using an R&D capital stock.

- market concentration (measured by the market share of the 4 largest sellers in a firm's branch of principal activity) did not have a significant influence on innovation output. It seems, however, that a high 'small business presence' (i.e. a high share of firms with less than 50 workers in the total number of firms in a firm's sector of principal activity) as an opposite measure seems to have a positive impact on the diffusion of imitative innovations;

- as expected, larger firms have a higher probability of being innovators than smaller firms, although this probability increases less than proportionately with firm size. Among the innovators, smaller firms tend to have higher shares in sales that are due to innovative products. As expected, the (absolute) number of new product announcements in manufacturing increases positively with firm size; however, this does not hold for services. One should mention here that the size structure of Dutch services is dominated by small and medium sized firms;

- location in central regions of the Netherlands has a positive impact on 'true' innovations and on new product announcements, pointing to the importance of regional knowledge spillovers. It is remarkable to note that the same does not hold for imitative innovations.

- the positive signs of demand growth underline the importance of Schmookler's (1966) 'demand-pull' effect. The finding that demand growth in the period 1990-92 has a positive impact on innovation output in 1992 points to an aspect of effective demand which has never been noticed by Keynes and the Keynesians;

- a remarkable outcome is that various forms of R&D collaboration and technology transfer have *no* significant impact on innovation output. In other words, holding constant the impact of variables such as R&D, firm size, etc., firms which collaborate on R&D or use external sources of technological knowledge do not perform better or worse than those who do it alone;

- other control variables (such as the average length of the product life cycle in a firm's sector of principal activity, or controls for firms that are dependent on their mother company) behave as expected.

Last but not least, we should give attention to differences between sectors. In earlier versions of our estimates, we experimented with finer sector dummies, finding, in most cases, little impact. Most of the tremendous differences in innovation intensity between sectors seem to be absorbed by R&D-intensity which we included as a control variable. In order to reduce the number of explanatory variables, we decided to use broader sector dummies along the lines of Pavitt's (1984) 'taxonomy' of sectors. We joined his categories of 'science-based', 'scale-intensive' and of 'specialised suppliers' industries in the category of high technological opportunity sectors and took, as a reference group, his 'supplier-dominated' industries as low technological opportunity sectors. As expected, we find that firms in high technological opportunity sectors are more successful innovators than are firms in 'supplier-dominated' industries. As compared to firms in the 'supplier-dominated' (manufacturing) sectors, service firms tend to have a lower innovation output, notably with respect to imitative innovations. In this point, our multivariate analysis deviates from the impression gained from the above descriptive tables. With respect to 'true' innovations and new product announcements, differences between services and supplier-dominated manufacturing sectors are less clear-cut.

9.5 CONCLUSIONS

R&D figures as a measure of innovation are misleading in several respects, which is due to at least three reasons:

- R&D efforts in smaller firms tend to be strongly undercounted in official R&D surveys implying that R&D figures are definitely not suitable for examination of the relationship of firm size, market structure and innovation, and even simple comparisons across sectors, regions or countries may be biased;

- in a number of countries, subsidising of small firm R&D in recent years gives a stimulus to report R&D that has hitherto not been reported in R&D surveys. As a consequence, observed changes in R&D time series are likely to be upward biased;

- the R&D share in total innovation expenditures is higher in manufacturing than in services. Therefore, a neglect of non-R&D innovation expenditure leads to underestimation of innovativeness in the service sector.

Patent figures are no good alternative to R&D. Recent empirical investigations of what F.M. Scherer once called 'the propensity to patent' give strong indications that the use of patent data as an innovation indicator will lead one to undercount the rate of innovation in:

- certain sectors, notably in low technological opportunity sectors;

- firms that do not collaborate on R&D.

- small firms: while undercounting *numbers* of small innovators, one will overestimate the innovation *intensity* of those small firms that do patent;

It is our general impression from the data collection process and from subsequent analyses that figures on both R&D and patents are biased towards 'hard' technical innovations in manufacturing, capturing insufficiently 'soft' service innovations. R&D and patents may well be inferior to the new output indicators.

Our summary of a multivariate analysis of new output indicators is encouraging since we find little differences between the indicator on new product announcements and the share in sales indicators from the CIS, notably when looking at 'true' innovations. This is remarkable, given the strong differences in percentages of innovating firms according to both measures. The two types of innovation output indicator are measured independently, and they are conceptually different: new product announcements measure the market introduction of innovations as far as they are published in journals. Some of these will fail to generate (substantial) sales. The share-in-sales indicators measure the sales generated by new products and are therefore confined to more or less successful innovations. The finding that the signs of coefficients of explanatory variables tend to be robust for such differences implies that insights into driving (or hampering) forces behind the innovation process (and policy implications) tend to be the same, whichever of the two types of indicators we happen to measure and to analyse.

REFERENCES

Acs, Z.J. and Audretsch, D.B. 'Analysing Innovation Output Indicators: The US Experience' in Kleinknecht, A. and Bain, D. (eds.), *New Concepts in Innovation Output Measurement*, London: Macmillan, 1993.

Arundel, A. and Kabla, I. 'What Percentage of Innovations is Patented?',*Research Policy*, 1998; Vol. 27, pp.127-141.

Bleichinger, D., Kleinknecht, A., Licht, G. and Pfeiffer, F. *The Impact of Innovation on Employment in Europe. An Analysis Using CIS Data*, Report to DG 13 of the European Commission, Luxembourg: EIMS Publication No. 46, 1997.

Brouwer E. and Kleinknecht, A.H. 'Technology and a Firm's Export Intensity: The Need for Adequate Innovation Measurement', *Konjunkturpolitik*, 1993; Vol. 39, pp.315-325.

Brouwer, E. and Kleinknecht, A.H. 'An Innovation Survey in Services, The Experience with the CIS Questionnaire in the Netherlands', *STI Review: Science Technology Industry* (OECD), 1995; No. 16, pp.141-148.

Brouwer, E. and Kleinknecht, A.H. 'Determinants of Innovation: A Microeconometric Analysis of Three Alternative Innovation Output Indicators' in Kleinknecht, A.H. (ed.), *Determinants of Innovation*, London: Macmillan Press, 1996, p.99-124.

Brouwer, E. and Kleinknecht, A.H. 'Measuring the Unmeasurable: A Country's Non-R&D Expenditure on Product and Service Innovation', *Research Policy*, 1997, Vol. 25, pp.1235-1242.

Brouwer, E. and Kleinknecht, A.H. 'Innovative Output and a Firm's Propensity to Patent',*Research Policy*, 1999; Vol. 27.

Brouwer, E., Kleinknecht, A.H. and Reijnen, J.O.N. 'Employment Growth and Innovation at the Firm Level: An Empirical Study', *Journal of Evolutionary Economics*, 1993; Vol. 3, pp.153-159.

Dosi, G. 'Sources, Procedures and Micro-Economic Effects of Innovation',*Journal of Economic Literature*, 1988, Vol. 26, pp.1120-1171.

Felder, J., Licht, G., Nerlinger, E. and Stahl, H. 'Factors Determining R&D and Innovation Expenditures in Germany Manufacturing Industries' in Kleinknecht, A.H. (ed.), *Determinants of Innovation: The Message from New Indicators*, London: Macmillan Press, 1996, pp.125-154.

Geroski, P., Machin, S. and Van Reenen, J. 'The Profitability of Innovating Firms',*Rand Journal of Economics*, 1983; Vol. 24, pp.198-211.

Hughes, K. *Exports and Technology*, Cambridge: Cambridge University Press, 1986.

Griliches, Z. (ed.). *R&D, Patents and Productivity*, Chicago: University of Chicago Press, 1984.

Griliches, Z. 'Patent Statistics as Economic Indicators', *Journal of Economic Literature*, 1990; Vol. 28, pp.1661-1707.

Kleinknecht, A.H. 'Measuring R&D in Small Firms: How Much Are We Missing?",*Journal of Industrial Economics*, 1987; Vol. 36, pp.253-256.

Kleinknecht, A.H. 'Testing Innovation Indicators for Postal Surveys' in Kleinknecht, A.H. and Bain , D. (eds.), *New Concepts in Innovation Output Measurement*, London: Macmillan, 1993.

Kleinknecht, A.H. and Reijnen, J.O.N. 'More Evidence on the Undercounting of Small Firm R&D', *Research Policy*, 1991; Vol. 20, pp.579-587.

Levin, R.C., Klevorick, A.K., Nelson, R.H. and Winter, S. 'Appropriating the Returns from Industrial Research and Development', *Brookings Papers on Economic Activity*, 1987; No. 3, pp.783-831.

OECD. *Proposed Standard Practice for Surveys of Research and Experimental Development, (Frascati Manual)*, Paris: OECD/DSTI, 1993.

Pavitt, K. 'Sectoral Patterns of Technological Change: Towards a Taxonomy and a Theory', *Research Policy*, 1984; Vol. 13, pp.343-73.

Scherer, F.M. 'The Propensity to Patent', *International Journal of Industrial Organization*, 1983; Vol. 1, pp.107-128.

Schmookler, J. *Invention and Economic Growth*, Cambridge: Harvard University Press, 1966.

Sterlacchini, A. 'Innovation, Formal Vs. Informal R&D and Firm Size: Some Evidence from Italian Manufacturing Firms', *Small Business Economics*, 1990; Vol. 2, pp.223-228.

Sterlacchini, A. 'Inputs and Outputs of Innovative Activities in Italian Manufacturing",*Quaderni di ricerca*, 1996; 78 (University of Ancona, Dept. of Economics).

---

[1]An earlier version of this chapter was presented at the CRIC workshop in May 1998 at Manchester and at the NIAS workshop in January 1999 at Wassenaar. This chapter summarises work carried out in collaboration with Erik Brouwer, Natasja Brouwer, Hana Budil, Tom Poot, Jeroen Reijnen and Wendy Smits at the University of Amsterdam (1988-94), at the Free University of Amsterdam (1994-97) and at Delft University of Technology (since 1997). The usual caveats apply.

[2]From unpublished sources of Statistics Netherlands I take that, in 1997, they estimated the number of firms performing R&D at about 2.900, while, in the same year, more than 10,000 firms received R&D subsidies.

[3]One should note that, in some market niches (with very small numbers of potential buyers), a firm may refrain from publication of new products in journals, other publication channels being more efficient.

Chapter 10
Structural Change and Technological Externalities in the Service Sector: Some Evidence from Italy

Gilberto Antonelli[1], Guilio Cainelli[2], Nicola De Liso[2] and Roberto Zoboli[2]

10.1 INTRODUCTION

The economic role of the tertiary sector is already predominant in all major economies. This is also true for the Italian economy, despite the fact that it is lagging behind the EU average. According to the last accounts of the Italian economy (ISTAT, 1999), in 1996 more than 60% of overall value added was produced in the tertiary sector. Around 2.6 million firms[3] and 7 million employees[4] were active in the market for services. On the whole, the tertiary sector engaged, in the same year, 2.94 million firms[5] and 9.69 million employees[6]. Moreover, the fact should also be taken into account that two sections of services, that is 'trade', 'hotels and restaurants' and 'other services'[7], are among the sectors in which the 'non directly observed economy' concentrates, if we consider the percentage of employees involved[8], while 'other services' are the leading section if we consider the percentage of self-employed workers[9]. All of this suggests that official figures tend to underestimate the relative weight of the tertiary sector in the Italian economy – whose overall performance of course depends on, and derives from, the interaction between all sectors of activity.

This being said, we have to consider the key role of the services both on the interpretative ground and in the policy sphere. Moreover, the process of market liberalisation and globalisation is fostering an economic transformation in which services *per se* and in connection with other sectors will acquire a new nature and play new roles[10].

If we fully appreciate this evidence, the well-known difficulties in defining, measuring and analysing the services sector are particularly worrying, since they could prevent us from understanding some of the most significant trends of structural change occurring in Italy as well as in the other advanced economies. In particular, the role of services in innovation processes, which are the main factor behind the changing structure of economies, is even more the subject of significant doubts and, sometimes, of misleading underestimation. In this respect, indirect approaches, which try to define the role of services through their contribution to the development of innovation in a systemic setting, can be very useful in overcoming the above mentioned difficulties. Moreover, the interpretation of the role of services in the framework of the theory of externalities can contribute in highlighting the role of services in both innovation and growth.

The purpose of the present work is to carry out an analysis of structural change and innovative externalities in the service sector in Italy, adopting a long-run and geographically-disaggregated perspective.

We will start by sketching some of the analytical difficulties still surrounding the interpretation of the role of services in structural change and innovation, and stressing some elements suggesting the need for a reappraisal of their role.

Secondly, we will examine some empirical evidence on the long-run evolution of the services sector in Italy during the period 1951-1991. In this examination, we will present some of the results of the research done by IDSE-CNR on long-run structural change and local systems of production in Italy. Data come from the IDSE-CNR data-set on services, which covers the whole post-war period and it is based on a geographical disaggregation at the *Province* level. In this framework, we will investigate the role of technological externalities also by means of econometric techniques.

Finally, we propose a new taxonomy of services which can be used for reinterpreting the dynamics of the service sectors at a high level of sectoral disaggregation in the decade 1981-1991.

10.2 SERVICES, INNOVATION AND PRODUCTIVITY: SOME ANALYTICAL PROBLEMS

The difficulties in defining and measuring services in the advanced economies are well known (OECD, 1998; Miles, 1996). These difficulties contribute to sustain a debate on tertiarisation and de-industrialisation in which the diverging positions are difficult to reconcile. Two extreme views can be sketched.

The first view considers the growth of services as a process which contributes to the productivity slow-down observed in the advanced economies during the last few decades. The alleged productivity-depressing effect of services' relative growth would be associated to some of their main features: the labour-intensive nature of most services; the laggardness characterising many non-business services; the limited competition allowed for in many service branches; the difficult submission of public services to cost-effectiveness controls; the outsourcing nature of most business services, which stresses the dependence of the services from manufacturing sectors[11]. This view is exemplified by the *cost desease* and the *productivity paradox* hypotheses (Baumol et al, 1989; OECD, 1991, 1996; Baily and Chakrabarty, 1988).

The second view, on the contrary, ascribes to the services and their growth an importance that could match the revolutionary changes engendered by the Industrial Revolution at the end of the 18th century. This second viewpoint stresses the qualitative as well as quantitative changes which are being imposed on the most advanced economies. Services are thus considered as the major users and promoters of technological and non-technological transfers between economic agents (Miles, 1998) and they can even achieve an autonomous capability to stimulate productivity, innovation and growth[12].

The heterogeneity of the service sector as a whole, which includes very traditional services together with the most advanced ones, is such that both views can be supported by some evidence, and the characterisation of services' growth as either a 'progressive' or 'regressive' force remains an open issue.

However, notwithstanding the above-mentioned fundamental divergence, in our opinion the analysis of innovation in services is still dominated by an 'industrial bias'. The idea that the main lever of technical progress and innovation *must* be looked for in manufacturing, while services play a role only as users of these innovations, still persists. This bias tends to obscure the proper nature and qualification of innovative processes in services. In fact, we have to derive all the implications from the growing awareness that: manufacturing sectors and services are more intertwined than is usually thought; services are characterised by their own specific innovative features; new products in new 'quasi markets' for services are developing, especially in the personal services and business services sphere.

10.2.1 Interdependence between manufacturing and services

A fact often neglected is that the growth of services is the result of the division of labour in production between manufacturing and services themselves. We can think of a firm whose core activity consists of the production of a manufactured good, but also performs internally accounting, software writing, marketing and post-sale assistance. If such a firm outsources the non-core activities in which its learning capabilities are more limited and concentrates on the core activity, it will be able to speed up its innovative capabilities. We can infer that such a firm is following a Smithian development path in terms of increased division of labour and increased *dexterity* or *learning-by-doing*. The providers of the services[13] will absorb those activities whose rate of progress might be considered a bottleneck from the point of view of the manufacturer, and they do not necessarily produce those services at increasing productivity rates. In this way, the provision of a 'stagnant' service can be a necessary prerequisite for a dynamic manufacturing. Put it another way, had we not this form of division of labour, the average systemic rate of technological change would be much slower, and the productivity gain in manufacturing can be partly attributed to stagnant services.

Another important fact to be considered, which emphasises a different form of complementarity between manufacturing industry and services, refers to the exchange sphere. The goods exchanged in the markets incorporate a growing information and service component, for instance in terms of quality certification and safety characteristics granted to the customers. *Learning-by-using* is relevant in this case. Cross fertilisation between improvements of both the material and immaterial components is at work. However, the information and service component seems to play a conditional role in the provision of the goods, fully apart from the degree of its improvement.

10.2.2 Specificity in innovation

Increased specialisation can give rise to accelerated learning processes in services as well as in manufacturing, even though they may be difficult to identify. Innovative processes in services, in fact, can often be assimilated to those of (very) small and medium-sized firms and enterprises (SMEs) so that many innovations are not recognised and classified as such (Hauknes and Miles, 1996). In particular, services and SMEs may share a series of characteristics such as: (i) the little relevance of formalised R&D activity; (ii) the measurement difficulties in relation to the sources of innovation; (iii) the dominance of incremental innovation; (iv) the limited capability to exclude others benefiting from successful innovations; (v) the high importance of human capital as innovative asset; (vi) the relevance of experience, tacit knowledge and different dimensions of knowledge (Loasby, 1996); (vii) the key role played by organisational innovations – which can prevail over technological innovations; (viii) the existence of strong externality effects in the transmission of intra- and inter-sectoral innovations; and, (ix) the negative influence of financial constraints. The concept of 'innovation without research' is therefore often associated to both services and SMEs, although it can be misleading to some extent[14].

10.2.3 Development of new products in new ;quasi markets'

There are two main forces driving the process of organisational and product innovation especially in the personal services and business services sphere.

The first one lies in the trend of globalisation and liberalisation already mentioned. Fields of activity that where previously inhibited to business enterprise are now opening themselves to customer-oriented approaches and cost-benefit calculations. Even if social capital and social costs play an extremely relevant role in this process (Antonelli and Pegoretti, 1995), we cannot underestimate the great potential of organisational and product innovation that they imply.

The second driving force depends on the fact that both the firm and the household are organisations whose boundaries with the markets are partially endogenous to the incentive structure of the economy. If the structure of the full cost of the 'basic goods' (Becker, 1965) consumed by the households[15] alters, in the sense that their opportunity cost increases relative to their direct cost, the provision of the basic goods will be gradually transferred to the market. Using Polanyi's framework (Polanyi, 1957), we are observing a deep transformation of the spheres of action of the 'exchange principle', the 'redistribution principle' and the 'reciprocity principle'. The exchange principle is gaining ground, but this seems true also for the reciprocity principle. This leads to the formation of new markets, new quasi markets and clubs for the production and exchange of new products.

10.2.4 The search for new conceptual tools

The limits of available analytical tools have stimulated the search for new conceptual frameworks capable of providing a unified theory of innovation in both manufacturing and services. An early attempt in this direction can be considered the theory of post-industrial society (Bell, 1973) and the problem of 'integrated approaches' has been reconsidered in terms of both sectoral and spatial dimensions (Gershuny and Miles, 1983). The focus of these approaches is on the characteristics of the 'service economy' – and thus of all the sectors with an increasing relative share of services – rather than on innovation in services *per se*. In the service economy, the increasingly extensive complementarities and interrelations between sectors guaranteed by service expansion can guarantee a wider diffusion and a greater effectiveness of innovations – independently from the origin of innovations themselves.

The view of the service economy as technology-driven, knowledge-intensive and skill-biased, with the pivotal role played by human capital, is becoming quite common in recent analyses of structural change (see OECD, 1996; Caracostas and Muldur, 1997). Furthermore, the information and communication technologies revolution puts dematerialised economic processes and the increasing tradeability of both goods and services at the centre of the service economy (Quah, 1996). The sound evidence of these processes at the micro level, however, is not paralleled by similar evidence at the macro level, where a large part of the 'potential' for change seems to remain unexploited. The reason can be traced to the fact that exploitation of the potential requires the establishment of a *new dominant style* in the economy involving a high degree of coherence between socio-technical and socio-economic components (Abramowitz, 1991).

The approach based on 'systems of innovation'[16], can supply useful insights in the understanding of the systemic dimension increasingly associated with innovative change in the service society, but so far it remains too focused on artefacts, formal research activity and industrial innovation. The central role of knowledge and human capital has started to receive its proper attention; as, in complex economic systems, knowledge and capabilities assume a critical importance in both broadening the possibilities of growth and change as well as helping to exploit them.

The new role of knowledge can be deduced from the increasing demand for knowledge-related skill which is associated with the growth of the service economy. Empirical evidence, based on various indicators and considering the demand for skills in the United States for the period 1950-1990 (Wolff, 1996), points out that both 'cognitive' and 'interactive' abilities showed a steady growth during the whole period (16.5% and 8.1%, respectively), while the demand for 'motive abilities' has decreased (-2.5%).

The interactive nature of the new information and communication technologies contributes to the process by changing the way in which economic agents interact, learn, work and get information. Flexibility, decentralization, outsourcing and downsizing are new structural dimensions which come with the increased importance of networking activities (Antonelli et al, 1998a, 1998b). The latter factors also point to the importance of the local dimension of the production systems. Local production systems – which are typical of many Italian areas – are characterised by strong advantages in terms of flexibility, effectiveness, presence of externalities and networking capabilities, and they can benefit from a development process based on knowledge, human capital and services.

10.3 STRUCTURAL CHANGE AND THE DYNAMICS OF SERVICES IN ITALY

Although the analytical limitations summarised above can hinder the possibility of interpreting the significance of services growth and of its implications on structural change, analyses of the growth in the service sectors in Italy are needed in order to reach deeper interpretations of its interaction with innovation. In the light of the limited empirical evidence available, IDSE-CNR created a database on services which includes homogenised data for nine services sectors at the *Province* level during the post-war period[17]. Starting from the description of some general trends of the service sectors growth in Italy, we present hereafter some stylised facts on both the long-run dimension and the geographical dimension of service sectors growth in Italy based on the IDSE-CNR data-set.

Table 10.1: Geographical distribution of labour standard units: 1980-1995

	1980	1985	1990	1995
	North-West			
Agriculture	6.71	6.23	5.10	4.36
Manufacturing	45.39	37.93	37.48	34.82
Services	47.90	55.84	57.42	60.82
Including:				
Trade	18.94	20.92	20.28	21.46
Transport	5.87	6.45	6.70	6.49
Credit and Insurance	1.94	2.19	2.26	2.45
Other services	8.32	12.15	14.04	15.42
Non-market services	12.83	14.13	14.15	14.99
	North-East			
Agriculture	12.65	10.91	8.76	7.19
Manufacturing	38.41	34.28	34.27	32.23
Services	48.94	54.81	56.97	60.58
Including:				
Trade	20.07	21.18	21.42	22.68
Transport	5.25	5.93	6.17	6.13
Credit and Insurance	1.44	1.61	1.68	1.82
Other services	6.68	9.52	11.32	12.93
Non-market services	15.49	16.58	16.39	17.03
	Centre			
Agriculture	11.17	9.20	7.81	6.37
Manufacturing	32.95	27.67	25.99	25.62
Services	55.88	63.13	66.20	68.01
Including:				
Trade	18.51	21.00	21.41	21.63
Transport	6.60	6.64	7.30	7.23
Credit and Insurance	2.11	2.16	2.29	2.41
Other services	8.56	12.39	13.64	14.66
Non-market services	20.10	20.94	21.56	22.08
	South			
Agriculture	22.88	18.23	15.71	13.88
Manufacturing	26.29	22.98	22.05	21.33
Services	50.83	58.79	62.24	64.79
Including:				
Trade	18.15	21.56	22.12	22.05
Transport	5.24	5.29	5.56	5.61
Credit and Insurance	0.97	1.05	1.15	1.28
Other services	7.80	10.70	12.17	12.61
Non-market services	18.67	20.20	21.24	23.24

Source: Golinelli (1997)

10.3.1 Services growth: some general trends

The relative share of the services sector in the Italian economy has grown steadily since 1911. According to a recent empirical work (Rossi et al, 1993) the share of labour input[18] of the private component of service sector has grown from 10% of 1911 to nearly 43% in 1990. The absolute values are 1,057,600 units in 1911 and 9,967,000 units in 1990. These figures can be compared with those of manufacturing industry - which grew from 2,981,000 to 6,940,000 – and agriculture – that decreased from 6,193,000 to 2,230,000 – to suggest that services have absorbed the amount of labour force leaving the other sectors and the whole increase of labour force itself. Public administration, with a share in total employment increasing from 5% in 1911 to more than 18% in 1990, had a great role in the growth of employment in the services sector.

However, the geographical distribution of growth in the service sectors has been far from homogeneous (Table 10.1 above). The growth of service employment in North-Western and North-Eastern Italy was mainly sustained by the 'other services', which includes 'business services' and 'personal services'.

A different situation can be observed in Central and Southern Italy, where the growth of service employment was mainly sustained by the public sector (as suggested by the increasing weight of the item 'non-market services').

A peculiar feature of the Italian service sector is the significant share of self-employed workers. The data shows a certain stability of the share of self-employed workers. However, the overall stability is the result of different dynamics: the strong growth in 'other services' and the decline in 'trade' and 'hotels and restaurants'.

10.3.2 The evidence from the IDSE-CNR data-set

As already mentioned, IDSE-CNR has created a new data set on services which is the result of a specific homogenisation and standardisation procedure of the Italian census data covering the period 1951-1991. Data are comparable over the whole period and highly disaggregated from the geographical point of view.

The period 1951-1991 can be divided into two sub-periods which corresponds to two different development stages of the Italian economy, and 1971 can be considered as a watershed in the process of structural change. While about two thirds of the increase in the number of firms and establishments concentrates in the period 1951-1971, only 40% of the increase in the number of employees occurred during the same period. During the period 1971-1991, instead, the above trend was almost reversed and a strong increase in the growth of employment in services occurred. The latter phenomenon may be interpreted as a sort of compensation effect because services partially absorbed the adverse employment impacts of changing macroeconomic conditions of the early 1970s.

Table 10.2: Share of self-employed workers in terms of standard labour units: 1951-1996

	51-55	56-60	61-65	66-70	71-75	76-80	81-85	86-90	91-96
Agriculture	77.23	77.43	71.45	69.87	66.44	67.71	69.52	68.79	67.45
Energy	0.15	0.15	0.15	0.12	0.13	0.21	0.18	0.15	0.25
Manufacturing	21.81	21.00	16.64	14.91	12.53	12.91	15.10	15.76	16.54
Construction	14.93	15.56	15.36	15.74	16.86	24.98	30.45	32.12	34.79
Services	44.92	44.92	42.92	42.78	42.23	42.96	45.25	45.49	44.34
Including:									
Trade	67.51	63.31	59.08	59.68	57.84	59.08	60.81	60.79	58.49
Hotels	51.23	48.54	45.53	43.52	42.32	42.18	45.31	46.26	45.77
Transport	11.36	14.10	16.57	17.07	16.97	17.40	19.99	20.39	21.33
Credit	1.42	1.38	1.22	1.14	1.05	0.90	0.67	0.61	0.57
Business services	28.08	29.93	33.35	35.40	37.11	39.34	42.29	43.14	42.68

Source: Golinelli (1997)

Table10.3: Firms, establishments and employees in the tertiary sector: 1951-1991

	1951			1961			1971		
	Firms	Establ.	Empl.	Firms	Establ.	Empl.	Firms	Establ.	Empl.
Transport	60.869	67.804	453.114	75.985	84.220	514.810	89.918	100.111	659.868
Communi-cation	1.003	8.597	120.859	1.029	9.653	177.228	712	15.471	230.562
Wholesale Trade	66.974	75.577	246.289	72.081	80.103	358.632	79.534	91.073	437.345
Retail Trade	488.018	502.596	971.912	782.208	813.622	1.551.941	887.756	922.545	1.732.778
Tourism	127.895	133.326	317.610	161.509	166.736	440.676	184.085	190.278	494.303
Other Serv. for Trade	33.336	34.319	55.952	17.994	18.947	55.339	30.543	33.126	55.773
Credit and banking	2.476	9.139	109.124	2.541	11.373	145.435	2.100	12.304	185.609
Insurance	6.411	15.360	49.614	8.564	17.728	75.336	15.756	27.208	112.294
Other Services	76.478	78.094	172.205	113.497	119.113	292.049	168.970	178.280	441.156
Total	**863.460**	**924.812**	**2.496.679**	**1.235.408**	**1.321.495**	**3.611.506**	**1.459.374**	**1.570.396**	**4.349.688**

Table 10.3 (continued)

	1981			1991		
	Firms	Establ.	Empl.	Firms	Establ.	Empl.
Transport	139.020	155.468	843.654	116.916	129.194	724.520
Communication	382	17.095	294.692	1.420	20.030	347.491
Wholesale Trade	115.186	130.925	666.847	141.753	163.804	814.048
Retail Trade	934.305	987.265	1.967.059	733.435	803.192	1.657.611
Tourism	218.895	231.635	680.115	222.882	243.165	764.490
Other Serv. for Trade	118.422	156.003	258.545	185.278	189.939	345.139
Credit and banking	3.141	16.728	315.428	7.839	26.954	389.160
Insurance	26.423	36.651	151.347	43.168	52.711	196.403
Other Services	202.386	352.469	950.696	319.262	392.882	1.422.892
Total	**1.758.160**	**2.084.239**	**6.128.383**	**1.771.953**	**2.021.871**	**6.661.754**

A negative trend emerged during the period 1981-1991 for 'transport' and 'retail trade', both experiencing a decrease in the number of firms, establishments and employees. Over the same period, a strong expansion of employment occurred in 'other services', 'other services for trade' and 'insurance', followed by 'credit and banking' and 'communication' and, to a lesser extent, by 'tourism'.

If we consider the whole period 1951-1991 and refer to the number of employees per firm as an indicator of firm size, the average size increased in four of the nine sub-sectors considered. The average size, however, decreased during the whole period in 'transport' and 'insurance', while it decreased only for a part of the period in 'other services for trade' and 'other services'. Finally, a dramatic fall in the average size of 'credit and banking' firms occurred during the 1980s.

Table 10.4: Average firm size

	1951	1961	1971	1981	1991
Transport	7,44	6,78	7,34	6,07	6,20
Communication	120,50	172,23	323,82	771,45	244,71
Wholesale Trade	3,68	4,98	5,50	5,79	5,74
Retail Trade	1,99	1,98	1,95	2,11	2,26
Tourism	2,48	2,73	2,69	3,11	3,43
Other Serv. for Trade	1,68	3,08	1,83	2,18	1,86
Credit and banking	44,07	57,24	88,39	100,42	49,64
Insurance	7,74	8,80	7,13	5,73	4,55
Other Services	2,25	2,57	2,61	4,70	4,46

In general, the structure of the services sector markedly changed over the period considered (Tables 10.5a and 10.5b). In particular, the 'other services' – which include some of the most innovative activities of the economic system – showed a rapid growth process and the share of the 'other services' continuously increased from 7% to 21%.

Table 10.5a: Structure of the tertiary sector (% in terms of employees)

	1951	1961	1971	1981	1991
Transport	18,15	14,25	15,17	13,77	10,88
Communication	4,84	4,91	5,30	4,81	5,22
Wholesale Trade	9,86	9,93	10,05	10,88	12,22
Retail Trade	38,93	42,97	39,84	32,10	24,88
Tourism	12,72	12,20	11,36	11,10	11,48
Other Services for Trade	2,24	1,53	1,28	4,22	5,18
Credit and banking	4,37	4,03	4,27	5,15	5,84
Insurance	1,99	2,09	2,58	2,47	2,95
Other Services	6,90	8,09	10,14	15,51	21,36

Table 10.5b: Structure of the tertiary sector (% in terms of firms)

	1951	1961	1971	1981	1991
Transport	7,05	6,15	18,15	6,15	6,37
Communication	0,12	0,08	4,84	0,08	0,73
Wholesale Trade	7,76	5,83	9,86	5,83	6,06
Retail Trade	56,52	63,32	38,93	63,32	61,57
Tourism	14,81	13,07	12,72	13,07	12,62
Other Services for Trade	3,86	1,46	2,24	1,46	1,43
Credit and banking	0,29	0,21	4,37	0,21	0,86
Insurance	0,74	0,69	1,99	0,69	1,34
Other Services	8,86	9,19	6,90	9,19	9,01

A positive trend also emerged for 'wholesale trade', 'credit and banking', 'other services for trade' and 'insurance'. Contrariwise, the weight of traditional sectors such as 'retail trade' steadily diminished – in terms of employees – between 1971 and 1991 from about 40% to 25%. Also 'transport' and 'tourism' showed a decreasing trend, while 'communications' did not experience any significant change.

Interesting hints can be derived from geographically disaggregated data using the usual partition in four macro-regions: North-West, North-East, Central and Southern Italy.

Two stylised facts can be recognised. First of all, some sectors, such as 'wholesale trade', 'other services for trade', 'credit and banking' and 'other services', showed a similar and positive trend at both national and macro-regional level – with the exception of 'transport' whose relative importance decreased in all macro-regions, excluding Southern Italy. Secondly, strong differences can be observed between the four macro-regions in the dynamics of the 'other services'.

Table 10.6: Tertiary sector in North-West Italy: 1951-1991

	1951			1961			1971		
	Firms	Establ.	Empl.	Firms	Establ.	Empl.	Firms	Establ.	Empl.
Transport	15.452	17.057	103.986	17.539	20.304	88.798	26.773	29.941	165.670
Communi-cation	262	358	9.690	134	407	17.118	143	706	56.178
Wholesale Trade	29.124	32.703	110.394	29.832	33.315	159.460	30.749	35.785	197.149
Retail Trade	141.585	151.830	324.999	223.270	231.920	503.719	244.731	254.222	534.800
Tourism	47.444	48.778	114.251	56.488	57.615	145.273	58.753	59.997	149.333
Other Serv. for Trade	12.912	13.277	22.169	7.210	7.615	25.013	10.742	11.560	22.692
Credit and banking	703	3.128	43.721	641	3.740	56.504	443	3.407	69.021
Insurance	1.497	3.200	6.983	2.862	4.193	23.338	5.260	9.244	43.790
Other Services	22.270	23.022	53.878	35.863	37.174	88.420	57.819	60.022	143.512
Total	**271.249**	**293.353**	**790.071**	**373.839**	**396.283**	**1.107.643**	**435.413**	**464.884**	**1.382.145**

Table 10.6 (continued)

	1981			1991		
	Firms	Establ.	Empl.	Firms	Establ.	Empl.
Transport	38.929	46.603	233.657	35.999	39.595	208.529
Communication	90	4.732	80.490	469	5.557	90.439
Wholesale Trade	40.337	45.523	258.218	46.635	53.319	297.696
Retail Trade	239.367	253.669	553.556	180.181	197.565	389.191
Tourism	60.204	63.848	192.342	59.050	64.743	213.855
Other Serv. for Trade	41.660	51.967	98.728	63.596	66.655	135.573
Credit and banking	980	5.431	116.728	2.765	8.744	137.012
Insurance	7.963	10.684	55.791	13.133	16.022	82.236
Other Services	64.682	107.539	315.843	94.332	114.621	446.526
Total	**494.212**	**589.996**	**1.905.353**	**496.160**	**566.821**	**2.001.057**

Table 10.7: Tertiary sector in North-East Italy: 1951-1991

	1951			1961			1971		
	Firms	Establ.	Empl.	Firms	Establ.	Empl.	Firms	Establ.	Empl.
Transport	15.705	16.516	55.224	20.785	21.927	79.141	24.853	26.699	95.681
Communication	231	263	3.219	131	252	4.452	84	85	351
Wholesale Trade	13.885	15.562	53.202	16.332	18.201	84.876	18.048	20.603	101.063
Retail Trade	82.084	87.943	192.659	143.752	150.538	307.594	161.415	169.372	342.024
Tourism	30.792	32.009	78.461	41.518	42.729	113.358	52.081	53.839	137.661
Other Serv. for Trade	9.724	9.887	14.463	3.401	3.547	8.973	9.021	9.487	14.280
Credit and banking	610	1.935	13.001	519	2.281	18.329	520	2.722	24.271
Insurance	1.216	2.100	10.072	1.654	2.748	14.438	3.128	5.116	23.784
Other Services	14.373	15.003	30.777	23.291	24.520	53.942	34.475	35.865	76.973
Total	**168.620**	**181.218**	**451.078**	**251.383**	**266.743**	**685.103**	**303.625**	**323.788**	**816.088**

Table 10. 7 (continued)

	1981			1991		
	Firms	Establ.	Empl.	Firms	Establ.	Empl.
Transport	50.873	55.195	231.305	36.660	39.822	163.280
Communication	79	3.457	50.890	305	4.166	61.123
Wholesale Trade	28.469	32.380	161.699	33.443	39.448	201.815
Retail Trade	168.033	180.533	398.663	129.459	144.607	354.281
Tourism	70.167	73.790	201.799	64.139	69.578	212.473
Other Serv. for Trade	42.619	50.628	74.233	54.144	56.178	90.049
Credit and banking	795	3.924	59.712	1.785	6.547	79.566
Insurance	5.296	7.144	27.365	9.037	11.374	37.468
Other Services	44.157	76.186	196.743	70.048	89.458	303.413
Total	**410.488**	**483.237**	**1.402.409**	**399.020**	**461.178**	**1.503.468**

Table 10.8: Tertiary sector in Central Italy: 1951-1991

	1951			1961			1971		
	Firms	Establ.	Empl.	Firms	Establ.	Empl.	Firms	Establ.	Empl.
Transport	11.896	15.801	236.354	17.922	21.335	267.477	20.756	24.842	320.097
Communication	204	7.637	105.437	350	8.293	148.880	261	14.456	173.647
Wholesale Trade	11.423	13.241	43.351	13.428	14.959	63.590	15.732	18.106	80.988
Retail Trade	97.386	107.000	202.087	173.345	181.789	333.749	202.803	211.571	397.664
Tourism	25.367	26.871	68.269	32.847	34.296	100.376	39.740	41.452	120.797
Other Serv. for Trade	4.778	4.993	9.469	3.327	3.522	10.098	6.774	7.476	9.644
Credit and banking	516	2.467	35.736	584	3.072	46.853	294	3.088	60.883
Insurance	1.689	7.008	24.094	2.007	7.891	28.657	3.853	8.413	33.928
Other Services	16.692	16.226	40.025	23.986	25.577	73.131	36.496	40.696	129.031
Total	**169.951**	**201.244**	**764.822**	**267.796**	**300.734**	**1.072.811**	**326.679**	**370.100**	**1.326.679**

Table 10.8 (continued)

	1981			1991		
	Firms	Establ.	Empl.	Firms	Establ.	Empl.
Transport	26.489	28.639	200.053	23.691	26.636	189.690
Communication	98	4.222	82.910	353	4.957	101.522
Wholesale Trade	22.209	25.945	128.638	29.548	34.451	163.398
Retail Trade	213.865	227.314	457.657	173.566	189.652	422.098
Tourism	46.722	50.101	159.147	50.179	54.824	188.592
Other Serv. for Trade	22.823	33.703	53.361	44.034	41.824	71.689
Credit and banking	457	3.704	81.357	1.692	6.073	99.214
Insurance	6.041	8.701	41.988	10.208	12.548	47.766
Other Services	45.627	85.051	244.099	75.055	92.599	321.358
Total	**384.331**	**467.380**	**1.449.210**	**408.326**	**463.564**	**1.605.327**

Table 10.9: Tertiary sector in South Italy: 1951-1991

	1951			1961			1971		
	Firms	Establ.	Empl.	Firms	Establ.	Empl.	Firms	Establ.	Empl.
Transport	17.816	18.430	57.550	19.739	20.654	79.394	17.536	18.629	78.420
Communication	306	339	2.513	414	701	6.778	224	224	386
Wholesale Trade	12.542	14.071	39.342	12.489	13.628	50.706	15.005	16.579	58.145
Retail Trade	166.963	155.823	252.167	241.841	249.375	406.879	278.807	287.380	458.290
Tourism	24.292	25.668	56.629	30.656	32.096	81.669	33.511	34.990	86.512
Other Serv. for Trade	5.922	6.162	9.851	4.056	4.263	11.315	4.036	4.603	9.157
Credit and banking	647	1.609	16.666	797	2.280	23.749	843	3.087	31.434
Insurance	2.009	3.052	8.465	2.041	2.896	8.903	3.515	4.435	10.792
Other Services	23.143	23.843	47.525	30.357	31.842	76.556	40.180	41.697	91.640
Total	**253.640**	**248.997**	**490.708**	**342.390**	**357.735**	**745.949**	**393.657**	**411.624**	**824.776**

Table 10.9 (continued)

	1981			1991		
	Firms	Establ.	Empl.	Firms	Establ.	Empl.
Transport	22.729	25.031	178.639	20.566	23.141	163.021
Communication	115	4.684	80.402	293	5.350	94.407
Wholesale Trade	24.171	27.077	118.292	32.127	36.586	151.139
Retail Trade	313.040	325.749	557.183	250.229	271.368	492.041
Tourism	41.802	43.896	126.827	49.514	54.020	149.570
Other Serv. for Trade	11.320	19.705	32.223	23.504	25.282	47.828
Credit and banking	909	3.669	57.631	1.597	5.590	73.368
Insurance	7.123	10.122	26.203	10.790	12.767	28.933
Other Services	47.920	83.693	194.011	79.204	96.204	351.595
Total	**469.129**	**543.626**	**1.371.411**	**468.447**	**530.308**	**1.551.902**

In particular, in the North-West 'retail trade' decreased, 'insurance' increased while 'tourism' declined after 1951 and then stabilised. A similar trend can be observed for 'retail trade' in the North-East, while 'communication' strongly increased in this same area. In Central Italy 'communication' fell, while tourism expanded markedly; 'retail trade' and 'insurance' remained stable. Finally, in the South 'retail trade' maintained the highest share – albeit decreasing – while 'communication' showed a strong growth. 'Transports', 'insurance' and 'wholesale trade' did not change significantly.

Table 10.10: Change in geographical distribution and structure of the Italian tertiary sector: 1951-1991

	North-West			North-East		
	1951	1971	1991	1951	1971	1991
Transport	13,16	11,99	10,42	12,24	11,72	10,86
Communication	1,23	4,06	4,52	0,71	0,04	4,07
Wholesale Trade	13,97	14,26	14,88	11,79	12,38	13,42
Retail Trade	41,14	38,69	19,45	42,71	41,91	23,56
Tourism	14,46	10,80	10,69	17,39	16,87	14,13
Other Serv. for Trade	2,81	1,64	6,78	3,21	1,75	5,99
Credit and banking	5,53	4,99	6,85	2,88	2,97	5,29
Insurance	0,88	3,17	4,11	2,23	2,91	2,49
Other Services	6,82	10,38	22,31	6,82	9,43	20,18

Table 10.10 (continued)

	Centre			South		
	1951	1971	1991	1951	1971	1991
Transport	30,90	24,13	11,82	11,73	9,51	10,50
Communication	13,79	13,09	6,32	0,51	0,05	6,08
Wholesale Trade	5,67	6,10	10,18	8,02	7,05	9,74
Retail Trade	26,42	29,97	26,29	51,39	55,57	31,71
Tourism	8,93	9,11	11,75	11,54	10,49	9,64
Other Serv. for Trade	1,24	0,73	4,47	2,01	1,11	3,08
Credit and banking	4,67	4,59	6,18	3,40	3,81	4,73
Insurance	3,15	2,56	2,98	1,73	1,31	1,86
Other Services	5,23	9,73	20,02	9,68	11,11	22,66

10.3.3 Technological externalities in services

By applying econometric techniques, the IDSE-CNR dataset on services can be used for investigating the potential role of technological externalities in explaining the long-run growth of the service sector in Italy described above. This analysis can contribute to the development of the systemic approaches considered in Section 10.2.4.

As it is now widely accepted, in particular in the literature on endogenous growth (Lucas, 1988; Romer, 1986, Aghion and Howitt, 1998), dynamic information externalities constitute a driving force of technological innovation and, thus, economic growth. The role of spatial proximity in facilitating the transmission of information, (new) ideas and tacit knowledge among firms and individuals has been recognised.

In particular, three forms of dynamic externalities have been identified in recent theoretical models: Marshall-Arrow-Romer externalities, usually referred to as MAR externalities; Porter externalities; Jacobs externalities (Glaeser et al, 1992; Henderson et al, 1995).

The three types of dynamic externality can be classified according to two criteria, i.e. knowldge spillovers and local market structure.

In MAR externalities knowledge spillovers occur among firms belonging to the same industry, while the internalisation of externalities is guaranteed by a high degree of market concentration.

In Porter's approach, externalities are maximised in regions characterised by a large number of specialised medium-sized firms; local competition accelerates imitation and improvement of the innovators' ideas.

Jacob's externalities identify the variety of the local industrial environment as the key variable in explaining the diffusion of knowledge spillovers, and thus the long-term economic development which mainly occurs through cross-fertilisation of ideas between firms belonging to different sectors.

Given these definitions of externality, we provide – following Glaeser's econometric methodology – a test for analysing the possible impact of each form of externality on long-run growth of the service sector in Italy[19]. The main results can be summarised as follows.

Table 10.11 : Determinants of employment growth in Italy (1951-91) – Dependent variable: Growth of employment between different periods

Independent Variables	*61-51*	*61-51*	*71-61*	*71-61*	*81-71*	*81-71*	*91-81*	*91-81*
Constant	-0.128	...	0.260	...	4.512**	....	0.953**	...
Log employment	0.030	-0.015	0.001	-0.004	-0.304**	-0.222**	-0.081**	-0.072**
Specialis-ation	-0.073	-0.046	-0.174	-0.132	-0.421**	-0.704**	-0.083	-0.068
Competition	-0.011	0.028	-0.083**	0.036	0.239**	0.052	-0.067	-0.038
Variety	0.879	0.477	0.535	-0.449	-2.835**	-1.684**	0.730	-0.373
S1	...	0.275	...	0.418	...	3.778**	...	0.871**
S2	...	-0.151	...	-0.884	...	5.949**	...	1.056**
S3	...	0.374	...	0.461	...	3.734**	...	1.161**
S4	...	0.556	...	0.376	...	3.751**	...	0.814**
S5	...	0.324	...	0.457	...	3.974**	...	1.067**
S6	...	-0.079	...	0.496	...	4.171**	...	1.082**
S7	...	0.339	...	0.546	...	3.722**	...	1.117
S8	...	0.473	...	0.548	...	3.358**	...	
S9	...	0.481	...	0.664	...	4.203**	...	
N. Obs.	171	171	171	171	171	171	171	171
Adjusted R^2	-0.05	0.102	0.077	0.224	0.807	0.902	0.170	0.397
SER	0.509	0.481	0.646	0.592	0.652	0.463	0.262	0.223
W.H.	27.7**	74.94**	11.92	115.9**	87.9**	142.0**	11.0	27.6
J.B.	400.8**	408.6**	4162.8**	962.3**	95.0**	202.4**	1664.3**	5489.9**

Note: Standard errors are computed with the White method in order to correct for the eventual presence of heteroscedasticity. W.H. is the test for heteroscedasticity proposed by White, while J.B. is the Jarque-Bera test for normality. S1, S2, …, S9 are dummy variables for each of the nine service sectors analyzed.
** significant at 5%, * significant at 10%.

(i) During the periods 1951-1961 and 1961-1971 the employment growth rate of the service sector in Italy did not depend on the existence of dynamic externalities; thus other explanations ought to be found.

(ii) Significant results emerge, instead, for the period 1971-1981: specialisation is statistically significant but it is negatively related to the growth of employment in the service sector – and this finding can be interpreted as evidence against the existence of inter-industry spillovers. A second result for the same period is the existence of convergence, in terms of growth rates, between sectors. Thirdly, and most important, our proxy for industrial variety – i.e. the Herfindahl concentration index – is statistically significant and has a negative sign. The higher is variety, the lower the Herfindahl index is, which means that variety matters; which, in turn, means that inter-industry spillovers affect employment growth. Finally, the sectoral dummies turn out to be quite significant.

(iii) For the last period taken into account (1981-1991) evidence of convergence and importance of sectoral specific effects emerges. This result can be interpreted as an index of modernisation of the Italian service sector.

10.4 A TAXONOMY FOR THE PERIOD 1981-1991

10.4.1 The taxonomy proposed

The evolution of the Italian service sector in the long-run has been outlined so far by using a rather aggregate version of the IDSE-CNR data set. In this section, we intend to go more in depth by using more disaggregated data – available only since 1981 – and analyse them through a specific taxonomy. Any taxonomy is inevitably plagued by shortcomings and this applies in particular to the services sector, where the various taxonomies available can all be criticised on some ground or another. However, the great variety of services and the very different paths followed by different branches in Italy suggest that a taxonomy can be helpful in providing a reference framework and beneficial in achieving a more coherent representation of stylised facts.

As a first step in designing the taxonomy we refer to the framework proposed by Petit and Soete (1997) and Pini (1998). These contributions underline that supply, demand and productivity in most services have been affected by information and communication technologies. The latter process, in turn, together with other determinants, has led to increased tradeability of the services themselves. Further to the tradeability and productivity dimension, we consider the 'networking' dimension of services as another criterion for classifying them. The resulting classification grid is the following:

(i) A first category includes those sub-sectors where mainly tradeable services are produced and productivity growth is high, as firms try to gain high market shares rather than to reach larger market; in this category are included sub-sectors that generate extended network interactions.

(ii) The second category includes sub-sectors where the production of tradeable services coexists together with low productivity gains, the latter being obtained mainly by introducing process information technologies; also in this case, the services considered generate extended network interactions.

(iii) In the third category services are included whose features do not allow easy tradeability – also because of their 'impermeability' to the adoption of new technologies – and which are characterised by a low level of productivity. Furthermore, these services tend to generate limited network interactions.

As a second step in the definition of classification criteria, we propose a distinction based on the infrastructural function of services. Services connect different economic activities, either inside individual economic organisations (firms, institutions, households) or outside them. We can thus distinguish: (a) services whose diffusion affects many organisations at the same time, thus generating systemic externalities; and, (b) services which affect organisations from inside, and have thus only 'internal' effects.

Tradeability, productivity, networking and infrastructuring effects of each service sub-sector will depend on supply and demand characteristics of the specific sub-sector – which overcomes the 'simple' statistical classification criterion. On the basis of this taxonomy we can attempt a reclassification and a reappraisal of the various sub-sectors – as defined according to ATECO 1991[20].

Table 10.12: A taxonomy of the service sectors

Tradeable services with high productivity (extended network interactions)		Tradeable services with low productivity (extended network interactions)		Non-tradeable services with low productivity (limited network interactions)	
Internal effects	**Systemic impact**	**Internal effects**	**Systemic impact**	**Internal effects**	**Systemic impact**
K74	I 61	G 51	J 65	G 50	I 60
	I 62	H 55	J 66	G 52	L 75
	I 63	J 67	M 80	K 70	O 90
	I 64	O 92	N 85	K 71	O 91
	K 72			O 93	
	K 73				

Note: We make use of the Ateco 1991 classification. The sections included are the following ones: K74 other business activities; I61 water transport; I62 air transport; I63 supporting and auxiliary transport activities; I64 post and telecommunications; K72 computer and related activities; K73 research and development; G51 wholesale trade and commission trade, except of motor vehicles and motorcycles; H55 hotels and restaurants; J67 activities auxiliary to financial intermediation; O92 recreational, cultural and sporting activities; J65 financial intermediation, except insurance and pension funding; J66 insurance and pension funding, except compulsory social security; M80 education; N85 health and social work; G50 sale, maintenance and repair of motor vehicles and motorcycles; G52 retail trade, except of motor vehicles and motorcycles; K70 real estate activities; K71 renting of machinery and equipment without operator and of personal and household goods; O93 other service activities; I60 land transport, transport via pipelines; L75 public administration, compulsory social security; O90 sewage and refuse disposal, sanitation and similar activities; O91 activities of membership organisations n.e.c.

10.4.2 A re-appraisal of specialisation processes in the Italian service sector

The reappraisal of services dynamics based on the proposed taxonomy highlights some additional features of the specialisation process at both the national and macro-regional level.

At the national level, the highest employment shares concentrate in the traditional sectors, i.e. 'retail trade', 'motor vehicles and motorcycles trade', 'hotels and restaurants', 'public administration', 'education' and 'health' (Table 10.13). Most sectors, however, show a decreasing trend in terms of employment.

The weight of 'other business activities' and 'computer and related activities' shows an evident increase during the period considered. On the other hand, 'R&D', 'banking and finance', 'insurance' and 'activities of membership organisations' are stagnant.

Table 10.13: Share of employees by service sub-sections in Italy (1981-1991)

Code*	Employees 1981	Share 1981	Employees 1991	Share 1991
50	468,167	5.3	491,148	4.4
51	760,068	8.6	902,672	8.2
52	1,868,031	21.2	1,913,442	17.3
55	654,880	7.4	734,372	6.6
60	606,634	6.9	539,124	4.9
61	27,209	0.3	21,415	0.2
62	20.156	0.2	23.922	0.2
63	199,798	2.3	188,919	1.7
64	294,692	3.3	347,511	3.1
65	327,948	3.7	410,109	3.7
66	45,560	0.5	48,938	0.4
67	66,004	0.7	110,488	1.0
70	45,899	0.5	83,189	0.8
71	15,908	0.2	19,634	0.2
72	51,882	0.6	180,952	1.6
73	29,430	0.3	43,458	0.4
74	508,682	5.8	878,637	7.9
75	0	0.0	912,846	8.3
80	1,359,540	15.4	1,398,069	12.6
85	832,812	9.5	1,125,587	10.2
90	67,841	0.8	78,218	0.7
91	115,537	1.3	110,624	1.0
92	190,218	2.2	201,283	1.8
93	255,695	2.9	289,743	2.6
Total	**8,812,591**	**100,0**	**11,054,300**	**100,0**

Notes: *50 Sale, maintenance and repair of motor vehicles and motorcycles; retail sale of automotive fuel; 51 Wholesale trade and commission trade, except of motor vehicles and motorcycles; 52 Retail trade, except of motor vehicles and motorcycles; repair of personal and household goods; 55 Hotels and restaurants; 60 Land transport; transport via pipelines; 61 Water transport; 62 Air transport; 63 Supporting and auxiliary transport activities; activities of travel agencies; 64 Post and telecommunications; 65 Financial intermediation, except insurance and pension funding ; 66 Insurance and pension funding, except compulsory social security; 67 Activities auxiliary to financial intermediation; 70 Real estate activities; 71 Renting of machinery and equipment without operator and of personal and household goods; 72 Computer and related activities; 73 Research and development; 74 Other business activities; 75 Public administration and defence; compulsory social security; 80 Education; 85 Health and social work; 90 Sewage and refuse disposal, sanitation and similar activities; 91 Activities of membership organisations n.e.c.; 92 Recreational, cultural and sporting activities; 93 Other service activities.

The picture is different when we look at the dynamics at the macro-regional level (Tables 10.14-10.18).

The North-West tends to specialise in sub-sectors with high productivity and tradeability as well as extended network interactions and systemic impact.

The North-East specialises in low productivity tradeable sub-sectors, with extended network interactions and systemic impact. This macro-region seems to have increased the potential of infrastructural services, even though not at the same pace as demand.

Central Italy seems to be half-way between low productivity, high tradeable sub-sectors with extended network interactions and low productivity, non-tradeable sub-sectors with limited network effects. Moreover, Central Italy shares with North-East a relevant weight of entrepreneurial organisations and trade unions, even though the weight of the latter is higher in the North-East.

In Southern Italy there is a strong connection between the supply of private and public services and a persistence of traditional services with limited network interaction. However, we can make a further distinction between South-East and South-West. The South-East – which includes Abruzzo, Basilicata, Lazio, Molise, Puglia – seems to be half-way between low productivity and high tradeability sub-sectors with extended network interactions and low productivity, non-tradeable sub-sectors with limited network effects. The impact of these services is mainly a systemic one. The South-West – which includes Calabria, Campania, Sardinia, Sicily – seems to keep a specialisation in non-tradeable, low productivity with limited network effects, and with internal effects.

Table 10.14: Specialisation in North-West Italy (1981-1991)

Code	Employees 1981	Share 1981	Spec. 1981	Employees 1991	Share 1991	Spec. 1991
50	136,799	5.2	0.977	139,853	4.4	0.988
51	286,169	10.9	1.258	333,902	10.5	1,283
52	525,956	19.9	0.941	536,815	16.8	0.973
55	184,125	7.0	0.940	203,343	6.4	0.960
60	187,768	7.1	1.035	153,092	4.8	0.985
61	8,440	0.3	1.037	4,566	0.1	0.740
62	960	0.0	0.159	2,396	0.1	0.347
63	73,998	2.8	1,238	66,564	2.1	1.222
64	80,490	3.1	0.913	90,439	2.8	0.903
65	121,250	4.6	1.236	146,641	4.6	1.240
66	19,654	0.7	1.442	21,922	0.7	1.554
67	23,869	0.9	1.209	39,076	1.2	1.227
70	18,575	0.7	1.353	36,396	1.1	1.517
71	5,894	0.2	1,238	6,345	0.2	1.121
72	22,414	0.9	1.444	69,683	2.2	1.336
73	9,380	0.4	1.065	13,468	0.4	1.075
74	188,397	7.1	1.238	312,521	9.8	1.234
75	0	0.0	...	177,775	5.6	0.675
80	327,267	12.4	0.805	328,131	10.3	0.814
85	237,637	9.0	0.954	315,665	9.9	0.973
90	18,013	0.7	0.888	19,331	0.6	0.857
91	28,907	1.1	0.836	27,958	0.9	0.877
92	50,700	1.9	0.891	53,872	1.7	0.928
93	79,825	3.0	1.044	87,376	2.7	1.046
Total	**2,636,487**	**100,0**	**1,000**	**3,187,130**	**100,0**	**1,000**

Table 10.15: Specialisation in North-East Italy (1981-1991)

Code	Employees 1981	Share 1981	Spec. 1981	Employees 1991	Share 1991	Spec. 1991
50	106,944	5.7	1.070	108,767	4.7	1.061
51	195,211	10.4	1.203	231,818	10.0	1.230
52	376,512	20.0	0.944	375,499	16.3	0.940
55	196,509	10.4	1.406	207,270	9.0	1.352
60	132,212	7.0	1.021	126,066	5.5	1.120
61	7,352	0.4	1.266	3,762	0.2	0.842
62	121	0.0	0.028	477	0.0	0.096
63	42,417	2.3	0.994	42,259	1.8	1.072
64	50,890	2.7	0.809	61,123	2.6	0.843
65	63,138	3.4	0.902	83,450	3.6	0.975
66	9,069	0.5	0.932	9,366	0.4	0.917
67	13,899	0.7	0.986	23,296	1.0	1.010
70	15,070	0.8	1.538	21,571	0.9	1.242
71	3,201	0.2	0.943	4.114	0.2	1.004
72	10,042	0.5	0.907	37,267	1.6	0.987
73	3,456	0.2	0.550	6,415	0.3	0.707
74	99,612	5.3	0.917	188,498	8.2	1.028
75	0	0.0	...	141,766	6.1	0.744
80	237,371	12.6	0.818	246,938	10.7	0.846
85	180,428	9.6	1.015	240,201	10.4	1.022
90	13,100	0.7	0.905	13,022	0.6	0.798
91	28,287	1.5	1.147	27,689	1.2	1.199
92	39,418	2.1	0.971	40,430	1.8	0.962
93	57,091	3.0	1.046	66,171	2,9	1.094
Total	1,881,350	100.0	1.000	2,307,235	100.0	1.000

Table 10.16: Specialisation in Central Italy (1981-1991)

Code	Employees 1981	Share 1981	Spec. 1981	Employees 1991	Share 1991	Spec. 1991
50	95,877	5.1	0.964	101,637	4.2	0.951
51	136,474	7.3	0.845	165,405	6.9	0.842
52	380,814	20.3	0.959	402,448	16.7	0.966
55	135,978	7.3	0.977	161,697	6.7	1.012
60	132,134	7.1	1.025	118,611	4.9	1.011
61	1,443	0.1	0.250	1,337	0.1	0.287
62	17,014	0.9	3.972	18,412	0.8	3.536
63	47,329	2.5	1.115	47,297	2.0	1.150
64	74,431	4.0	1.189	90,974	3.8	1.203
65	81,788	4.4	1.174	97,137	4.0	1.088
66	12,807	0.7	1.323	13,103	0.5	1.230
67	12,966	0.7	0.924	23,305	1.0	0.969
70	8,543	0.5	0.876	18,240	0.8	1.007
71	3,900	0.2	1.154	4,850	0.2	1.135
72	13,537	0.7	1.228	46,798	1.9	1.188
73	11,646	0.6	1.862	15,941	0.7	1.685
74	110,171	5.9	1.019	179,762	7.5	0.940
75	0	0.0	...	241,423	10.0	1.215
80	269,595	14.4	0.933	269,067	11.2	0.884
85	176,696	9.4	0.998	226,896	9.4	0.926
90	14,033	0.7	0.973	16,304	0.7	0.958
91	28,915	1.5	1.178	28,111	1.2	1.168
92	51,101	2.7	1.264	55,865	2.3	1.275
93	55,473	3.0	1.021	61,276	2.5	0.972
Total	**1,872,665**	**100.0**	**1.000**	**2,405,896**	**100.0**	**1.000**

Table 10.17: Specialisation in South-East Italy(1981-1991)

Code	Employees 1981	Share 1981	Spec. 1981	Employees 1991	Share 1991	Spec. 1991
50	43,430	5.8	1.092	45,264	4.7	1.062
51	45,380	6.1	0.703	57,880	6.0	0.739
52	183,760	24.5	1.158	184,197	19.2	1.110
55	41,120	5.5	0.739	47,705	5.0	0.749
60	45,095	6.0	0.875	42,016	4.4	0.898
61	168	0.0	0.073	373	0.0	0.201
62	89	0.0	0.052	195	0.0	0.094
63	8,725	1.2	0.514	8,412	0.9	0.513
64	24,164	3.2	0.965	30,823	3.2	1.023
65	17,039	2.3	0.612	25,107	2.6	0.706
66	984	0.1	0.254	1,221	0.1	0.288
67	5,024	0.7	0.896	8.120	0.8	0.847
70	996	0.1	0.255	2,175	0.2	0.301
71	672	0.1	0.497	1,106	0.1	0.649
72	1,963	0.3	0.445	8,840	0.9	0.563
73	1,194	0.2	0.478	2,382	0.2	0.632
74	38,397	5.1	0.889	67,450	7.0	0.885
75	0	0.0	...	96,279	10.0	1.216
80	154,907	20.7	1.341	167,302	17.4	1.380
85	78,784	10.5	1.114	104,151	10.9	1.067
90	8,400	1.1	1.458	8,063	0.8	1.188
91	11,676	1.6	1.190	9,304	1.0	0.970
92	16,042	2.1	0.993	15,319	1.6	0.877
93	20,627	2.8	0.950	25,207	2.6	1.003
Total	**748,636**	**100.0**	**1.000**	**958,891**	**100.0**	**1.000**

Table 10.18: Specialisation in South-West Italy (1981-1991)

Code	Employees 1981	Share 1981	Spec. 1981	Employees 1991	Share 1991	Spec. 1991
50	85,117	5.1	0.957	95,627	4.4	0.980
51	96,834	5.8	0.671	113,667	5.2	0.634
52	400,989	24.0	1.130	414,483	18.9	1.091
55	97,148	5.8	0.781	114,357	5.2	0.784
60	109,425	6.5	0.950	99,339	4.5	0.928
61	9,806	0.6	1.898	11,377	0.5	2.675
62	1,972	0.1	0.515	2,442	0.1	0.514
63	27,329	1.6	0.720	24,387	1.1	0.650
64	64,717	3.9	1.156	74,152	3.4	1.075
65	44,733	2.7	0.718	57,774	2.6	0.709
66	3,046	0.2	0.352	3,326	0.2	0.342
67	10,246	0.6	0.817	16,691	0.8	0.761
70	2,715	0.2	0.311	4,807	0.2	0.291
71	2,241	0.1	0.742	3,219	0.1	0.826
72	3,926	0.2	0.398	18,364	0.8	0.511
73	3,754	0.2	0.672	5,252	0.2	0.609
74	72,105	4.3	0.746	130,406	5.9	0.747
75	0	0.0	...	255,603	11.6	1.410
80	370,400	22.1	1.435	386,631	17.1	1.393
85	159,267	9.5	1.007	238,674	10.9	1.068
90	14,295	0.9	1.110	21,498	1.0	1.384
91	17,752	1.1	0.809	17,562	0.8	0.799
92	32,957	2.0	0.912	35,797	1.6	0.896
93	42,679	2.6	0.879	49,713	2.3	0.864
Total	**1,673,453**	**100.0**	**1.000**	**2,195,148**	**100,0**	**1.000**

10.5 CONCLUSIONS

From the analysis carried out the first conclusion we can draw is that a satisfactory interaction between economic theory, economic history and innovation measurement has not yet been reached. Therefore the collection of more evidence on the main characteristics of structural change seems useful in order to focus the appropriate context of analysis for the measurement of service innovation.

In particular, further research is needed to assess the relative impact of specialisation and variety on economic performance at the local level. This is especially true with respect to:

(i) the nature and characteristics of knowledge involved;
(ii) the trade-off between path dependence and discretionary action in innovative processes.

We have also to take into account that a significant part of the impact of innovation operates through social and individual capabilities, which implies that there may exist greater emphasis on potential for growth, rather than on actual growth. This, in turn, implies inevitable 'redundancy' effects and social costs in knowledge and technology accumulation.

Moreover, in the new reference framework mapped by liberalisation and market unification we can point out that:

(a) the linkage between innovation and competitiveness becomes more direct, innovation is conceived more and more as an actual exploitation of past investments, and finalised goals dominate over the progress of basic knowledge in research activity – and, we could ask: in this framework who is in charge of knowledge accumulation?

(b) the innovation in the organisation of economic agents, also by means of the formation of 'new quasi markets', is a kind of compensation, a way of responding to the relative scarcity of basic knowledge accumulation, which helps to exploit existing redundancies of knowledge and social capital.

Despite the theoretical and analytical limitations in defining the role of services at both the macro and micro level just highlighted, we have analysed the evolution of the service sector in the long run in Italy, pointing out its sectoral and geographical differentiation as well as its possible links with innovation.

The main suggestion emerging from our work is that the characteristics of service sectors are such that we cannot necessarily measure the interactions with the innovative process only by means of *direct* indicators of change. In fact, given the quantitative importance of traditional 'non-innovative' services, it is important to identify those forms of change – which do not necessarily occur along traditional channels – taking place in the form of externalities. Externalities consist of transmission of competencies, ideas and knowledge, and can be captured only through models based on *indirect* effects, as those tested at the empirical level in Section 10.3.3. The possible limitations of such an approach do not prevent further attempts to push forward new conceptual frameworks and models to overcome the loose vision still prevailing on the role of services in structural change and innovation in advanced economies.

Appendix

An econometric specification allowing for a test of the likely impact of each type of externality on long-run growth of the services sector is given by the following equation (Glaeser et al, 1992):

$$\log\left(\frac{l_{i,t}^{k}}{l_{i,t-1}^{k}}\right) = \alpha + \beta \log(l_{i,t-1}^{k}) + \delta S_{i,t-1}^{k} + \varsigma C_{i,t-1}^{k} + \eta V_{i,t-1}^{k} + \varepsilon_{i,k}$$

The dependent variable in this equation is the growth rate of employment of the service sector i localised in region k, between period t and period t-1: that is, for example, between 1961 and 1951.

In order to measure the dynamic externalities we employ different indices. In particular, following Glaeser et al (1992) the specialisation of the service sector i localised in region k is measured by means of the following expression:

$$S_{i}^{k} = \frac{\dfrac{l_{i,k}}{l_{k}}}{\dfrac{l_{i,IT}}{l_{IT}}}$$

where $l_{i,k}$ denotes the employment of the service sector i in region k, l_k the total employment of region k, $l_{i,IT}$ the employment of service sector i in Italy, and, finally, l_{IT} the total employment in Italy.

In a very simple way, this index – known in the literature as the Balassa index – relates the share of the service sector i localised in a given geographical area with the same share calculated at the national level. The economic meaning of this index is quite simple: values of the index greater than *1* must be interpreted as evidence of specialisation, whereas values smaller than *1* show the presence of a despecialised sector.

In order to test the role played by variety, that is the importance of diversity in the industrial and economic environment, a simple measure has been defined in the recent literature (Henderson et al, 1995), employing an Herfindahl index of the following type:

$$V_{i}^{k} = \sum_{j \neq i} s_{jk}^{2}$$

where s_{jk} is the share in region k of the service sector j over all other service sectors, with the exception of sector i. An increase in V_{i}^{k} therefore should be interpreted as an increase in the local concentration and hence as a reduction in the variety of the region. On the contrary, a low value of this measure means that the region is characterised by a high level of diversification and thus by some variety of its productive structure.

We wish to investigate the effect of these kinds of externalities, controlling as well for the market structure of the local service sectors: that is specialisation-*cum*-competition and specialisation-*cum*-monopoly. Our measure of this phenomenon is therefore given by the following relationship:

$$C_{i,k} = \frac{\dfrac{u_{i,k}}{l_{i,k}}}{\dfrac{u_{i,IT}}{l_{i,IT}}}$$

where $u_{i,k}$ denotes the number of establishments of service sector i in region k, $l_{i,k}$ the number of employees in the same local sector, while the value of the denominator show the same magnitudes at the national level.

As in the case of the specialisation index, if $C_{i,k}$ has a value greater than *1* means that the local market of this sector i is more competitive than the same industry at a national level.

Finally, $\varepsilon_{i,k}$ is the error term with the usual properties.

REFERENCES

Abramowitz, M. 'The Post-War Productivity Spurt and Slowdown: Factors of Potential and Realisation' in OECD, *Technology and Productivity*, Paris: OECD, 1991.

Aghion, Ph. and Howitt, P. *Endogenous Growth Theory*, Cambridge, Mass.: MIT Press, 1998.

Antonelli, G. and Pegoretti, G. 'Paths of Technological Change, Markets for Production Factors, Irreversibility', *Quaderni Dynamis*, 1995; No. 6/95, Milan, IDSE-CNR.

Antonelli, G., Guidetti, G., Leoncini, R., Nosvelli, M., Pombeni, L. and Zamparini, L. *Apertura dei mercati locali del lavoro e fabbisogno di risorse umane da parte delle imprese*, Milan: Franco Angeli, 1998a.

Antonelli, G., Cainelli, G. and Notari, C. 'Externalities, Structural Change and Service Sectors in Italy. Some Empirical Evidence from a New Data Set', paper presented at the workshop "Conceptualising and Measuring Service Innovation", 20-21 May, CRIC, University of Manchester, 1998b.

Antonelli, G. and Zoboli, R. 'Ricerca e innovazione in una economia di servizi', paper presented at the XXXIX Scientific Meeting of the Italian Economic Association, October 1998 (to be published).

Baily, M.N. and Chakrabarti, A.K. *Innovation and the Productivity Crisis*, Washington: The Brookings Institution, 1988.

Baumol, W.J., Batey Blackman, S.A. and Wolff, E. *Productivity and the American Leadership. The Long View*, Cambridge: MIT Press, 1989.

Becker, G. 'A Theory of the Allocation of Time', *Economic Journal*, 1965; 75, pp.493-517.

Bell, D. *The Coming of Post-Industrial Society*, London: Heinemann, 1973.

Caracostas, P. and Muldur, U. *Society, The Endless Frontier. A European Vision of Research and Innovation Policies for the 21st Century*, Brussels: European Commission, 1997.

Carlsson, B. (ed.). *Technological Systems and Economic Performance: The Case of Factory Automation*, Dordrecht: Kluwer, 1995.

Dahmén, E. ' "Development Blocks" in Industrial Transformation' in Carlsson B. and Henriksson G.H. (eds.), *Development Blocks and Industrial Transformation: The Dahménian Approach to Economic Development*, Stockholm: Almqvist & Wicksell, 1991, also in Carlsson, B., *Industrial Dynamics*, Boston: Kluwer, 1989, pp.109-121.

De Liso, N. and Metcalfe, J.S. 'On Technological Systems and Technological Paradigms', in Helmstädter, E. and Perlman, M. (eds.), *Behavioral Norms, Technological Progress and Economic Dynamics*, Ann Arbor: University of Michigan Press, 1996, pp.71-95.

Dosi, G., Freeman, C., Nelson, R., Silverberg, G. and Soete, L. (eds.). *Technical Change and Economic Theory*, London: Pinter, 1988.

Gershuny, J.I. and Miles, I. *The New Service Economy: the Transformation of Employment in Industrial Societies*, London: Frances Printer, 1983.

Glaeser, E., Kallal, H., Scheinkman, J. and Schleifer, A. 'Growth in Cities', *Journal of Political Economy*, 1992; Vol. 100, No. 6, pp.1126-1152.

Hauknes, J. and Miles, I. 'Services in European Innovation Systems. A Review of Issues', *STEP Report*, n. 6/96, Oslo, 1996.

Henderson, V., Kuncoro, A. and Turner, M. 'Industrial Development in Cities',*Journal of Political Economy*, 1995; Vol. 103, No. 5, pp.1067-1090.

IDSE-CNR. *Trasformazioni Strutturali e Competitività dei Sistemi Locali di Produzione*, Milan: Franco Angeli, 1999.

ISTAT. *Rapporto annuale. La situazione del Paese nel 1998*, Rome: ISTAT, 1999.

Loasby, B. 'On the Definition and the Organization of Capabilities', paper presented at the IDSE-CNR seminar, June, mimeo, 1996.

Lucas, R.E. 'On the Mechanics of Economic Development', *Journal of Monetary Economics*, 1988; Vol. 22, pp.3-42.

Miles, I. 'Innovation in Services: Services in Innovation', WP of the Manchester Statistical Society, Feb 1996.

Miles, I. *Services, Systems and Innovation. Cinderella Comes to the Ball*, Manchester, CRIC, mimeo, 1998.

Momigliano, F. and Siniscalco, D. 'Mutamenti nella struttura del sistema produttivo e integrazione fra industria e servizi' in Pasinetti, L. (ed.), *Mutamenti strutturali del sistema produttivo: integrazione fra industria e settore terziario*, Bologna: il Mulino, 1986.

OECD. *Technology and Productivity*, Paris: OECD, 1991.

OECD. *Technology, Productivity and Job Creation: Analytical Report*, DSTI/IND/STP/ ICCP(95)14/ FINAL, Paris: OECD, 1996.

OECD. *Services: Measuring Real Annual Value Added*, Paris: OECD, 1988 (http://www.oecd.org/ std/servhome.html).

Petit, P. and Soete, L. 'Technical Change and Employment Growth in Services: Analytical and Policy Challenges', *Nota di lavoro n. 46.97*, Milano: Fondazione Enrico Mattei, 1997.

Pini, P. 'Occupazione, tecnologia e crescita: modelli interpretativi ed evidenze empiriche a livello macroeconomico' in *Sviluppo tecnologico e disoccupazione: trasformazione della società*, Roma: Accademia Nazionale dei Lincei, 1998.

Polanyi, K. 'The Economy as an Instituted Process' in Polanyi, K. (ed.),*Trade and Market in the Early Empires*, Free Press, 1957.

Quadrio Curzio, A. 'Globalizzazione: profili economici', lecture held at the Faculty of Law, University of Bologna, April 1999.

Quah, D.T. 'The Invisible Hand and the Weightless Economy',*Cepr Occasional Papers*, 1996; No. 12.

Romer, P.M. 'Increasing Returns and Long Run Growth',*Journal of Political Economy*, 1986; Vol. 94, pp.1002-1037.

Rossi, N., Sorgato, A. and Toniolo, G. 'I conti economici italiani: una ricostruzione statistica, 1890-1990', *Rivista di Storia Economica*, 1993; Vol. 10, No. 1.

Wolff, E. 'Technology and the Demand for Skills', *STI Review*, 1996; No. 18.

---

[1] Department of Economics, University of Bologna and IDSE-CNR, Milan, Italy.
[2] IDSE-CNR, Milan, Italy.
[3] Around 75% of the total number of firms active in the industrial and tertiary sectors taken together.
[4] More than 50% of the total employees active in the industrial and tertiary sectors taken together.
[5] Which amounts to 84% of the total number of firms active in the industrial and tertiary sectors taken together.
[6] Around 62% of the total number of firms active in the industrial and tertiary sectors taken together.
[7] Which, in this case, includes personal and business services as well as services supplied by the public administration.
[8] Respectively 26.9% and 18.7% of the total.
[9] 14.1% of the total.
[10] According to some estimates (Quadrio Curzio, 1999), the service sectors are dominant both in the 15% world PIL share in which globalisation is accelerating and in the 50% world PIL share in which globalisation is just starting.
[11] See for instance Momigliano and Siniscalco (1986).
[12] For instance, the infrastructural character of some activities has been underlined in IDSE-CNR (1999) and ISTAT (1999).
[13] Such as accounting, software writing and marketing.
[14] See Antonelli and Pegoretti (1995).
[15] Whose service content is very substantial.
[16] See, among the many contributions, Carlsson (1995), Dosi et al (1988), Dahmèn (1991), De Liso and Metcalfe (1996).
[17] This is part of a more general research work on structural change in Italy over the very long run developed by IDSE-CNR (1999).
[18] Labour input measured in terms of the 'standard labour unit'.
[19] As for the econometric specification see the appendix.
[20] Ateco is the Italian standard classification of economic activities.

PART III:
CASE STUDIES

Chapter 11
Information Technologies in Non-Knowledge Services: Innovations on the Margin?[1]

Ken Ducatel

11.1 INTRODUCTION

It has often been remarked that services are relatively neglected in research, especially given that they constitute the lion's share of the economy and their centrality to job creation. Partly, this neglect relates to an assumption that services are not really driving the economy. In particular, services are often viewed as less innovative. Recently, such a view is in the process of being overturned with, as this volume testifies, a mini-boom of interest in innovation in services. Most of the attention has fallen onto two sorts of services, in which innovation seems likely to be more prominent. These are scale intensive or branch-network informational services (especially the 'FIRE' group of services represented by financial, insurance and real estate services) and especially 'knowledge-intensive business services' (KIBS) such as professional consultancy and design or software services (Soete and Miozzi, 1989). Some services such as telecommunications are in both categories.

The result of all these efforts is to create a second level of neglect of service innovation in which the non-informational or non-knowledge intensive services (non-KIS) activities such as: retailing, distribution, cleaning, tourism have been largely left out (although see Sundbo, 1995, 1998; and Gallouj 1994). Is such relative neglect justified? Certainly not on the grounds that these services are unimportant in themselves, they are major contributors to employment and employment growth. Sometimes, these sectors are also significant investors in ICTs - across the board, not just in informational or technology-based sub-sectors (Table 11.1).

Table 11.1: European IT expenditure trends by vertical markets (1998)

	Relative market size	Annual Growth Rate (%)
Discrete manufacturing	100	10
Banking	90	16
Process manufacturing	75	13
Retail & Wholesale	70	10
Central Government	55	6
Insurance & other finance	45	11
Business services	39	11
Local Government	36	8
Communications	34	13
Education	19	8
Agriculture, mining & construction	18	7
Transport	14	10
Health care	13	9
Utilities	11	11
Other	5	7

Source: Data adapted from EITO (1999)

However, does investment in ICTs in itself mean that innovation is taking place in these services? It is arguable that these services are of marginal importance to innovation studies. Perhaps they are merely applying technologies that have been developed elsewhere. For instance, they could be investing in turnkey technologies to perform generic administrative functions such as payroll services, word and document processing, automated billing or electronic data interchange. Are such changes really 'innovative', do all firms undertake such innovations? They might represent the application of techniques which have been developed by a computer service company (i.e. the real innovation work is undertaken by a KIBS not the 'non-KIS'). Also, do such innovations affect the *core* of the service? Surely, the technologies that affect these services are really just incremental improvements to the internal operations, with no real innovation taking place in terms of the character of the service, which remain low tech and based on direct personal service delivery?

In this chapter, I attempt to answer these questions by drawing on published evidence from, mostly qualitative, studies of technical change in these services. The material is organised in three further sections. First, it seeks a conceptual basis for the work by reviewing some recent writings on service innovation. Second, it examines on a sector by sector basis some of the key innovations taking place in these services. Third, on the basis of this discussion, it tries to categorise the innovations that are taking place in order to assess the significance of the changes, and to draw out the effects on work structures related to the different types of innovation.

11.2 SERVICE CONCEPTS: INNOVATION AND KNOWLEDGE

The recent growth of attention to innovation in services has largely focused innovations in informational and/or knowledge-based services. A closer look at these focuses reveals certain predispositions in the analysis of innovation in services. For example, Soete and Miozzi (1989) suggest the following taxonomy:

- Specialised technology suppliers and science-based suppliers (this category includes Knowledge Intensive Business Services such as engineering consultancy or computer programming).

- Production and scale intensive services and network services. This includes large data processing services, banks and insurance services that are dependent on elaborate informational systems and networks. It also covers those that are dependent on physical investments such as transport and distribution.

- Supplier dominated services such as retail trade, personal services, maintenance and repair.

Each cluster of services has a different relationship to innovation. In the first cluster, KIBS do not just apply technological knowledge but may actually generate new (technical) knowledge[2]. The distinction between firms that are developers of innovations and those that merely apply innovations from elsewhere is critical to our evolving understanding of service innovation. The interest in KIBS is at least partly related to an apparent professionalisation of innovation in services. The staff of KIBS is more likely to be highly educated and may have a more or less explicit research or development bias in their work, in areas such as marketing, design or strategy. In technology based KIBS (such as software services or telematics) there may even be a cadre of science or technology trained personnel. These people provide a direct port of entry for technologies, and in some cases may actually be undertaking the innovations that will later be absorbed into routine technological systems of technology providers[3].

It is the case that within even some highly professionalised KIBS (for example, accounting) innovation practices are less often identified as such by the actors involved and are often undertaken on a short-term project or ad hoc basis. Overall, there are fewer official R&D departments or workers and a weaker linkage to the innovation system in the form or technology or science-based research in institutes or universities. Despite these caveats it is possible in the studies of innovation in services to see verisimilitude to the industrial model of 'R&D' or 'innovation'. Studies of KIBS have found that evidence of professional and/or technologically trained staff, departments with a remit to undertake development and signs of separation from line-operations through access to specific development budgets. In the context of the current volume, these characteristics are salient because they imply that KIBS are more likely to be explicitly plugged into a formal innovation system than other forms of service which lack this body of formally trained and assigned personnel.

In the second cluster of scale intensive or network-based firms, the interest of innovation researchers has focused on informational services, especially the financial services and telecommunications and audio-visual services. Telecommunications and audio-visual are inherently technology-based, often with formalised research departments and staff. The interest in banks and insurance companies, meanwhile, undoubtedly reflects their pioneering role in the commercial application of computers and telecommunications (see, for example, Barras and Swann, 1983; Fincham et al, 1994; or Knights and Morgan, 1995). Many of the applications have been in so called horizontal applications such as the administration of payrolls, accounting and bookkeeping functions, transaction processing. These are of course generic problems encountered by all large-scale businesses that have to move around a lot of data. But the core business of banks and insurance firms are also transactional-intense and informational by nature. This makes them natural candidates to use information and communication technologies. It means that they exhibit interesting innovative behaviour that affects both processes and the character of the end service. They are also distributed organisations with a network of branches, sales offices, and thus were early users and indeed developers of computer networks and telematics. Innovation studies in these industries reveal therefore, that despite the traditional view of services as simply applying innovations developed elsewhere, service organisations have led to the development and transfer of new knowledge computer systems and services. Such changes can lead to radical innovations in the form of services delivered (for example, branchless telephone banking or ATMs). Also, many informational services have built-up rather large and formalised departments for the implementation of new technologies. The effects of introducing such workers can have substantial ramifications for the wider corporate culture. For example, wide ranging changes in banking career structures can be associated with the recruitment of computer specialists. These new technological workers not only challenged existing patterns of career expectations, but introduced new models of 'legitimate' banking behaviour and new approaches to searching for business opportunity based on formal models, mathematical techniques and so on.

However, for a firm to be classed as 'innovative', is it necessary that it involves technical competences and a formal institutionalisation of innovation? Recent conceptualisations of innovation emphasise the difficulty of transferring technology as a discrete and ready formed package. In fact, most innovation involves a close interface and interaction between the innovator and the user and subsequent adaptation. Indeed, most innovation is now seen as recursive involving extended interactions between producer and user, involving learning on both sides, the codification of tacit knowledge and the creation of new tacit knowledge. Here is a broader conceptualisation of innovation which permits non-technical acts and non-technical staff to be brought into the ambit of innovation.

A very useful example of such theories is in the recent work on service innovation by Gallouj and Weinstein (1997). They generalise earlier work by Saviotti and Metcalfe (1984) which suggested that innovation can be thought of in terms of a change affecting one or more aspect of the set of characteristics of a good. For services, the overall set of characteristics includes a number of sub-sets including:

- The final use characteristics (Y) – which are determined by how the user sees the service in terms of its price, level of performance, technical complexity and so on.

- The technical characteristics (X) – which are made up of the technical systems necessary to deliver the service (including tangible aspects such as particular machines or chemicals and intangibles but codified procedures or routines such as formal models or legal procedures).

- The complementary sets of competences (C) needed by the producer of the service and the clients in order to deliver and use the service as it is currently configured.

In a service the distinctions between these categories are perhaps more fluid than would be the case with a traditional industrial product. In particular, there are important interactions at the customer-producer interface in which the technical characteristics intervene. Also both the technical characteristics and the competences of the firm interact to affect the ability to deliver final use characteristics. We can imagine that a vector in which the elements are levels of the different attributes involved can describe each set of characteristics (as follows, note that producer competences are represented as C and customer competences as C′.).

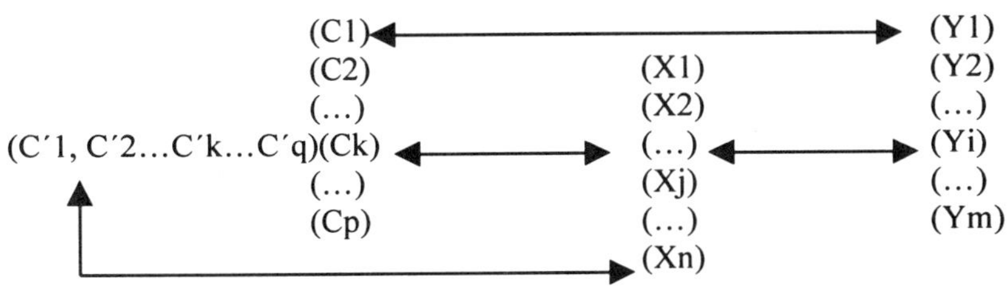

In brief, a service innovation is a change in level or a substitution of any of the elements in each of the vectors represented above. In their analysis, Gallouj and Weinstein make a number of important and careful distinctions between the tangible and intangible elements of the technical characteristics.

However (for simplicity of exposition one must suppose) in their general form of innovation they collapse together the internal technical characteristics and the external ones. The internal technical characteristics are the technical attributes aspects of the service which are not explicit in the client interface (i.e. they are in the 'back office'). (We might imagine another vector of characteristics 'x' which interacts with the external 'front office' characteristics (X), the firm's competences (C) and the final characteristics (Y), but perhaps not the customer competences (C′).) For many types of service (and indeed industrial) innovation this distinction is likely to be important. Think for instance of the difference between branchless telebanking and branch network automation. Both the internal and external technical characteristics of course have tangible and intangible components. They also make a distinction between the intangible technical characteristics and the competences of the firm. Intangible technical characteristics are codified and standardised as routines, whereas competences are more tacit, relying on human capabilities at an individual or group level.

The utility of Gallouj and Weinstein's framework is that it provides a systematic way of disaggregating different types of innovation, in order to see how the different elements vary in different cases. Thus:

1. In the strictest case of *radical innovations* the before and after situations should have no common elements at all. Thus, the structure of competences of client and provider are destroyed and there are completely new internal and external technical characteristics and user characteristics. In a less extreme case there may be quite a lot of continuity of elements. For instance in an innovation the use characteristics (Y) may be unchanged (for example the technical performance is identical), but it requires different skills on the part of the provider and the customer and the technical characteristics are quite different in the front and back offices.

2. *Improvement innovations* can be defined as having the same elements as before but with different levels. Thus, perhaps Y changes because of a change in C or X. Or, an internal technical productivity gain (x) may cause a lower price and thus the levels of Y.

3. *Incremental innovations* are analytically easy to distinguish between improvement innovations because they involve removal, new or substitutions of elements rather than changes of level. However, a very large increase in level may be empirically very hard to distinguish from a substitution.

4. *Ad hoc innovations* are non-programmed improvements, perhaps resulting from customer-producer interactions that lead to a reproducible improvement in performance because a higher level of knowledge is achieved (elements of C or C′ are introduced).

5. *Formalisation or standardisation innovations* are where improvements are achieved through the codification of improvements, perhaps as a result of learning by doing (represented by increases in elements of C). This may lead to higher levels of the intangible elements of X or x. In these cases an improvement is routinised and so rendered easier to understand and to transfer.

6. *Recombinant or architectural innovations* occur when the service characteristics are split apart (unbundled) or combined with other services in order to create a new service structure (bundling). This, it is claimed, is an extremely common form of service innovation. What it involves often is the destruction of competences which relate to the old structure (or 'architecture') or the service but not those which relate to individual modules or sub-components. The fact that innovation in services can happen by changes in the mix of service characteristics also provides an clue as to why service innovation is so hard to detect. Namely, the innovations are often rather unspectacular, hard to evaluate, low cost, involving a lack of explicit development, lacking prototypes and difficult to protect from being copied.

In addition to providing a useful framework for analyses, which we will use below to look at non-KIS innovations, the approach of Gallouj and Weinstein offers a way to discriminate between different types of innovation. On a superficial level any change in any of the elements of the service characteristics is an 'innovation'. This means that changes in the service performance can be seen as an innovation. Of course, changes in the tangible technical characteristic sets are technological innovations in the traditional sense of hardware improvements. But, changes in intangible, codified, technical characteristics are also innovations. Changes in either internal or external technical sets are also innovations. So are changes in the competences required of either producers or suppliers are innovations.

Using this framework the extent of innovation can be interrogated. For instance, in principle it is possible to trace the extent to which the introduction of new machinery in the back office leads to a change in the service characteristics, the producer competences or the customer competences. More importantly, given our current interest in the relationship between services and systems of innovation, how do the immediate changes in the service characteristics associated with a particular innovation ramify through the other sets of characteristics? How do different innovatory set-ups (or innovation systems) in terms of different customer-supplier interactions, levels of competences on both sides, starting points for technical characteristics and so on affect the speed and path of an innovation in a particular industry? Finally, again relating to our interest in systems of innovation, to what extent does the firm exhibit purposive behaviour in relation to innovation? What forms of transfer from tacit to codified behaviour can we identify? Or, is the propensity to innovate related to different levels of formalisation of competences?

It will be very useful to seek answers to these questions as they provide for a much more structured inquiry into the scope of innovation than has been possible until now. In this case, I hope to show that it allows a broader approach to innovation than the tradition one stemming from the science or formal technology base, which is still rather dominant. At the same time it provides a discipline for enquiry, because not all behaviour leading to change is given equal weight in the analysis. This broad but discriminating tool should allow us to look at the non-KIS innovation analytically, and this task is begun in the next section.

11.3 SECTORS BY SECTOR – KEY ICT INNOVATIONS[4]

The services chosen for analysis here are normally in the private sector. As we can see from Table 11.2, these services are certainly worth examining if only because their contribution to employment and their growth rates in recent years. However, the question here is what types of innovation are taking place in these sectors (their significance) and whether these sectors are marked by certain types of innovation typified by and whether ICTs are leading to substantial or only small-scale innovations in these services. As a first step towards this goal, we provide a listing of the major ICT innovations we have been able to detect in these services (Table 11.3). In some of these sectors (for example, catering and cleaning) the major innovations relate more to other technologies than ICTs. However, in all the sectors there is a substantial amount of new ICT equipment being deployed. Also, in all of these services the adoption of new technologies is accompanying structural shifts in organisational patterns.

Table 11.2: Services employment shares and changes in Europe

	Contribution to employment in Europe (% of total EU salaried employment)	Net contribution to employment 1985-95 (000s)
Construction	6.4	-6.87
Inland transport	3.0	2.47
Retail and wholesale	13.9	30.88
Hotels and restaurants	3.8	17.68
Total services	72.6	195.72

(Source: EU Panorama, 1997)

Table 11.3: ICT innovations by sector

Service	Key ICT-innovations
Construction	CAD design - virtualisation of project descriptions Construction project control systems Bar-coding of stock Smart buildings (data cabling, opto-electronics, computer control, security systems, environmental controls) Smart roads/traffic systems, etc. (*non ICT: off-site production/on-site assembly; new materials*)
Security services	Integrated services/facilities management Access control, portable alarm systems, intruder detectors, fire sensors, etc Smart badges, Automatic fault notification Portable communication systems
Cleaning services	Environmental sensors bar coding of facilities, etc *(non-ICT: new cleaning products, machines and techniques; specialised cleaning services e.g. toxic waste management)*
Retail and wholesale	Supply chain integration Efficient Consumer Response (automatic stock replenishment) Electronic Data Interchange EPoS/EFTPoS, Bar coding, Portable Data Capture Customer profiling, loyalty programmes, etc Added value services (customer advice and information) On-line shopping Outlet performance tracking Flexible Staff Rostering
Transport services	Supply chain integration (hub automation, robotised warehouses, logistical services, etc) Smart vehicles and intelligent highways (route guidance, convoy driving, geographic information systems etc) Tracking and tracing, bar coding, portable data capture units Mobile communication systems Customer profiling, service customisation Quality/performance indicators
Hotels and restaurants	EPoS, Ordering pads, touch screen terminals, bill preparation Automatic reservation, ordering, check-in, etc Central reservation systems (GDS) Stock control Performance tracking (outlets and individuals) Modular customisation (yield management, menu pricing, etc.) On-line ordering (home delivery) Customer profiling Added value services (cybercafes?, electronic room services) (*non-ICT: food preparation, cook-chill, portion control*)

11.3.1 Construction[5]

Construction's image as a low-technology industry belies a reality in which there is substantial technological development (Senker et al, 1998). Smart buildings provide an example of a clear radical innovation that can affect all service characteristic sets. Very often the extra costs associated in constructing a smart building in itself will imply changes in the final use characteristics (Y) for example high security for banks, superior environmental performance, ease of access for disabled people. Some smart building technologies are also very much in the user's interface (X). Doors may open automatically or, otherwise, prevent unauthorised access. Lights may turn on and off depending on whether the room is in use. Modular construction can rapidly accommodate changes of activity according to need.

Internal technical characteristics in the production of (x) are also fundamentally transformed in smart buildings to accommodate sophisticated electronic control equipment. Changes here affect the competences demanded of users and, especially of service workers. The new competence demands occur at all stages of construction, design, erection of the structure, installation of equipment, maintenance and operation and finally refurbishment. Traditional tradesmen, especially electricians and plumbers, as well as specialist technicians have to work with (or at least avoid damaging) the electronic nervous systems that make up smart buildings. In some cases the new competences (C) must be added over the top of existing skills. There will also be a proportionally greater growth of computer and network support technicians for installing and troubleshooting.

The introduction of new materials, methods and so on also affects the industrial structure of the construction trade. A good example is the way that DIY methods can undermine some traditional processes and skills. Maintenance free materials, easy fit pipes all reduce the demand for certain trades in both the construction and the maintenance/renovation phases. Most of the impact here, however, is probably not coming from ICT, but from new materials and from the organisational pressures which result.

As regards the industrial level innovation system, therefore, smart buildings mean that there will be a shift towards formal knowledge-based activities. There will be an increase in demand for formally qualified technical staff. One can expect also a growth of specialist (KIBS) service suppliers and sub-contractors. At the same time there are important shifts in the intangible technical characteristics of construction, driven by customer demands for higher quality and greater responsiveness to client demands as well as the need for construction services to respond to innovations in materials and methods and in the furnishings of buildings. Construction is importing management techniques from complex manufacturing environments (aerospace, and automobiles) such as concurrent engineering to increase project co-ordination or 'lean construction' to reduce reworking of defects. Here we see important changes in the levels of internal technical intangibles and a supporting role for ICTs as technical tangibles in the form of CAD systems or project management systems. These process innovations are very evident in high cost 'smart buildings', where the building process is much more complex and has new elements which must be controlled and where there is a higher level of formal sophistication in project management. But, these management process innovations are also evident as improvement innovations (changes in levels of X and C) in conventional building projects. Even small scale building projects can use such methods as virtualisation techniques for project design and costing to allow customers to achieve a closer specification of their needs. A very simple example could be the widespread use of computerised kitchen designs and the use of project management tools to price and plan the tasks of the different tradesmen involved.

Smart buildings are clearly important as radical innovations, but they are still rather rare. There is a very high level of inertia in the construction industry, because of the low rate of replacement of the building stock. Also, many construction sector firms are small and low-tech, with traditionally skilled employees having low levels of formal training. All these factors imply that the receptiveness to new technologies will be rather low in the sector.

Certainly, technological innovation will chip away at the traditional industrial structure but in a direction that is driven by economic and organisational drivers. New markets, firms and occupations will emerge around these new technologies. However, Gallouj and Weinstein suggest that '*recombinant innovation*' will turn out to be the most important. This is clearly seen in construction where market pressures are driving a trend towards integrated 'facilities management' services, in which construction is just the first phase of a on-stop service contract spanning maintenance, cleaning and servicing.

11.3.2 Services to buildings and their residents (cleaning, security, etc)

The trend towards integrated facilities management in larger construction projects also affects physical building services such as cleaning, security and grounds-maintenance. All these services centre on the repair, control and protection of buildings[6]. In addition, these services also involve looking after or protecting people such as caretaking or guardianship. Technological innovation in this area awaits the installation of electronic infrastructures, which is currently a rather piecemeal affair. Nevertheless, there could be an important role for ICT, especially in security services with the high level of automatic devices available to monitor and control buildings and their environments (see Table 11.4). Such developments can change quite dramatically the performance of building services and the markets in which they are applied. For example, the introduction of high technology security systems are known to change the experience of high rise living in public housing projects, by raising the security of individuals. Electronic monitoring systems can be used to allow elderly people or those with chronic health problems to continue to lead independent lives (Rantanen and Lehtinen, 1999).

Table 11.4: Security services market Netherlands (Mecu)

	Market	Market share (%)
Installation/Electronic security systems	425.2	37
Manned Guarding	264.3	23
Locks, etc	195.4	17
Security equipment	172.4	15
Transport of valuable	57.5	5
Alarm monitoring stations	23.0	2
Council/advice	11.5	1
Total	1149.3	

(Source: adapted from EU Panorama, 1997)

These innovations are clearly radical innovations as they affect all the service characteristic sets. But, they arise from (or perhaps require) recombinant innovations, in that, very often, previously separate service functions are integrated to form new service structures. To see this more clearly, it is interesting to note an analysis of domotics in French public housing (HLMs). In this study, Toussaint and Lozier (1997) distinguish between 'services for buildings' and 'services for living'. The driver in the first case is perceived to be in the provision of an efficient infrastructure for the building (security, control and fault identification/rectification in machines such as lifts and entry phones, heating and cooling control, remote metering of use, logging of faults and clear-up rates, etc.). In the case of services for living the dominant driver is the quality of the social environment, care and support for residents, personal contact with the caretaker, etc.

These are two quite distinct service philosophies and there are very important lessons to be learnt here about the relationship between social and technological drivers of innovation. For the purposes of the present chapter, however, the important point to note is the way in which user and service provider competences interacted in the introduction of the domotic systems. In many cases the ICTs were ignored by end users – the people living in the housing projects. Why? Because the residents had satisfied their needs (or expectations) of housing with having a safe, dry and warm roof over their heads. In terms of the innovation theory, we can see this as a failure to address the customer-service interface. In addition, that the residents did not accept the technologies will in part reflect unobtainable calls on their competences of the clients (C'). Here we might need an extended concept of 'competences'. As Haddon and Silverstone (1999) note the acceptance of new technologies by such groups requires a close match between previous experiences and their assessment of needs. Thus, technologies that require a conscious extra effort to learn are only likely to be of interest if they have an explicit pay-off for the end user.

More generally, it should also be pointed out that large parts of the task profile of these services are very hard if not impossible to automate. Not only are caring and personal service elements more difficult to substitute, but some of the basic tasks, such as cleaning, gardening, servicing and monitoring of the condition of facilities, involve activities which are relatively simple for humans but tremendously difficult for machines. For this reason many of the innovations in the sector are achieved through formalisation, standardisation or modularisation, i.e. by converting competences into routines (Sundbo, 1994). Thus, in introducing new ICTs in these sectors, effects on the client-service interface are likely to be rather muted. Rather, the main effects of innovation are likely to be internal, with the raising of the technical know-how of building service operatives. (This raises serious questions about how retraining will take place.)

The fact that end users will often have to be shielded from direct interaction with the technology means that service operatives may also have to be trained to intermediate between the technologies and the clients. This means that the 'hard' ICT infrastructure investments in buildings will require quite a lot of innovation in tacit and codified human competences (C and x). For this reason we may expect to see a high growth in the use of informatics tools (especially portable and wearable communication devices) as electronic aides to building services workers in treating problems, rather than direct automation of the service itself.

11.3.3 Distribution (retail, wholesale and transport services)

As with building services recombinant innovation in the form of supply chain integration features strongly in the distribution sector. Here ICTs have been exploited on a broad base in recent years to achieve the disintermediation of wholesalers by retailers (integrating backwards up the supply chain) and by professional logistical services which offer complete solutions to distribution issues.

Retailers have pioneered the use of ICTs in services first from the point of view of stock control and rationalisation. Technologies such as EPoS, EFTPoS, bar coding and portable data capture units have allowed larger retailers to develop a unique capacity to control the supply chain, because they can construct a 'demand pull' inventory system, which allows the matching of stock levels very closely to purchasing events. Such strategies, for instance, mean that much of the stockroom in stores can be eliminated and turned over to revenue generating shopfloor space, without risking stock outs on key lines. This often requires that they can operate their distribution and transport infrastructures to much tighter specifications than could be offered by traditional warehousing services. Many retailers have set up their own warehousing facilities in order to gain flexibility and control over the supply to their stores.

Wholesalers, then, are subject to a process of ICT-led 'disintermediation' in which retailers (and, in some cases, manufacturers) have taken over the distribution function. Meanwhile, the physical distribution, and sometimes the storage, functions are allocated to transport specialists who bid for contracts against a pre-specified level of service. For these reasons, the physical transport business is increasingly becoming an information-logistical service (i.e. KIBS). Of course, ICTs are central to this transformation particularly to increase the timeliness of transport services, fleet management, fleet planning, automatic fee collection, driver decision aides such as route guidance, automated transaction processing through EDI and the tracking of vehicles (see Hepworth and Ducatel (1992) for a review of the implications). The result is that a key area of innovation is in the planning and design of goods handling facilities: moves towards hub automation, automated loading/ unloading, storage, picking and packing, etc.

The most visible expression of technology intensive transportation is in express courier services. This sector is dominated by fast growing private operators, who have invested heavily in new information technologies in order to achieve the level of service required (speed, reliability, guaranteed delivery, home collection and delivery, real-time consignment tracking, delivery confirmation, etc). The very high levels of technological investment raise substantial barriers to entry to new operators, especially in the integrated couriers that offer overnight door-to-door services across on an intercontinental scale. In fact, the international market is dominated by a few leading express agents (DHL, UPS, FedEx, GD Express Worldwide on a global level and Khune & Nagel, Jet Services, TAT Express and Securicor in Europe).

Clearly, here there is a mutually reinforcing cluster of innovations at different levels. Overall, supply chain integration is a recombinant innovation that, in the extreme, eliminates certain competences (wholesalers). The driver of the process is undoubtedly the search for incremental innovations by retailers, with increases in store size, more accurate planning of inventory to match consumer demand, and faster fulfilment of orders. We can also see the emergence of new competences and markets (integrated logistical services). Probably, the most far-reaching effects are that there is a systematic and purposive search, particularly by the retailers, to use the data in their information systems to add value and raise business performance. This is nothing other than a codification of existing ad hoc and tacit processes. In other words, a formalisation of innovative behaviour has taken place. Often, the process is associated with the establishment of specialised workers and departments involved in learning from the operations of the informatics systems.

It is impossible to look at ICT innovations in distribution without considering teleshopping, which is clearly a radical innovation as it involves the automation of the retail function. However, the level of development of teleshopping remains very low, despite great claims for the future of e-commerce (see OECD, 1999). More radical distribution applications, such as teleshopping, electronic kiosks, electronic merchandising, and new interactive TV, and virtual malls, have been widely touted in the past two years. But their impact on overall sales and employment remains quite slight and there is some doubt that it will ever significantly displace traditional retailing. First of all the vast majority of e-commerce is expected to be inter-business (70-80 per cent). Second, even on the highest projections of growth for <u>total</u> e-commerce (for example, US$1.3 trillion by 2003) the turnover is small compared to overall retail sales. The likelihood is that, like catalogue shopping, on-line shopping will be restricted to a relatively small if significant share of the retail market. Just as catalogue shopping never grew much beyond 5% in the UK. It is also with a concentration in certain product lines - computer equipment and services, travel (see below) and leisure goods (books, music, videos (and pornography))[7], and amongst certain types of consumers - mostly 'white male middle-class suburb' - of the Internet. Retail e-commerce is interesting here because it once again emphasises the important of the client-service interface in determining the rate and direction of innovations. That, these pioneer teleshoppers have certain (and rather unusual) levels of technical and social competence is crucial to understanding why acceptance has been slow so far. One of the interesting questions that result is the extent to which it will be possible change this competence profile through further innovations in technical interface and/or the service support. One example of the former is the fact that full graphics browsers (i.e. the world-wide web) were needed before a wide acceptance of home computer network began even to be thinkable. On the latter, telephone-computer integration may play an important role, where a friendly voice is available to potential consumers as the click of a button on a web page (see Cornford et al, 1999).

11.3.4 Hotels and restaurants

Hotels and restaurants represent large services in terms of turnover and employment, but they are fragmented with starkly different demands, sub-sectors, patterns of ownership and control, and forms of employment. As in other consumer-based industries, much of the recent growth has followed the trend towards differentiation following the increased sophistication of tastes in tourism.

Restaurants have shown very strong growth, especially in the fast food sector where standardisation type innovations in the form of chains and franchise operations have made a strong impact. Nevertheless, the industry is still very much dominated by small, local orientated family enterprises, especially in restaurants such as pizzerias or crêperies, which offer lower cost meals but do not require high levels of skill or start-up investment. Technological innovation in this part of the tourist industry tends to be concentrated in the larger firms and is focused on food preparation such as: semi-prepared foods, cook-chill and portion control procedures. Such techniques allow greater control and predictability and lower levels of waste. Batches of food prepared in central production units also contribute to lower costs and greater flexibility and may permit the operation of an outlet over longer hours, with fewer or lower skilled personnel.

By contrast, ICTs are entering the business rather slowly and at the administrative margins, mainly through order processing and the monitoring of the performance of outlets. Improvement innovations are being achieved through technologies such as electronic ordering pads, touch screen terminals, automatic menu pricing and magnetic strip readers which are integrated into bill preparation and transaction processing technologies. These technologies are most important in increasing the speed and reliability of the service by eliminating some steps in service processing. Using such technologies orders can be sent to the production station allowing very fast fulfilment of orders. "In some cases, appetisers and drinks may be ready to serve virtually seconds after a server has finished entering the order and has left the table" (ILO, 1997, p.41). Automatic menu pricing and touch key screens also save time and can reduce errors by serving staff by removing the need to calculate prices, thus perhaps changing the competence base of restaurant staff and changing the character of the service to the end user. Where the restaurants are part of a larger facility (a hotel or leisure centre, for example) immediate up-dating of a client's bill increasingly takes place, thus adding value to the service (also effecting changes in Y).

Such ICTs may also aid the formalisation and codification of businesses by allowing managers to monitor the service. Magnetic strip readers are principally used in inventory control in food and drink preparation areas and to provide a check on pilferage and waste, to identify the origin of disputes (with clients or between departments). Such features may be very important in the operation of chain restaurants where throughput and reliability of service and food are key competitive demands. However, the fundamental applications of ICTs here are to improve organisational performance. Overall, what is taking place is a significant industrialisation of the restaurant business. Arguably the core of the service itself still rests on the quality of the food, the ambience and the personal service and there is a generalised expectation of higher standards of food and service. Thus, consumers choose to eat out in fast food restaurants some of the time to get the benefits of high levels of predictability or speed, etc. The same people, however, will increasingly demand personal service and variety of cuisine. This is good news for traditional restaurants (those that compete on the quality, creative cooking and service) but will also affect the restaurant chains, leading to 'mass customisation' strategies based on ICT tracking of performance, customer profiling and so on. Here the resulting vector of Y will reflect changes in all the other characteristic sets. So, for instance, to offer highly individualised services to clients might call on the same technical hardware (tangible x) but completely different sets of interface technologies (X), routines (intangible x) and above all different competences.

Hotels are subject to rather similar processes of ICT-mediated industrialisation as restaurants, only more so. The key trend is the integration of customer sale (point of sale) and property management systems, which have evolved separately and are orientated to the different needs of different departments (Rowe, 1995). Smaller hotels still tend to use computers for billing, accounting and so on. Large hotels and chains have made progress towards integrating all the service elements of the hotel business: room booking and pricing (yield management systems which produce prices against forecasts of occupancy rates); room status (is the room clean and ready, have faults been dealt with by maintenance, etc.); guest accounts (remote check-in for and/or pre-registration; automated updating the ledger by integrating bar and restaurant checks, credit card clearance, etc.); personnel management (rostering, payroll, performance); and, property management (energy control systems, remote air conditioning, security systems, etc.).

The developments in hotel services have strong parallels, here, with those discussed above in the building services industry. There are choices to be made between automation and personal service elements. For most hotels the main driver of technological innovation is service quality and guest satisfaction (Y), but personal service comes at a price. For instance, Piketty (1997) suggests that totally automated hotels, where reservations, check-in and payment have the absolute minimum of human intervention are growing much more rapidly in France than in the United States, where customers expect and are willing to pay for a high level of personal service[8]. This illustrates how important are social influences on technological choice. Also, it shows how closely interrelated are the different characteristic sets – in this case we can see strong trade-offs between C, C' and X in the specification of Y variables. Moreover, the 'high-level personal service' model does not exclude the use of ICTs, rather they are used as electronic 'aides' to add value to the service. For instance by establishing databases of consumer preferences from electronic 'guest preference pads' that are built up by front of house staff and managers through interactions with guests: a clear case where ICTs are engaged to codify ad hoc innovations.

In another parallel with the construction/building services sectors, hotels have always led the field in the development of smart buildings. Today, rooms are often equipped with ICT: client interface systems (automated room service with touch screen terminals, automatic wake-up, remote check-in and checkout, on-line comment cards, etc.); added value services (in-room pay per view, modem connections, smartphones and videophones, multimedia consoles, on-line local information services, fax machines, voice mail, electronic safes, etc.); and, room control systems (electronic door locks, energy management, security devices).

In addition, the growth of business, conference and exhibition related stays has strongly increased the importance of offering additional facilities and services; on-line computers, faxes and modem equipped phones, business centres, call centres, secretarial support, interpreting, videoconferencing and so on. All these features affect the service characteristics, although arguably in an incremental way rather than through radical transformations of the basic service offer (i.e. room and breakfast). In the case of conference and exhibition orientated services, however, at stake is a significant case of, recombinant innovations, in order to develop a new and strongly growing market area.

The technical skills to service such technologies, however, are not likely to be held in-house but will lie with the agents of the equipment vendor. Thus, the technologies are not deeply ingrained into the internal technical characteristics or competence sets of the firm. Although, of course the hotel management will, perhaps, have to be able to interpret clients needs in a reliable manner and be able locate competent suppliers, to get equipment repaired.

Also fundamental to hotel innovation are Central Reservation Systems (CRS) linked to telematic systems, such as Minitel, or into the computer networks of larger travel industry organisations in complementary sectors (airlines, car rental firms, travel agencies, tour operators, etc.). The consequence of hotel information coming on-line, however, has tended to reinforce still further the position of the airlines in tourist reservations because they own and operate the systems. In this sense we see the airlines using the asset of their electronic databases (in the form of competences and intangible and tangible internal technical characteristics) to gain competitive advantage. With the increasing technological intensity of the industry, the growing range of demands and the increasing standards expected it is not surprising that there has been a widespread decline in the small independent hotels. However, in some countries, we see partnership and co-operation, with voluntary chains having a significant presence in countries such as France and Germany. For instance, Thisco (The Hotel Industry Switch Company) is an association of smaller hotels which provides global electronic access to allow their members to provide direct booking facilities to travel agents[9].

In some cases small hotel groups are linked into the networks of larger operators (especially via airline computer reservation systems) or mutual networks (such as Thisco's Travelweb Internet site). In a few cases they appear in their own web sites. However, although it is expected to be important in the future, the hotel industry has been quite slow to take to the web. Bookings through the Internet were estimated at only 73,000 in 1995, rising to 450,000 in 1998. Nevertheless the potential global reach of the medium was seen as important in the recent International Hotel Association White Paper (IHA, 1996) on the future of the industry.

It can be argued, then, the fundamental service characteristic of hotels is largely unaffected by ICTs. The key services are comfortable beds, clean rooms, efficient and friendly staff and so on: the traditional values. Thus there is no real scope for radical innovation. Also, although in the case of automated hotels there are changes demanded in client competences, most of the time the clients are not required to change their skill base – why should they. As we have noted some clients may well demand new services (especially, perhaps, emailing, net surfing road warriors). For non-business users probably more important have been non-ICT related changes, such as demands for: 'short-breaks', second annual holidays, cultural tourism, the growth of environmentally aware/sensitive tourism, and the emerging demands of older tourists.

However, ICTs are in the picture, especially at the technical interface with client reservation, billing and room technologies. These technologies are also very important in changing the internal technical characteristics of the hotel industry, the competences demanded of personnel and organisation. First, by capturing information, hotels can achieve a degree of mass customisation. Second, ICT-related internal efficiencies can make the service more cost effective and appear seamless to the client. Again ICTs have played an important role in permitting architectural innovations that blur the boundaries of the hotel service with other areas of travel and tourism services - such as car rental, add-on tours and conference facilities. Such changes are very significant in developing new markets and areas of profit. Third, the innovations in the infrastructure of the hotel are important both in changing the interaction between guests and staff and in differentiating between hotels in terms of their emphasis on different aspects of the overall service offer. Some hotels will clearly be able to trade on offering more ICTs (for example, on-line access in all rooms) than others (perhaps in the future some hotels will advertise that they are 'unplugged' - even from MTV).

11.4 ARE NON-KNOWLEDGE SERVICES SIGNIFICANT INNOVATORS?

We can now return to our more general concerns about the nature of innovation in these sectors. The first point to note is that the relative neglect of these industries is unwise - there is a lot of innovation taking place with a wide variation in its depth of impact.

Moreover, despite heterogeneity of the non-KIS services, we can see some interesting commonalties. In most cases the business proposition of the industry centres on the client-provider interface. It is probably inevitable therefore that radical innovations, which transform this interface, are problematic. In particular, they are hard to achieve where a change in the competences of the client is required. We saw this strongly in the cases of domotics and teleshopping. Nevertheless, there are always clear cases where certain clients are prepared to acquire these new skills. A question for innovating firms arises, therefore, about how the transfer of the vanguard user's experience takes place so as to broaden the market. A cognate, but rather different, innovation effect is where services require extra work from the consumer. The obvious examples are superstores and fast-food outlets, where the innovation is to rationalise the service in order to offer self-choice, speed and lower prices. Such innovations are unlikely to cover all types of consumer demand. For instance compare the French automated hotels with the US experience. But as we have seen with superstores, price can provide a powerful rationale.

The most significant changes in the service sectors considered here are likely to be associated with innovations that mainly centre on incremental, improvement and the very important category of recombinant innovations. In many cases these latter innovations are associated with disintermediation and the streamlining of internal production processes. That is to say, the integration of previously separate service activities and roles into an integrated package to meet client demands. Here ICTs play a very important role in breaking down the old divisions between the service categories and allowing centralisation of the service offer. We can see this very clearly in all the services under consideration:

- the centralised co-ordination of many different contractors to deliver a complex construction project on time and to cost.

- the growth of integrated approaches to building and managing property which span construction, maintenance, cleaning, security, etc.

- the development of integrated carriers (express couriers) which have total ownership of the logistical chain including from collection to delivery.

- backward integration by the retail industry in order to gain greater control over time-dependent goods.

- backward integration by fast food chains and hotel groups in order to guarantee the supply (timing, quality, consistency) of food and packaging inputs.

- partnership arrangements in the travel and tour trade spanning booking, flights, accommodation, food and car rental.

These types of processes are based on integrating the information systems of previously different actors or departments in order to track performance, to develop information on customer usage of services, and to add value to the services (or at least identify ways to extend the service offer). Often, the process operates through the emergence of a dominant player for instance: airlines in the CRS; retailers in Efficient Consumer Response (ECR) systems. A pivotal ownership of the information transactional data systems, in other words the intangible internal technical assets, undoubtedly reinforces this dominance.

Again, the result is service restructuring. The drivers are business imperatives not the technologies themselves. The technologies are merely the essential media for the change. However, it is clear that the result is a quite different service offer to the consumer. Even if the direct acts of consumption or service processes do not inherently change, the nature of the search, the processes of comparison, the cost structures, the level of flexibility are all subject to changes. And of course, many firms and roles (competences) will disappear in the process.

Because it is not the ICTs which are driving change it is equally the case that ICTs permit a process of re-intermediation which counter balances (to some extent) the major processes of centralisation and streamlining just described. Here we can see the emergence of value added services which exploit the information impactedness of consumers in the 'digital economy'. Very often the trick here is to use computer power plus extended staff competences to offer solutions (rather than just more information). In some cases these competences are heavily codified (for example the digital scripts followed by call-centre staff). And, as we have seen, very often the matching of the client to the service can depend on the development of a client profile that is generated through logging transactions. But, then the service offer specifically used the human touch to fill the 'service vacuums' which consumers are unwilling or unable to fill themselves. In many cases these vacuums may have been created by disintermediation and streamlining. For instance, these might be the provision of direct personalised care and contact covering the parts of the service activity that ICTs cannot reach.

Although these basic trend are the same across the sectors we are looking the precise form seems to vary according to the service concerned. Thus, we have:

- Professionalised services offering construction project design and co-ordination services.

- Specialised facilities management services which do operate services themselves but sit at the centre of a web of sub-contractors covering the various disciplines concerned.

- Guardianship, care and counselling services providing advice to consumers, residents, travellers, etc.

- Specialist shops, hotels and eateries with a high knowledge of their clientele generated through personal contact and customer profiling.

- Specialised travel brokers dealing with specific types of clients or needs.

The implication of these developments seems to be fourfold. First there is a professionalisation by adding value to services which historically tend to be quite low and status-skill. Second, the ICTs are used to reinforce traditional notions of personalised service (i.e. X+C). Third the ICTs will be embedded in the client service worker interface, even in highly personalised interactions (X+C). Fourth, we might anticipate that new divisions of service labour might be opening up between professionalised value-added co-ordination functions and the (client care orientated) client/point of sale functions[10].

All these examples of innovatory characteristics, however, do not allow us to escape from some of the key problems of understanding innovation. As we argued above, the main reason why these non-knowledge services have been relatively neglected in studies of service innovation is probably due to their lack of institutionalised innovative activity. In practical terms, this means that they do not employ many technologists, there are few research departments or research budgets, and they do not identify with the word innovation. These views are not mere prejudice (see the companion chapters in this volume for empirical evidence, especially Preissl and Kleinknecht). But it may be that the times they are a-changing. As we have seen above there are clear trends in the institution of purposive innovation activities in sectors such as construction and distribution. Especially, with the introduction of ICTs, we see much more extensive and deliberate efforts on the part of restaurant and hotel industries to codify and learn from their expertise. There are trends to hire more graduate level staff for line management and planning activities. Increasingly technical skills are needed in order to install and maintain electronic infrastructures in buildings and hotels. The overall message that emerges is that *non-knowledge services are becoming more knowledge based.* They are doing this themselves and through sub-contracting and through the entrance of KIBS, as re-intermediaries in the service vacuums created by service rationalisations elsewhere.

One implication, of course, is that they will become more like technology-based firms in terms of their innovative behaviour. In so doing such industries will also become more highly integrated into the science-based thinking that conventionally defines the innovation systems. They will invest more in technological developments, they will recruit more technologists and in so doing will connect more closely with the formal knowledge base of the economy. But there are two further ideas that may be worth investigating concerning the interaction between different actors in the service innovation system. Whilst Gallouj and Weinstein's framework has proved useful to diffract the different dimensions of firm level innovation, it gives us few tools to understand the co-evolutionary effects of interactions between firms in the partnerships and alliances which are so common now in business and innovation. How does learning pass from one non-knowledge firm to another and is it different to the relationships between firms traditionally regarded as innovative? This is a key area of conceptual development. Finally, as with other services non-knowledge intensive services are close to the client interface. As we have seen this often defines their business propositions and makes radical innovation rather difficult. It also means that these services tend to innovate in a more diffuse manner, by codifying experience into reproducible innovatory steps. As various writers (for example, Miles, 1996 and Sundbo, 1998) have recently suggested, with a general trend towards end-user involvement in innovation, all firms may have something to learn about how to innovate from the service sector. As we have seen, this may even be true for non-knowledge intensive services.

REFERENCES

Barras, R. 'Towards a Theory of Innovation in Services', *Research Policy*, 1986; 15(4): pp.161-173.

Barras, R. and Swann, J. *The Adaption and Impact of Information Technology in the UK Insurance Industry*, London: Technical Change Centre, 1983.

Bilderbeek, R., den Hertog, P., Huntink, W., Bouman, M., Kastrinos, N. and Flanagan, K. 'Case Studies in Innovative and Knowledge Intensive Business Services', *TNO-report*, 1994; STB/94/041, Apeldororn.

Bilderbeek, R., den Hertog, P. and Meihuizen, H. 'Innovation In and Through Knowledge Intensive Business Services in the Netherlands. National Profile Report', *TNO-report*, 1998, STB-98-03, Apeldoorn.

Cornford, J., Gillespie, A. and Richardson, R. 'Regional Development in the Information Society' in Ducatel, K., Webster, J. and Herrmann, W. (eds.), *The Information Society in Europe: Life and Work in an Age of Globalization*, Boulder CO: Rowman and Littlefield, 1999.

Ducatel, K. *Unexpected Innovations: The Future of Work in Non-Technology Based Services*, Seville: IPTS, 1998.

EITO. *European Information Technology Observatory 1999*, Frankfurt/Main, EITO, 1999.

EU Panorama. *Panorama of European Industry, 1997*, Commission of the European Communities, DGIII, Luxembourg, 1997.

Fincham, R., Fleck, J., Procter, R., Scarbrough, H., Tierney, M. and Williams, R. *Expertise and Innovation*, Oxford, 1994.

Gallouj, F. *Économie de l'Innovation dans les Services*, Paris: Harmattan Logique Economique, 1994.

Gallouj, F and Weinstein, O. 'Innovation in Services', *Research Policy*, 1997; 26(4/5): pp.537-556.

Haddon, L. and Silverstone, R. 'Information and Communication Technologies and Everyday Life: Individual and Social Dimensions' in Ducatel, K., Webster, J. and Herrmann, W. (eds.), *The Information Society in Europe: Life and Work in an Age of Globalization*, Boulder CO: Rowman and Littlefield, 1999.

Hepworth, M. and Ducatel, K. *Transport in the Information Age: Wheels and Wires*, London: Belhaven, 1994.

IHA. *Into the New Millennium, White Paper on the Global Hospitality Industry*, Paris: IHA, 1996.

ILO. *New technologies and working conditions in the hotel, catering and tourism sector*, Geneva: ILO, 1997.

Knights, D. and Morgan, G. 'Strategy under the Microscope: Strategic Management and IT in Financial Services', *Journal of Management Studies*, 1995; 32(2), pp.191-214.

Miles, I. *Innovation in Services: Services in Innovation*, Manchester: Manchester Statistical Society, 1996.

Miles, I., Kastrinos, N., Flanagan, K., Bilderbeek, R., den Hertog, P., Huntink, W. and Bouman, M. *Knowledge Intensive Business Services; Their Role as Users, Carriers and Sources of Innovation*, Commission of the European Communities, DGXIII, SPRINT-EIMS, Luxembourg, 1995.

OECD. *The Economic and Social Impacts of Electronic Commerce: Preliminary Findings and Research Agenda*, Paris: OECD, 1999.

Piketty, T. 'Les creations d'emplois en France et aux Etats Unis: <<services de proximité>> contre <<petits boulots>>?', *La Revue de la CFDT*, 1997; pp.3-31.

Rantanen, J. and Lehtinen, S. 'Health and the Information Society' in Ducatel, K., Webster, J. and Herrmann, W. (eds.), *The Information Society in Europe: Life and Work in an Age of Globalization*, Boulder CO: Rowman and Littlefield, 1999.

Rowe, I. 'The Impact of Technology on the Travel Industry: Developments and Trends', *FT Management Reports*, London, 1995.

Saviotti, P. and Metcalfe, J.S. 'A Theoretical Approach to the Construction of Technological Output Indicators', *Research Policy*, 1984; 13: pp.141-151.

Senker, P., Gann, D., Clarke, L. and Wall, C. (with Moody, D. and Trill, J.). *Strategic Review of Construction Skills Training*, London: Construction Industry Board/Thomas Telford, 1998.

Soete, L. and Miozzi, M. 'Trade and Development in Services a Technological Perspective', *MERIT Working Paper*, 1989; No. 89-031, Maastricht: University of Limburg.

Sundbo, J. 'Modulisation of Service Production', *Scandinavian Journal of Management*, 1994; 10(3): pp.245-246.

Sundbo, J. 'Development of a Service System in a Manual Service Firm. A Case Study of the Danish ISS' in Swartz, T., Bowen, D. and Brown, S. (eds.) *Advances in Services Marketing and Management*, Vol. 5, Greenwich – CA, 1995, pp.161-191.

Sundbo, J. *The Organisation of Innovation in Services*, Roskilde: Roskilde University Press, 1998.

Toussaint, Y. and Lozier, F. 'Home Automation, Employment and Services", R*eport to IPTS*, Seville, November, 1997.

Windrum, P., Flanagan, K., Tomlinson, M., Boden, M., Miles, I., Bolisani, E. and McKinlay, C. *Innovation in UK Services: Recent trends and patterns*, Manchester: PREST, University of Manchester, 1998.

---

[1]The views expressed in this chapter are the authors and are not necessarily those of the European Commission.
[2]For a review of work on KIBS see the outputs of the European Community funded EIMS and "SI4S" studies in reports such as Miles et al, 1994, Bilderbeek et al, 1994 and 1998, and Windrum et al, 1998.
[3]This suggests a cross-sectoral generalisation of the notion of a Reverse Product Cycle in services where technologies are the concrete manifestation of earlier process innovations in the service sector (Barras, 1986).
[4]The information in this section is based on an as yet unpublished report by the author: Ducatel (1998)
[5]Construction is included here, even though it is not strictly in the service sector, because much of the work is non-knowledge based, it is a big sector which is often ignored in innovation studies, and is increasingly linked to other 'building services' such as repair, security, cleaning and facilities management.
[6]Obviously, some of these service activities lie outside 'building services' an obvious example is the transportation of valuables which represents a large part of the security industry.

---

[7]Examples in these areas include on-line bookstores such as Amazon (based in Seattle, US) and Oxford based Bookshop (http://www.bookshop.co.ac); insurance with UK Underwriters Eagle Star; the French based 'Eurodat' virtual music shop which allows the digital downloading of music.
[8]Piketty (1997) makes a similar analysis of differences in the retail sector.
[9]It would be an interesting to ask how such alliances fit within the model suggested byGallouj and Weinstein (1997), and this indeed is a question which directly addresses the relationship between service innovation and the innovation system within which they are situated.
[10]A clear example of this might be in the development of call centres (see Cornford et al, 1999).

Chapter 12
Innovation in Services: The Dynamics of Control Systems in Investment Banking[1]

Paul Nightingale and Rob Poll

12.1 INTRODUCTION

This chapter has the following aims. First, to show that a pattern of innovation in a number of service sectors can be conceptualised within a slightly modified version of Chandler's 'Scale and Scope' framework. The second aim is to show how twentieth century service firms that process and authenticate information differ from manufacturing firms in the nineteenth century that processed physical materials. Processing information rather than physical materials reduces the dependence on manufacturing technology and therefore increases the emphasis within the Chandlerian framework on issues of control.

How does innovation take place in high-tech service sectors? In manufacturing industries innovative activity concentrates on the product and the process technologies used in production. Services, on the other hand, are traditionally defined in terms of the intangible nature of their 'products[2]. As a consequencc, services have tended to be thought of as fairly un-innovative 'residual' sectors that are not manufacturing or primary production.

Recent research has tended to move away from any clear-cut distinction between services and manufacturing and has highlighted their interdependence, showing how manufacturing firms are increasingly stressing the value-added service based features of their products (Miles, 1996; Uchpakinan, 1997). The distinction between services and manufacturing is now more commonly understood as a continuum rather than a dichotomy. Thus, the intangible aspect is dominant in sectors like teaching, and the tangible aspect is dominant in sectors like detergents, but there is also a large middle ground, such as fast food, where the tangible and intangible aspects of service provision overlap (Uchpakinan, 1997). This overlap between services and manufacturing would indicate that there might be a similar overlap in how innovation takes place. If this is the case, it should be possible to understand service innovation using slightly modified versions of theoretical frameworks developed for understanding innovation in manufacturing.

In this chapter we conceptualise the continuum between service and manufacturing firms in terms of the tangibility of functions, so that service firms are paid to *provide a function* while manufacturing companies produce goods that are bought to *perform a function.* As a consequence 'getting a function performed' is the fundamental construct, and manufacturing represents a subset where this takes a tangible form[3].

To understand how service firms use technology to provide intangible functions, we have used Chandler's 'Scale and Scope' framework to explore innovation in investment banking. The Chandlerian framework explains how firms grew by achieving economies of scale and scope in the physical processing of materials. Looking at the large service firm of the late twentieth century, firm dynamics are more closely tied to the processing and authentication of information than to the production of physical goods. This necessitates a shift in emphasis because the same technology can be used to process both large and small amounts of information. Since scale and scope economies can be achieved without major changes to the technology the importance of 'controlling' throughput becomes relatively more important[4].

Control can be improved by the application of technology and organisational changes (Hopwood and Miller, 1994; Power, 1993; Kaplan and Johnson, 1987). While Chandler (1992; 1977, pp.110-112), Yates (1989), and Pavitt (1997) have stressed its importance to the large firm, and many authors have done work on the organisational side of control in the accountancy literature, the link between technology and organisational control is under researched. The subject has however received increasing emphasis through the work of Beniger (1986), Davies (1996) and Hughes (1990). The chapter follows on from their work by examining the qualitative changes that occur as control systems increase in complexity, and consequently how the dynamics of the control systems act as a technical limit to increases in economies of scale and scope. We will argue that investment banks invest in technology in an attempt to control the risks of financial losses associated with increases in the size (scale) and complexity (scope) of trades. The control systems co-evolve with internal and external regulations and undergo qualitative changes as they increase in size.

Limitations of the Data

The chapter uses a case study methodology which is limited in its generalisability and creates the danger of producing results that are time, sector, country and technology specific. We attempt to alleviate this problem by linking the case study to an established framework, that developed by Chandler to explain the dynamics of the large manufacturing firm. While this increases the potential generalisability of the results, a number of minor modifications have had to be made. These changes, in particular the emphasis on regulation, are specific to banking. Moreover, investment banking is more than just trading. The majority of investment banking activity takes place in similar settings to commercial banking, and commercial banking follows a fundamentally different technology trajectory than the one highlighted here (Crede, 1997).

The evidence of the case study should therefore be interpreted as suggesting that Chandler's framework is surprisingly robust and requires a few minor changes to deal with innovation in investment banking, but the modifications themselves are probably very sector specific. On the other hand, the qualitative changes in the control system are determined by its increasing complexity, rather than the specifics of the investment banking sector, and in the conclusion we tentatively suggest that similar phenomena can be found in other service sectors.

The following section (Section 12.2) discusses Chandler's explanation of the growth of the firm, stressing the importance of controllingthrough-put in order to achieve economies of scale and scope. Section 12.3 outlines the distinction between commercial and investment banking. Section 12.4 explains the dynamics of investment banking innovation and shows how the ability to exploit scale and scope eventually hits constraints based on the regulation of risk-exposure. These constraints focus innovation on the risk control system, a different level of analysis. As the control system expands in size, power and complexity it goes through a series of qualitative changes we have called: craft; mass standardisation and complex. Table 12.1 highlights the key features of each type of control system.

Table 12.1

Feature	Craft	Mass standardisation	Complex
No of servers	1-100	100-1000	1000+
Maintenance	craft based	standardised	very complex
Architecture	Decentralised	centralised	Centralised
Advantage	limited	scale & scope	+ system
Key factor	automation	change function	risk & reliability
Problems	few	predictable	1-in-a-million
Importance	limited	business cost	business critical
Prod/process	process	product	bundled
Make/buy	Both	buy & customise	in-house
Risk analysis	Limited	end of day	quicker and wider

Section 12.5 explores the experiences of the case study bank and stresses how increases in the size and complexity of the control system allowed the bank to exploit new economies of scale and scope. As a consequence bottlenecks in the evolution of the control system act as limits on corporate

dynamics. The chapter concludes in Section 12.6 with a discussion of the issues raised and their potential application to other service sectors.

12.2 THE EVOLUTION OF THE CHANDLERIAN FIRM

Classical economists such as Babbage and Marx explained the industrial revolution by emphasising how co-ordinated machines in factories replaced uncoordinated human hands. Chandler writing a hundred years later emphasised the importance of technical managers who have the ability to understand, analyse and modify production and distribution. Chandler's concern moved on from the factory of the industrial revolution, to the large firm that was for him the 'central institution of managerial capitalism'.

For Chandler what was new about these institutions, and what drove their growth and determined their final form, was their ability firstly to achieve potential scale and scope advantages in production, and, secondly, to organise and co-ordinate the resources required to turn potential cost advantages into actual advantages. In old industries productivity increases come from adding machinery and labour, but in the new Chandlerian firms, productivity increases came from 'drastic changes in labour capital ratios' producing greater economies of scale (1990, p.22). Chandler defines economies of scale as 'those that result when the increased size of a single operating unit producing or distributing a single product reduces the unit cost of production or distribution' (1990, p.17). If the production processes could be operated at their 'minimum efficient scale' they would have 'an impressive cost advantage over smaller plants that did not reach that scale' (1990, p.24). Similarly, reductions in unit costs could be produced by 'making a number of products in the same production unit from much the same raw andsemi-finished materials and by the same intermediate processes...' (1990, p.24). Thus, economies of scope are 'those resulting from the use of processes within a single operating unit to produce or distribute more than one product' (1990, p.17).

These cost advantages required co-ordination of the flow of materials through the production process, as 'costs per unit rose rapidly' if 'the realised volume of flow fell below capacity' (1990, p.24). The determinants of profit and costs became 'the actual amount processed within a specific time' thus potential economies of scale became dependent on 'both size-rated capacity-and speed-the intensity at which the capacity is utilised' (1990, p.24). This emphasis on co-ordinating the speed and size of processing created a distinction between the 'potential economies of scale and scope, as measured by rated capacity, [which] are the physical characteristics of the production facilities' and the 'actual economies of scale and scope, as determined by throughput, [which] are organisational' (1990, p.24). These organisational economies depend on 'the organised human capabilities essential to exploit the potential of technological processes' which are 'knowledge, skill, experience and teamwork' (1990, p.24).

The ability to control and co-ordinate resources originated in the railroads of the 1870s. Chandler notes '...unless the movement of trains and the flow of goods were carefully monitored and co-ordinated, accidents occurred, lives were lost and goods moved slowly and with uncertainty' (1992, p.264). The control problems meant that rather than exploiting increased economies of scale, the railroads of the 1850s saw per-mile operating costs actually increase with size (Beniger, 1986, p.227). This is an important point, the exploitation of economies of scale and scope can be limited, slowed down and eventually halted by lack of control.

The Chandlerian framework for understanding the 'central dynamic' (1990, p.596) of the large firm consists of four mutually dependent and interacting 'technological imperatives':

- Firstly, the initial reductions in unit cost are closely tied to production technologies that can exploit economies of scale. This linkage to production technologies explains why large firms were formed in steel and chemicals but not in shoes.

- Secondly, the increased capital costs mean that there is a potential for losses if co-ordination of the production processes fails. This creates an emphasis on controlling the production processes, to make them efficient and to ensure that they are directed to profitable markets. The difference between potential and actual reductions in unit cost is a management and organisational issue.

- Thirdly, the co-ordination of resources through the production process requires control along the 'value chain' into marketing and distribution.

- Fourthly, the interaction of the technical ability to produce at low unit cost, and the managerial ability to control and co-ordinate resources, allows firms to exploit economies of scope between functionally specialised technologies and departments, creating distinct organisational forms.

Chandler (1977, 1990) argues that if firms can respond to new market or technological opportunities and make the 'three pronged set of investments' in manufacturing, marketing (distribution) and management when markets are growing rapidly, they can consolidate economies of scale and scope. These investments gave firms first-mover advantages in building up the production and distribution networks needed to bring down unit costs. They also allow firms to compete through functional and strategic efficiency by improving products and processes, using R&D, finding better supply sources, improving marketing and product differentiation and strategically moving 'more quickly into expanding markets and out of declining ones' (1990, p.36).

From Physical to Information Processing

Chandler's framework has been very successful in explaining the evolution of large manufacturing and transport firms where economies of scale and scope are closely tied to 'physical processing'. While manufacturing still matters, the service sector has recently increased in relative importance for two reasons. Firstly, relative differences in productivity have shifted employment from manufacturing to services. Today more than 5.4 million people are employed in the US banking sector, more than twice the combined number of employees in automobiles, computers, pharmaceuticals, steel and clothing manufacture (Frei et al, 1998, p.1). Secondly, there is an increasing recognition that the service sector, and financial services in particular, is in the vanguard of the introduction of new information technologies (Barras 1990, 1986). The case study bank, for example, spends some $1 billion a year on IT.

This raises two questions; if these sectors are so important in terms of productivity, employment and innovation what are the dynamics of firm growth in these sectors, and can Chandler's framework be applied to firms that produce intangible outputs where scale and scope are no longer tied to 'physical processing'?

We argue that Chandler's framework is remarkably robust and requires a minor shift of emphasis rather than a reworking. Of the four technological imperatives listed above only the first, 'reductions in unit costs closely tied to production technologies' is out of place. The others, namely co-ordination of production processes, control along the value chain, and managerial ability to co-ordinate resources between functional groups and specialised technologies, are all applicable to 'information processing' sectors. The shift is a shift in emphasis away from production and towards control.

Control in the Firm

Beniger (1986) used the Chandlerian framework to understand why the processing of information 'has begun to overshadow the processing of matter and energy' (1986, p.1). Looking at the period 1870-1910 from a Chandlerian perspective Beniger notes that 'the American economy had become distinctively more purposive... during these decades' (1986, p.vii) and improved control was required to ensure the through-put of materials through the production processes described by Chandler. Beniger links control to both communication between the processes and a description of their intended behaviour, and secondly, to a feedback mechanism that guides and redirects future action[5]. Increases in the amount, speed, and quality of information allow a closer comparison between intended and actual procedures, increasing the quality of control.

As the various technologies of production and distribution evolved at different rates, mismatches caused unpredictable features to emerge. These 'crises of control' prevented potential cost reductions being turned into actual cost reductions creating new loci of innovation. New control systems were then developed to bring processes 'back into line'. Thus, each innovation created a potential lack of control, as unintended consequences emerged within the technological system; but control innovations in organisations and technology allowed purpose to be re-imposed, creating both a radical change and a return to a transformed previous state (1986, p.9). For example, the reorganisation of the railways after train crashes, not only transformed their operation, but also returned train travel to its original purpose of providing safe transportation. Beniger's work is important as he shows that lack of control can act as a limit on the ability of firms to achieve economies of scale and scope. Consequently, innovations in control can remove these bottlenecks in firm dynamics.

Another author who has adapted the Chandlerian approach is Davies (1996) in his explanation of innovation in telecommunications. Davies argues that 'an explanation of growth in large technical systems has to account for the economic drive to realise cost-saving economies of scale and scope' (1996, pp.1144-1145). However, the Chandlerian framework needs to be slightly modified to incorporate the economic advantages of controlling and co-ordinating the routing of traffic through a telecommunications network or other large technical system. These 'economies of system' improve the utilisation of installed capacity which increases its capacity producing economies of scale and can 'open up opportunities for supplying new services', increasing economies of scope (Davies, 1996, p.1164)[6]. Economies of system reduce costs by increasing the utilisation of capacity without expanding the size or scope of production. Since control systems allow new production economies they shift from having a monitoring role to playing an increasingly business-critical role in delivering goods and services. In effect they become 'bundled in' with service provision.

Chandler's four stage framework can be modified to deal with 'information processing' industries by shifting emphasis from production to control, incorporating Beniger's idea that control problems provide both limits and opportunities for firms, and Davies' notion that they can be used to achieve economies of system. In doing so, a framework of three interacting stages can be produced to explain firm dynamics.

1. In the first stage economies of scale and scope are exploited until the firms hit control limits. These focus innovation on systems to control the problems of co-ordination associated with the increased complexity of service provision.

2. In the second stage, the introduction of control systems has re-established control and the firm can achieve new economies of scale and scope. In doing so the control-system becomes essential to the ability to compete and the technology becomes 'bundled-in' with service provision. Thus the control system shifts from a monitoring role to a business-critical role putting an increasing amount of emphasis on reliability.

3. In the third stage, the control system can be used to achieve additional economies of system. The control system can be used to monitor and better manage the production of value added services and ensure the reliability of the business critical network. This allows firms to quickly move into new markets

This framework will be used to explain the dynamics of innovation in investment banking. In the next sections we will apply Chandler's framework to explain the dynamics of innovation in trading technologies. As exploiting economies of scale and scope is dependent on control, the control system itself becomes a focus of innovative activity and its internal dynamics, that are the result of linking nodes in a network, provide limits and opportunities for change. The following section will provide a brief introduction to the banking sector before explaining the dynamics of trading technologies.

12.3 THE NATURE OF BANKING

Banks represent some of the largest, most profitable, stable and influential firms in developed economies. In most countries, banks represent around 5-10% of GDP and a similar proportion of employment. While the traditional lending activities of banks may be shrinking, the market for financial services is rapidly expanding (Berger, 1995; Greenbaum, 1996). Within banking, a key distinction needs to be made between commercial and investment banks. While both are involved in savings intermediation and help move surplus funds from investors to borrowers, they differ fundamentally in how this is done. 'In the case of investment banking, the rechannelling [of surplus savings] takes place through the underwriting of securities for sale to other investors, while in the case of commercial banking, financial intermediation is carried out by the commercial banks holding assets until their maturity....' (Crede, 1997, p.5). That is, commercial banks loan assets they hold, while investment banks will make a 'loan' by creating a security, for example a bond issue, and selling the loan as a contract in the market place.

A typical investment bank structure might divide into Asset Gathering, Mergers and Acquisitions, and Trading.

- Asset Gathering involves Fund Management where the bank runs financial services like Unit Trusts for private clients or large institutions like pension funds. Investment Banks also play a role as 'Custodians' where they act as a third party holding the real money and financial instruments while another bank manages funds on behalf of a pension fund.

- Investment banks also deal with corporate finance and might help issue and underwrite the bonds for firms that want to raise money. This also includes services like Mergers and Acquisitions and privatisations.

- Trading involves buying and selling in the secondary markets for Equities, Fixed Income (bonds), Foreign Exchange and Commodities, and trading derivatives of these products.

This study has been concerned with innovation in the trading environment and is limited insofar as it has not attempted to explore innovation in the rest of the investment banking industry where in contrast to the trading environment, banks provide services based on the 'tacit authentication' of privileged information (Crede, 1997).

12.4 INNOVATION IN INVESTMENT BANKING

Investment banking is a typical service sector where profitability is linked to the efficient contextualisation and processing of information rather than being tied to capital machinery as in manufacturing industries. For example, the profitability contribution of a worker on a machine is determined to a large extent by the technology itself, and a worker would be hard pressed to increase profits substantially. A trader, by contrast, who was smarter or had better analysis than others in the market, could very easily make significantly more profits with very little restriction from technology. This ability of a trader to have large effects on profits is called leverage. The flip side of high leverage is that while profits are almost 'unlimited' so are losses, creating an incentive on understanding and managing the extent and likelihood of any losses occurring, namely risk.

Risk in banking is a slightly different concept than risk in the economic literature: Market Risk refers to the potential losses that will occur in a portfolio or position in the market if there are changes in market prices, rates, indices, etc.; Credit Risk refers to exposure to losses that arise if a legal counter-party to a transaction fails to perform; Concentration Risk refers to exposure to losses from holding onto proportionally large positions in any single type of security, for example from one government, or one geographical area, or particular industry. The risk associated with any transaction will have a negative effect on its value, that takes into account the likelihood and distribution of potential losses.

However, risk is also dependent on the position of the bank in other markets and can be actively manipulated so that a trader can balance a trade that might make a loss if the markets move, with one that will make an increased profit, transferring the risk away. Similarly, if a position has concentrated exposure in one sector or market, the bank can swap that exposure with other banks to dilute the risk. As a consequence each transaction will have a safety-margin where the potential risk of losses is acceptable; but unless the position of the firm and the markets are well understood and analysed, the safety margin surrounding a trade may be largely unknown. Computer systems are used to gather and analyse information on risk positions to try and ensure that the calculated risk position corresponds to the real risk positions so that actual profits come close to the potential profits. The analysis of the calculated risk distribution is used by the dealers to help judge how large a prudent investment should be. This judgement depends on the dealer having the tacit knowledge to judge and authenticate the information in the context of changing markets. As a consequence there are large differences between dealers in performance which is reflected in large remuneration packages.

Risk-exposure is therefore controlled by technologies that gather data, and by formalised control mechanisms that ensure that information is acted upon. Breakdowns in systems of control can lead to erroneous risk information, which have resulted in substantial recent losses for a number of international investment banks. Because the banking industry is deeply embedded in systems of trust, control of risk is not only regulated internally but is externally audited and regulated at an international level to limit any damage to stakeholders (Conroy, 1995; Merton, 1995). Thus, risk-exposure regulations act as a limit on potential profitability within the firm, unlike manufacturing where profitability is tied to rated capacity and the co-ordination of through-put.

The focus on risk reduction centres on a strategy of collecting, analysing and auditing data on risk exposure (Simons, 1996; Santomero, 1997; Kimball, 1997; Powers, 1993). Within investment banks risk is managed by the banks most senior officers, who monitor and review the current practices. They will in turn be supported by Risk Committees who will be responsible for specific aspects of control and liaise with the departments responsible for Law, Government Affairs, Treasury and Controllers who are functionally separate from the business units. The implementation of risk management is monitored in most investment banks by Internal Audit departments who ensure that the processes for measuring, monitoring and reporting risk positions are in place. These trading limits are approved for each trading division of the company and assigned to trading desks which are in turn monitored by desk risk managers. These departments rely on quantitative and qualitative risk measurements and analysis, such as Value-at-Risk, that are based on data coming from the IT network.

The quality of this input data can be improved in two ways. Firstly, risk can be mitigated by globalisation of analysis and managerial control. If risk exposure is analysed separately in London, New York and the Far East and all three find themselves 'close to the edge' on a particular trading position in the same market, then even though each local risk exposure is within regulatory limits, the global exposure is such that the bank will make losses. If risk can be analysed on a global scale, not only can these problems be avoided, but the risk can also be managed by passing on exposure from operations in London, to New York and the Far East. They then adjust positions in the markets while London markets sleep, by selling or swapping exposure with other banks or by various forms of financial engineering. Globalisation therefore focuses innovation on expanding the network and linking together trading within the trading floor and between international trading rooms.

Secondly, risk can be mitigated by increasing the speed of analysis. The risk exposure is constantly changing as trades are made and the markets move. The longer data collection and its analysis calculations take, the more the calculated and actual risk positions will diverge. As a consequence, the uncertainty band around a risk-exposure probability-distribution will be dependent on the speed of analysis; real time analysis will be more accurate than day old analysis. Any improvements in speed and accuracy allows traders to safely trade 'closer to the edge' and increase profits.

Knowing the nature of exposure enables trades to be appropriately priced, increasing the profitability of trades, and allows previously inaccessible products to be produced. In short, it generates economies of scale and scope and as it moves from a primary function of analysing pre-existing products, to a secondary function of making new products possible. This shift, from a regulatory, to a business-critical function creates another focusing device on innovation with its increased concern for reliability. The more banks understand risk, the more dependent they are on understanding exposure for their business and the consequences of unreliability increase.

The three main focusing devices for innovation are globalisation, speed, and reliability. These push innovation in risk control-systems towards globalised networks that are faster, more reliable, more powerful, with greater function and wider scope. These desires are not necessarily compatible, as increasing the complexity of the system does not increase reliability. Understanding the dynamics of investment banking requires firstly recognition that the high leverage of trading and heavy regulation focuses innovation on control, secondly, that there is innovative pressure on the technology of the control systems to increase in size and function, and thirdly that as the control system expands it opens up new opportunities for the bank to exploit economies of scale and scope. Thus a shift in unit of analysis is required from the dynamics of the firm to the dynamics of the control system technology. The next section discusses the experiences of the case study bank as it has introduced control systems and they have grown in complexity.

12.5 THE EXPERIENCES OF THE CASE STUDY BANK

The previous section highlighted how regulations about risk-exposure prevent traders from exploiting economies of scale and scope in investment-banks. These limitations can be reduced by analysing and managing any risk in the trading portfolio; if risk can be controlled then banks can safely make higher profits. The data for this analysis is improved by including a larger sample of traders, (what we have called globalisation), and by increasing the speed of analysis, which reduces the difference between the real exposure and the calculated exposure. This section will highlight the experiences of the case study bank to show how attempts to control risk by expanding the size and power of the IT infrastructure produced qualitative changes in the architecture of the technology. These qualitatively different architectures have been called Craft, Mass Standardisation, and Complex; their main features are listed in Table 12.1. The organisational and technical problems associated with transitions between each architecture act as 'crises of control' that the bank must overcome to achieve new levels of productivity.

Craft

In the 1980s, traders in the financial markets were limited in the size of trades they could make by predetermined limits imposed by the bank. Their ability to increase the scale of trades depended on their ability to hedge risk through financial engineering. The data and computer systems used for this engineering were 'local' to the trader's Front Office system, while the separate Back Office applications were automated on mainframes. The Front Office systems were normally 'dumb', and simply displayed raw data or contained a limited amount of simple analysis software that could be bought off-the-shelf from specialised software houses. Typical Front Office technology at the time consisted of Reuters dumb 'green screen' boxes that displayed raw 'tick' data.

The initial introduction of software to increase the power of analysis was localised and small scale. Brady and Target (1996) analysed the adoption of IT in investment banks in the late 1980s and found that they could be considered almost a competitive burden that offered little sustainable competitive advantage.

In systems of this kind, the number of computers was small (between1 and 100) which allowed systems maintenance, software updates and innovation to be carried out locally, as it is feasible for someone to move from computer to computer installing new software. If problems occur on one machine maintenance was reactive and localised with limited additional benefits. The benefits of globalising the system and increasing its power required an expansion from these local, craft maintained systems to a wider architecture.

The late 1980s represented stage one of the modified framework, where the ability to exploit economies of scale and scope reached limits imposed by risk regulation. The initial introduction of information technology was not intended to transcend this 'crisis of control' and in the late 1980s any advances were minor and limited. The next stage came with a radical change in architecture from a small scale, local, craft maintained system to a far larger more powerful system.

Mass Standardisation

In the late 1980s, a competitor bank decided to move to the UNIX platform. At the time there were no applications and very little understanding of what the technology could do. In response to this, the case study bank followed suit and introduced a UNIX network into its Fixed Income Trading Area. The people involved in the project had come from academia and decided that the UNIX environment should be globally consistent. The project was initiated as a strategic 'blue sky' research project in response to the actions of other banks rather than as part of a strategy to produce a then unknown global risk environment.

Once the desk-top systems were in place they could begin to perform some analysis. The bank stopped taking 'end-to-end' data feeds from suppliers and installed their own analysis and display systems in between the feed-input and the traders' screens. For example the new systems allowed the bank to compute £:$ ratios from £:¥ and $:¥ ratios and then look for arbitrage possibilities. These simple applications were easily copied and the profits limited.

Once 'real' applications became available, the number of machines increased quickly and machines were installed globally. By 1992 the bank had installed well over 100 machines. The introduction of the larger Unix system provided the bank with more powerful, global analysis of risk exposure that allowed the bank to monitor and then manage larger and more complex trades. This transcended the previous limits on scale and scope. As a consequence the control system shifted from having a business monitoring function, and became essential to service provision; a process referred to as 'bundling in'[7].

As the control system becomes more business-critical there is an increased emphasis on reliability to ensure that traders do not loose money buying or selling on false information. As noted above, increased reliability is not necessarily compatible with the increased complexity, scale and power of the control-system which makes maintenance more difficult.

Once the number of machines goes beyond 100, it becomes increasingly difficult to up-date software by moving to each machine in turn. Advantage comes from standardising the process by producing solutions at one point and installing them throughout the system. This required a separation within the software between the infrastructure and the applications programmes. By centralising the maintenance and upgrading of software the network architecture underwent a qualitative change and moved from a local craft base - where problems were solved in isolation at the local level - to a centralised standardised one where solutions could be mass produced in a globally consistent way.

The standardisation ensures that local parts of the system do not diverge on potentially incompatible, path dependent trajectories. This allows new functions to be easily added through new software and software patches. Standardised software has the added advantage of providing a familiar environment for banking staff as they move around the world.

'Mass standardised' networks allow information from a number of different systems to be integrated and analysed, which gives banks the advantage of better analysis from wider sources. This gives traders a more accurate understanding of their exposures, allowing them to trade safely 'closer to the edge' by trading more of the same product, (an economy of scale), and/or integrating different products to produce a more customised service, (an economy of scope). This moved the bank to stage two of the modified framework where the control system allows new economies of scale and scope and becomes 'bundled in' with the provision of the service creating an increasing emphasis on reliability.

Complex

Transcending the limits on risk exposure required the bank to expand the size and power of the network further. This involved an integration of several existing systems into one larger system. The Equities division had acquired a large, global UNIX system in 1992. They had originally intended to go for a non-UNIX environment but had encountered scalability problems. Other divisions such as Commodities, Foreign Exchange and Asset Management followed a similar route and acquired their own systems and built their own architectures. However, it soon became apparent that these business units would benefit from running each other's applications. The bank decided to integrate these systems (with the exception of the Fixed Income business unit which was run as a relatively separate entity) into a single environment. This was done to allow different divisions to share applications, rather than because the firm thought that integrated technologies would allow the technology to perform new functions. A software engineer was hired from another bankwho had previously integrated their systems for the same reason. The integration project took some 2-3 years which highlights the technical difficulty of the process.

This left the case study bank with two complex (1000+ system), global environments, one in Fixed Income and one in Equities, Commodities, etc. These two environments have subsequently been integrated as part of a four year project. The project was initially intended to allow a greater sharing of applications and to upgrade the outdated operating systems that both environments were built with. The control system was initially designed to allow applications to be shared but turned out to have unintended *latent functions* that allowed risk analysis to be globalised.

This latent function began to emerge in the early 1990s when risk analysis was transferred from the Back Office mainframes to the front-desk Unix systems. Traditionally risk analysis had been done at the end of the day as part of the Back Office routine. The mainframes in the New York office would take data from the trading desk systems and use it to compute the risk position overnight. This was done as part of the process that provides traders with their morning position and information on cancelled and unconfirmed trades. With the advent of the desktop UNIX boxes, risk analysis can be done on local data within the UNIX environment to give the traders a better understanding of their real time risk exposure. As one moves to more complicated derivatives this becomes more complex and computationally expensive.

With the shift in risk analysis from the Back Office mainframe environment to the Front Office Unix environment, the bank moved into the third stage of the modified framework; where increased control allowed the traders to achieve new economies of scale, scope and system where control systems allow risk-exposure and maintenance to be actively managed.

When a trading room performs risk-analysis at the end of the day and has the results ready in the morning, any changes in the markets during the night will cause the calculated position to diverge from the real position. This produces a band of uncertainty around the result that is dependent on how much the markets have changed during the time of the calculation. If the risk could possibly be analysed in real time, traders would see the band of uncertainty compress dramatically. Shifting risk analysis onto the UNIX environment reduces the time taken for analysis, which reduces the uncertainty surrounding the calculation and allows traders to trade safely 'closer to the edge'. This increases their leverage as it increases the size and scope of safe trades. Moreover, if the results of risk-analysis can be made available by the end of the day, rather than by the beginning of the next day, they can be passed from London to New York. This give the traders in New York a more 'global' understanding of the bank's exposure, which they can actively manage while the London market shuts down overnight. This ability to understand a more 'global' scenario and actively manage global risk positions with control systems, is analogous to the management of capacity utilisation in telecommunications highlighted by Davies (1996).

The advantages of increased speed, and therefore accuracy, of real-time risk calculations and the ability to manage risk on a global rather than local basis, are direct consequences of shifting from a local to global IT architecture. But this shift in the architecture of the control system comes at a cost. While the benefits of integration increase with complexity, so do the problems (Simon, 1969; Hobday, 1998). By the time the system contains 1000+ machines it has made a qualitative shift from the mass standardisation system. Problems that would traditionally only occur once in a million days on one machine, happen far more frequently (often daily) on a system of 7000 interconnected machines all running large numbers of interacting programmes. Since the large complex control system is even more business-critical than the mass standardisation system *reliability* and *maintenance* become more important.

While it is possible to find UNIX engineers who can quickly and easily solve mass-standardisation problems, in the 'complex' stage most of the easy problems have been solved. The new problems require software engineers who can understand the evolution of a system made up of sub-systems advancing at different rates, for different purposes, with new functions emerging and periodic consolidations and re-configurations of systems architecture. The complexity of the task necessitated a shift towards a more 'organic' flexible organisational structure within the software team, rather than the hierarchical structures that were used to ensure consistency in the mass-standardisation phase.

The combination of high complexity and increased business importance changes the nature of the innovation process. Software engineers cannot predict all the interactions between programmes and therefore have to incrementally build on previous changes. This is complicated because the control system is in use 24 hours a day and cannot be simply 'turned off' to update software or hardware. The implementation process must not affect the overall functioning of the system and must be able to return sub-systems to their previous states if anything unplanned happens; a highly uncertain process likened to 'changing the foundations of a house with the house still standing'.

The business importance of the network means that changes must be made to maintenance as well. Instead of having each office solve its problems locally, which would create an inconsistent environment, the global scope of the network allows a centralisation of some of the maintenance function. A centralised 24-hour 'systems monitoring' base in the data centre allows the bank to see problems in the system as they emerge and *move from reactive maintenance to proactive maintenance.* If a problem is found with a system in London it is fixed locally and then as a matter of course the London team can log-in to the rest of the world and make changes so that the same problem does not confront New York when they start work. This can only be done with a globally consistent environment monitored and maintained from a centralised position, (in the bank's case this monitoring and maintenance can be anywhere in the world). Moreover this reduces the risk of catastrophic failure, as work on 'mission-critical software' has shown one of the main causes of catastrophic failure is local maintenance corrections to programmes that compromise 'system integrity' (Leveson, 1994).

The development of the 'complex' control system allowed the bank to exploit new economies of system as well as larger economies of scale and scope and moved the bank into stage three of the modified framework. New economies of scale and scope come from faster, more global analysis of risk exposure. The economies of system emerge because the architecture of the network allows centralised monitoring and pre-emptive management of unexpected risk exposures. This, in turn, allows traders to actively manage economies of scale and scope in their portfolios. Economies of system also emerge within the network itself, where centralised control allows proactive maintenance to ensure the reliability of the business-critical system. Because the sub-systems have not been allowed to diverge, the bank can quickly and safely roll-out applications and co-ordinate resources to move in and out of new markets. These benefits come at the cost of increased complexity in the control system, which in turn, changes the nature of the innovation process and the qualities of software engineers required to understand and solve systemic problems. This increased uncertainty, complexity and business importance in turn requires a change in the management structure of the software engineers towards a more decentralised, devolved structure.

12.6 DISCUSSION AND CONCLUSION

This chapter has shown that a slightly modified version of Chandler's framework can be successfully extended from nineteenth century manufacturing firms to explain innovation in a modern high-tech service firm. Despite processing information rather than materials, a familiar Chandlerian pattern of organisational and technological integration can be found in the experiences of the case study bank where increasingly large economies of scale and scope required control to be re-established and focused through technological and organisational innovation.

Chandler (1992) and Beniger (1986) have shown that a firm's dynamics are governed by an iterative process where firms expand scale and scope until control bottlenecks are reached and then re-establish control. In this chapter we have explored the qualitative changes in control systems themselves.

Moving to more powerful, faster, reliable and wider risk analysis is not trivial, as software and hardware systems do not scale in a linear way (Boehm, 1997; Hobday, 1998). As the complexity increases, new problems emerge that were unforeseen when the initial architecture was designed (Simon, 1969). As a consequence of this 'tight coupling' the larger and more complex a system, the harder it is to maintain (Perrow, 1988) and increasing emphasis must be placed on getting the systems architecture right (Boehm, 1997; Brooks, 1995; Hobday, 1988). A historical look at the evolution of UNIX systems within banks identified three qualitatively different stages: 'Craft', 'Mass Standardisation' and 'Complex'; whose key features are highlighted in Table 12.1.

The control system within the bank went through three qualitatively different stages. These were:

1. Small scale introduction of control systems to allow the bank to achieve new scale and scope economies. The control systems are simple and maintained on a craft basis.

2. The control system expands and allows the bank to achieve more scale and scope economies. In doing so it becomes 'bundled in' with the provision of the service and reliability becomes more important. The increasing scale of the control system limits further expansion until a qualitative change centralises a standardised innovation and maintenance process.

3. The control system continues to expand and opens up new scale and scope opportunities. The new centralised infrastructure allows new economies of system that can be used to actively manage risk-exposure for the traders and pro-actively maintain the infrastructure. The complexity of the system allows new latent functions to emerge, primarily the ability to analyse risk in close to real time. These advantages come at a cost, as the same complexity that enables the system to deliver new services, also makes maintenance and innovation increasingly complicated, necessitating a qualitatively different style of software engineering. Having the right systems architecture allows the bank to quickly roll out applications and move into expanding markets.

Extrapolating the framework to other Service Sectors

Since many of the features of these quantitatively different stages are dependent on changes in the control system it is feasible that they might be repeated in other instances. In further work on innovation in control systems we have developed a tentative model of innovation in information intensive services based on the qualitative changes in the control systems, the main features of which are summarised in Tables 12.2 and 12.3. In common with Barras (1986, 1990) we find that the first stage concentrates on process innovations, and the second stage concentrates on product innovations. However, in the third stage we emphasise the importance of control, economies of system and enhanced reliability. In a fourth stage we highlight the importance of consolidating advantages.

Table 12.2

	Standardise and Process	**Expand Function**	**Economies of System**	**Consolidate**
Innovation Focus	Automate and Monitor **Processes**	New **Products** and Services & Bundling IT with service provision	**Control**, Enlargement and Maintenance	**Control** and Maintenance
Control System	Simple & local	Mass produce	Complex	Very Complex
Economies	Scale (speed)	Scope	System	Complexity Barriers
Advantages	Limited	Small	Large	Very Large
Problems	Limited	Architecture and Expansion	Reliability and Expansion	Reliability and Change

In the first stage, innovation focuses on controlling service performance and bespoke IT systems are applied to automate *processes*. The centralisation of control requires the separation of the IT infrastructure from the applications and control system. This allows the function of the modularised applications software to be extended and increasingly integrated with service provision. Because this increases the complexity of the control system, advantages can be had from automating changes and standardising the applications used in the firm. This transforms the nature of software innovation from a craft based form to a more standardised process, automated from a central point.

In the second stage, the introduction of control systems has allowed the firm to achieve new economies of scale and scope and has enabled the development of new functions (which can possibly be products). In doing so the control-system becomes essential to the ability to compete and the technology becomes 'bundled-in' with service provision. The shift from a monitoring to a business-critical role puts an increasing amount of emphasis on reliability.

In the third stage, the control system can be used to achieve additional economies of system. The control system can be used to monitor and better manage the production of value added services and ensure the reliability of the business critical network. This is vital for firms that deliver continuous services like water, electricity, telecoms, banking, retailing and transport. By gathering information on how the system is behaving, maintenance can be preventative rather than reactive, and sub-systems can be brought on-line and off-line without affecting the overall performance. At this stage the system can be customised to meet particular customer requirements, allowing new functions and economies of scope. Because very complex, software based systems that have evolved over many years are very hard to produce, they act as significant barriers to entry and make it difficult for new competitors to enter the market[8].

As a consequence, the ability to project manage any changes to the control system becomes more important to corporate success at this level.

Table 12.3

Sector	Automate	Increase function	Economies of System and Complexity Barriers	Consolidate Competitive Advantage
Investment Banking	Early 1980s back office systems	Risk analysis moved from back to front office and the power of risk analysis increased.	Perform global real time risk analysis and maintain the business critical system from a central point.	Exploit advantages globally, consolidate in London, New York and Tokyo.
Retailing	EPoS used to automate ordering and stock control (late 1970s early 1980s)	EPoS data used to understand customer requirements. Larger scale needed for better analysis	Network shifted from stores to the entire supply chain to control manufactures	Larger supermarkets dominate market and small players fall out. Attempt to move international
Water	IT monitors used to find local leaks (early 1980s)	Network put in place to understand water movements and losses (late 1980s).	Centralised control system put in place in for entire region. Used for preventive maintenance to ensure supply to priority users (i.e. hospitals)	Infrastructure monopoly.
Lifts	Control panels within lifts electronified, (1980s).	Data from lifts passed to control room within building, (late 1980s).	Control rooms for entire cities set up to deal with emergencies and move from reactive to proactive maintenance	Industry shaken up by changes in markets and technology.
Postal (UK)	Large customer bar-coding for low level automation.	Shift to Integrated Mail Processing (IMP) and to exploit economies of scale and scope	£150 million distribution network centres set up to exploit data from IMP and match resources to managed flows. Move to provide logistics for other firms	Regulated Monopoly lobbies for privatisation in attempt to get access to lucrative European Markets
Telecom	Digitisation of analogue lines.	Function of telecoms expands and scale and scope economies exploited	Controlling infrastructure becomes vital. Competitive advantage to ensure reliability and effective customer costing.	Firms move to manage control for other networks in globalised market.

In the fourth stage, the technological advantages derived from the control system require organisational changes to produce potential competitive advantage. In industries that can compete globally, competition and innovation produce absolute competitive advantages within national systems of innovation. While firms compete away their advantages with their local competitors through an arms war of technological and organisational innovation, that same process means that they develop exploitable absolute advantages on an international scale and the control systems allow them to expand into foreign markets at lower cost.

While this evidence is anecdotal and at an early stage it does suggest that, like manufacturing firms, service firms are driven by a 'technological imperative' to achieve new economies of scale and scope. In doing so, they seem to undergo a series of quantitative changes that require corresponding changes in the nature of co-ordination and control of service provision. If, as we are tentatively suggesting, the qualitative changes in the technological control system are largely determined by its complexity then the four stage model we have abstracted from our initial case studies could be applicable in a number of other service sectors. The insights Chandler made about the railroads of the 1880s may still be very pertinent to understanding service led industrialisation in the next century.

REFERENCES

Barras, R. 'Towards a Theory of Innovation in Services: The Vanguard of the Services',*Research Policy*, 1986; 15, pp.161-173.

Barras, R. 'Interactive Innovation in Financial and Business Services: The Vanguard of the Services Revolution', *Research Policy*, 1990; 19, 3, pp.215-37.

Beniger, J. *The Control Revolution: Technological and Economic Origins of the Information Society*, Cambridge, Mass: Harvard University Press, 1986.

Berger A.N., Kashyap, A.K. and Scalise, J.M. 'The Transformation of the US Banking Industry: What a Long Strange Trip Its Been', *Brookings Papers on Economic Activity*, 1995; 2, pp.55-218.

Boehm, B. *Large Software Projects*, Photocopy Mimeo, 1997.

Brady, T. and Target, D. 'Strategic Information Systems in the Banking Sector - Holy Grail or Poison Chalice', *Technology and Strategic Management*, 1996; Vol. 7, No. 4, pp.387-406.

Brooks, F.P. *The Mythical Man Month, Essays on Software Engineering*, 25th Anniversary Edition, Addison-Wesley, 1995.

Chandler, A.D. *The Visible Hand: The Managerial Revolution in American Business*, Belknap Press, 1977.

Chandler, A.D. *Scale and Scope: the Dynamics of Industrial Capitalism*, Belknap Press, 1990.

Chandler, A.D. 'Corporate Strategy, Structure and Control Methods in the United States During the 20th Century', *Industrial and Corporate Change*, 1992; 1, 2, pp.263-284.

Conroy. 'Risk based Capital Adequacy Guidelines - Sound Regulatory Policy or a Symptom of Regulatory Inadequacy', *Fordham Law Review*, 1995; Vol. 63, No. 6, pp.2395-2443.

Crede, A. *Technological Change and the Information Society: An Analysis of Credit Risk Assessment and Cash Handling Procedures in Commercial Banking*, University of Sussex, D. Phil., SPRU, 1997.

Davies, A. 'Innovation in Large Technical Systems: The Case of Telecommunications',*Industrial and Corporate Change*, 1996; Vol. 5, Number 4, pp.1143-1180.

Freeman, C. and Soete, L. *The Economics of Industrial Innovation*, 3rd Edition, London: Pinter, 1998.

Frei, F.X., Harker, P.T and Hunter L.W. *Innovation in Retail Banking*, Wharton Financial Institutions Centre Working Paper, 97-48-B, 1997.

Greenbaun. '25 Years of Banking Research', *Financial Management*, 1996;Vol. 25, No. 2, p.86.

Hobday, M. 'Product Complexity, Innovation and Industrial Organisation',*Research Policy*, 1998; 26, pp.689-710.

Hopwood, A. and Miller, P. *Accounting as Social and Institutional Practice*, Cambridge University Press, 1994.

Hughes, T. *The System Builders*, Cambridge: Cambridge University Press, 1990.

Kaplan, S. and Johnson, T. *Relevance Lost - The Rise and Fall of Management Accounting*, Cambridge Mass: Harvard Business School Press, 1987.

Kimball, R.C. 'Specialisation, Risk, and Capital in Banking', *New England Economic Review*, 1997; p.51.

Leveson, N. *Safeware: System Safety and Computers*, Reading Mass: Addison Wesley, 1994.

Miles, I. *Innovation in Services: Services in Innovation*, Manchester: Manchester Statistical Society, 1996.

Pavitt K. 'Technologies, Products and Organisation in the Innovating Firm: What Adam Smith Tells Us and Joseph Schumpeter Doesn't', *Industrial and Corporate Change*, 1997; 7, 3, pp.433-452.

Perrow, C. *Normal Accidents*, Basic Books, 1988.

Porter, M. *The Competitive Advantage of Nations*, New York Free Press, 1990.

Power, M. 'Auditing and the Politics of Control in the UK Financial Services Sector' inMcCahery, J., Picciotto, S. and Scott, C. (eds.), *Corporate Control and Accountability*, Oxford: Claredon Press, 1993, pp.187-202.

Santomero, A.M. 'Commercial Bank Risk Management: An Analysis of the Process',*Journal of Financial Services Research*, 1996; Vol. 12, No. 2-3, pp.83-115.

Searle, J. *The Construction of Social Reality*, Allen Lane, 1995.

Simon, H. *The Sciences of the Artificial*, MIT Press, 1969.

Simons, K. 'Value at Risk, New Approaches to Risk Management', *New England Economic Review*, 1996; p.3.

Uchpakinan, K. *Dynamics of Competitive Strategy and IT-based Produce/Process Innovation in Financial Services: The Development of Electronic Banking Services in Thailand*, Unpublished D.Phil. Thesis, University of Sussex, Brighton, UK, 1997.

Yates. *Control Through Communication: The Rise of System in American Management*, Baltimore: John Hopkins University Press, 1989.

---

[1]This chapter is based on a paper presented to the CRIC 'Service Innovation' workshop. Some of the sections are taken from a second paper based on the original work 'Innovation in Investment Banking: Control Systems in the Chandlerian Firm' *Industrial and Corporate Change* (forthcoming). We would like to extend our thanks to Dr. Andrew Davies, and discussants at the CRIC 'Service Innovation' workshop for their help in this research project. This work was funded by the ESRC, as part of the Complex Product System Innovation Centre, held jointly between the Universities of Sussex and Brighton.

[2] Freeman and Soete, for example, define services as 'those activities (sectors) where output is essentially consumed when produced' (1997, p.403). Miles (1996) defines services in terms of the transformation of materials, people, or information to highlight the variety, intangibility and client-intensity of services. Porter (1990, p.240) similarity highlights the intangible-function aspect of services: 'The term services encompasses a wide range of industries that perform various functions for buyers but do not involve, or only involve incidentally the sale of tangible products' (Porter, 1990, p.240).

---

[3] The distinction between manufacturing and services is further complicated because there exists a further distinction within the category of 'performing a function' between services that exist by the brute physical fact of them happening and services that exist 'only by way of collective acceptance' of a certain status (Searle, 1995, p.124). For example a judge can clean windows at the weekend, but a window cleaner cannot sit in a court. Our concern is only areas of service innovation where the function is determined by brute physical facts.

[4] Where control is defined as 'purposive influence towards a predetermined goal' (Beniger, 1986, p.7).

[5] The word 'Control' itself comes from the Latin verb *contrarotulare,* meaning to compare something 'against the rolls' of paper used as official records (Beniger, 1986, p.8).

[6] Economies of system while related to economies of speed are distinct from them. Improvements in the control of a fibre optic network - where the traffic is already flowing at the speed of light - can improve capacity utilisation and provide economies of scale without increasing the size of the network or the speed of the traffic.

[7] We are grateful to Andrew Davies for discussions on this point.

[8] This is consistent with firms making low profits.

Chapter 13
Research and Technology Outsourcing and Systems of Innovation[1]

Jeremy Howells

13.1 INTRODUCTION

This chapter seeks to locate the process of research and technology outsourcing in relation to the wider innovation process and to investigate the growth of the phenomenon within a system of innovation framework. To reflect this, the chapter is in four main parts. The first main section outlines the factors behind the process of research and technology outsourcing process and explanatory models that have been used to conceptualise the process. The next section then charts the growth of research and technology outsourcing and puts in an historical perspective, linking up with the explanatory and conceptual frameworks presented in the preceding section. Developing some of these findings, the chapter then places the process of research and technology outsourcing within the wider systemic and market context before reviewing the implications of research and technological outsourcing more specifically within the 'system of innovation' approach to innovation. These approaches centre on the 'systems of innovation' model based originally on the 'national systems of innovation' (NSI) studies developed by Freeman (1987, 1988), Lundvall (1988, 1992) and Nelson (1992; Nelson and Rosenberg, 1993), but also includes the 'technology system' approach developed by Carlsson (1995).

Much of the recent discussion and analysis concerning research externalisation has concentrated upon inter-firm collaborations between manufacturing companies (manufacturing-manufacturing relations) operating in the same or related sectors. Such analysis has been within the context of long term dynamic competitive and technological trends. It has stressed that firms, even large multinational corporations, can no longer expect to be totally dependent on their own in-house research and technical resources to maintain their innovative performance. This chapter seeks to outline the implications of these dynamic trends for the conduct of the innovative activity by firms in relation to research and technology outsourcing and place it within the wider context of a systems of innovation framework. The growth in the external sourcing of R&D and other design and technical activities by firms has played an important role in the creation and development of the research and technology (R&T) 'market'[2], in terms of the commercial purchasing and trading of R&D and technology. This chapter will seek to explore the development of this market and in turn relate its development to the growth of organisations serving this R&T 'market'. This includes Contract Research and Technology Organisations (CRTOs), as well as a wider group of companies and organisations with some involvement in the R&T market.

The main focus of this analysis is therefore on the outsourcing of R&D and technical activity formerly undertaken by manufacturing and service firms and now outsourced to firms and organisations which supply the R&T market. These are primarily private CRTOs, but also includes a range of private non-profit and related hybrid organisations. The focus is, therefore, not on what has become termed as 'inter-firm' research and technical collaboration, involving both horizontal co-operation, with firms from the same or related manufacturing sector (sometimes direct competitors) and vertical co-operation, associated with collaboration along the supply chain between suppliers and their customers. Nor does the analysis specifically cover 'university-industry' or 'government-industry' collaborations, although certain hybrid organisations such as university-owned spin-off companies and government laboratories which are trading companies are increasingly important in the research and technical contract market (see later). However it is acknowledged that all these forms of research and technical collaboration make up part of the external research and technical profile of the firm, and are of relevance to the wider analysis of R&D outsourcing and externalisation. In turn the emergence of these different actors within the R&T market and system has important implications for the development of the innovation support infrastructure in national and local systems of innovation. These wider systemic issues will be explored in more detail in the latter half of this chapter.

13.2 FACTORS ASSOCIATED WITH RESEARCH AND TECHNOLOGY OUTSOURCING

A range of studies has sought to explain the growth in the use of external research and technical resources by firms and why firms should seek to collaborate or contract out their R&D, design and engineering needs (see, for example, Pisano et al, 1988; Thomas, 1988; Pisano, 1990; Arora and Gambardella, 1990; Hagerdoorn, 1990; Hagerdoorn and Schakenraad, 1990; Charles and Howells, 1992; Häusler et al, 1994; Katz and Martin, 1997; Chen, 1997). Some of these studies highlight the transaction-cost model developed by Williamson (1975, 1979; see also Ouchi, 1980) as a means of explaining why there may be a shift from an in-house vertical and hierarchical provision of such research needs towards a more market mediated approach. In brief, Williamson (1975) sought to distinguish between 'markets', based on contracts, and 'hierarchies', based on transactions undertaken within the firm as alternative governance mechanisms for activities undertaken within or between the firm. The lower the transaction cost the more likely the activity is to be undertaken outside the firm via market mechanisms. These types of transactions are likely to be simple, more certain and predictable in their outcome involving more standard, non-specific assets. Ostensibly the transaction cost model explains on one level why there should be no contracting-out of R&D activities of the firm at all. This is because the high degree of complexity and uncertainty associated with R&D makes contractual relationships difficult or impossible to negotiate (Brockhoff, 1992). Thus under conditions of bounded rationality, together with a high degree of complexity and uncertainty, bargainers are unable to define complete contractual agreements because neither all alternative features nor their corresponding prices can be determined (Williamson, 1975, p.25; see also Ouchi and Bolton, 1988, p.11). However, not all R&D has high degrees of uncertainty and complexity attached to it (Freeman andSoete, 1997). There are forms of R&D which have low levels of uncertainty and complexity where market-mediated contracts can function. By contrast, more complex and uncertain research requirements would preclude market-based contracts (Mowery, 1994, p.581). The transaction cost model therefore suggests that contracted out research and technology would be restricted to these shorter term, simpler, less uncertain technology activities.

However, Williamson has more recently sought to develop and highlight different contractual arrangements under classical, neo-classical and relational categories of contract law (Williamson, 1986; using concepts originally outlined by MacNeil, 1978) which allow for a wider range of contractual relations. Thus, Williamson (1996) has highlighted hybrid forms of governance structures which are between the 'pure' forms of markets and hierarchies and which can help account for greater complexity in contractual and organisational forms in the real world. Others have also sought to stress networks and other forms of collaborative organisation as governance structures between markets and hierarchies (see, for example, Håkansson, 1987; Powell, 1990). Despite this, researchers, such as Robins (1987) and Englander (1988), believe that is not possible to adequately operationalise transaction-cost model because of inherent logical and empirical problems in the approach.

Notwithstanding this, a study by Croisier (1998), using the Co-operative Agreements and Technology Indicators (CATI) database developed by the Maastricht Economic Research Institute on Information and Technology (MERIT) at the University of Limburg, has indeed operationalised the transaction cost economics model in relation to external research. According to three main governance forms he was able to use ranging from ('market' oriented) R&D subcontracting, through to co-operative R&D without equity participation and on to co-operative R&D with equity participation (a more 'hierarchy' oriented structure) certain assumptions about what form external research and technology will take were confirmed. Thus Croisier (1998, p.295) noted "the more the project is located at the beginning of the research process, the higher the number of partners and the technologies implied, the larger the geographical scope of the project, the more it is likely that market modes will be replaced by more hierarchical modes, i.e. subcontracting by co-operation and co-operation without equity participation by co-operation with equity participation". The study provides a good set of parameters on which projects are likely to be conducted, but provides very little explanation behind the dynamics of such processes. Above all it provides little or no explanation as to why there are shifts in the degree of research and technology outsourcing undertaken by firms (see later). It is maintained here that the extension of the transaction cost economics framework by Williamson and the reinterpretation of his work by others, although valuable, can only go so far in helping to explain and understand the phenomenon of technology outsourcing. The other main conceptual framework that has been used in analysing research and technology outsourcing is that provided by Winter (1987) and Prahalad and Hamel (1990) in relation to their respective and related notions of 'strategic assets' and 'core competences' of the firm. Both approaches emphasise that firms have key assets or competencies that have resulted from previous rounds of investment and from learning-by-doing. These core competences can be seen as 'resources' (Hall, 1992) as well as capabilities which are accumulated over the long term which firms seek to both develop and deploy to gain competitive advantage (Penrose, 1959; Richardson, 1972). Coombs (1996) has sought to employ the 'core competence' model as a lens through which to analyse changes in the strategic management of R&D over time. The core competences model is a valuable conceptual approach in that it provides a more holistic and structured framework from which managers can make a priori judgements regarding the internal/external technology boundaries of the firm and a posteriori make evaluations about their success.

However, care must be taken when emphasising both the benefits to firms of such trends towards R&D and technical outsourcing (Coombs, 1996) and the implications of such a shift in terms of in-house research and technical effort. Thus Coombs (1996, p.354) suggests why firms may be over-reaching themselves in their desire to decentralise and outsource their R&D portfolios, ultimately weakening their core technological competences (although this view is somewhat at variance with researchers who have sought to highlight the 'virtual organisation' (Chesbrough and Teece, 1996) and more especially the virtual organisation of R&D activities (Chiesa and Manzini, 1997)). Häusler et al (1994) have also emphasised that current literature tends to give a rather positive image of external research links, even though there are good reasons against co-operation. MacPherson (1997) in his study of producer service outsourcing more generally (including research and technical services), has noted that the 'necessity effect' (where a firm has an urgent need but no in-house resources to satisfy that requirement) was the most single important factor motivating outsourcing for firms in New York State (see also Oliver and Blakeborough, 1998). There may be firm size effects here, with more smaller firms experiencing this problem (see Rothwell, 1991).

Historical insights into the process are also valuable here. Mowery (1984), for example, has highlighted the British reliance on the external contracting of R&D and 'market' mechanisms as a key factor hampering innovation and export performance amongst British firms before 1950. In particular the use of consulting engineers in the British electrical industry in the late nineteenth and early twentieth centuries hindered innovation and the design process in the industry (Byatt, 1979). In the latter context, the increase in external outsourcing of research and technical activity does not imply that there is necessarily a contingent decline in-house R&D activity. Thus, Mowery (1983, p.369), in the context of independent research organisations during the period of 1900-1940, has noted that "rather than functioning as substitutes, the independent and in-house research laboratories were complements during this period, exhibiting a division of labor in the performance of research tasks". Similarly, Arora and Gambardella (1990, pp.373-4) have suggested external collaboration of research is complementary to, rather than a replacement for, in-house research activity. Such views have also been supported by Lowe and Taylor (1998) in the relation to the purchase of existing technology through licensing from other firms.

13.3 GROWTH AND HISTORICAL PERSPECTIVES

Given these factors behind R&T outsourcing it is not surprising that there has been significant growth in the amount of extra-mural or contracted out research in advanced industrialised economies since the 1970s (Whittington, 1990). Thus, in Canada an estimated 7% of industrial R&D expenditure was extra-mural in 1993 (Rose, 1997, pp.10-11). Similarly in the UK some 10% of Business Expenditure on Research and Development (BERD) was extra-mural in 1995. Veugelers (1997, p.308) in a study of Flemish companies also noted that 10% of the total R&D budget of such firms was spent extra-murus. Moreover there has been a substantial growth in external research work. In Germany an increasing proportion of company R&D budgets are spent externally, from a figure of 3.5% in 1969 to 9.8% in 1983 (although levelling off to 9.2% in 1989; Häusler et al., 1994). In the UK there has been a doubling in the amount of extra-mural BERD in the UK between 1985 and 1995, from £450 million (in constant 1995 prices; 5.5% of the total) to some £935 million (10.0%) in 1995. In some sectors this level was even higher, with an estimated 15.5% of UK pharmaceutical R&D being extramural in 1995 (Jones, 1997, p.13).

However, although this growth in outsourcing has been rapid in recent years it should not be considered new. Thus, as Teece (1988, p.258) notes in relation to the United States "During the late nineteenth century and the first half of the twentieth century... practically all of it [research] had been conducted outside of the firm in stand-alone research organizations". Up until the inter-war period, in-house R&D laboratories were considered novel and most firms that needed to undertake research would contract it out to universities or independent research scientists when necessary. Even in sectors with a relatively long scientific tradition, such as the pharmaceutical industry, this was still the most usual method of conducting research up until the First World War (see, for example, Liebenau, 1984; Swann, 1988). However, during the 1920s, 1930s and 1940s there appears to have been a significant decrease in R&D contracted out to independent research organisations. This can be seen in the decline in the proportion of scientific professionals working in such establishments to all scientific professionals working in all (in-house and independent) research laboratories in the US (Table 13.1); a trend which is symptomatic of the wider 'professionalisation' of research (Freeman 1982, p.10) at this time.

In Britain, the emergence and development of in-house R&D departments came much later and was less complete in important respects when compared to that occurring in the United States (Mowery, 1983; 1984; Mowery and Rosenberg, 1988) and more especially Germany. Indeed, Germany was more advanced than the United States in terms of not only the formation of industrially-based research (Meyer-Thurow, 1982; Homburg, 1992; Marsch, 1994), but also in relation to the development of industry-academic links (see, for example, comparisons made by Sanderson, 1972).

It was therefore only effectively after the inter-war period (for the United States) and more especially after the Second World War (for most leading European countries) that the 'tradition' of large, centralised R&D laboratories became commonplace. The post-war period up until the mid 1960s represented a phase where most firms used only limited external R&D and technical sources, however by the late 1960s this had changed as firms started to experiment with using multiple outside sources of technology (Graham, 1985, p.87). By the mid-1980s there appears to have been a more marked rise in the use of external sources of R&T, although still from a relatively low levels. External sourcing of research and technology is still growing strongly, but how far this growth will continue remains to be seen as a number of firms and sectors seem to be overreaching themselves in terms of over-dependence of external resources for innovation (Coombs, 1996). Firms cannot become over-reliant on external research and technology as in-house R&T capacity is still required to effectively scan, purchase and absorb new technologies from outside.

Table 13.1: Employment of Scientific Professionals in Independent Research Organizations as a Fraction of Employment of Scientific Professionals in all In-House and Independent Research Laboratories in the United States, 1921-46

1921	15.2
1927	12.9
1933	10.9
1940	8.7
1946	6.9

Source: Mowery (1983)

However a new era in terms of research and technical competences and collaborative patterns for firms may now be emerging as companies seek a more balanced and holistic approach to their research and technical requirements. Firms have increasingly come to accept that they need in-house capacity to adequately appraise, select and then use 'brought in' research and technical elements. They also recognise the need to retain certain core technological competences within the firm because they can develop them more effectively on a hierarchical, non-contractual basis, including here reasons associated with secrecy, trust and appropriability.

Table 13.2 outlines these different stylised phases drawing on a range of different studies which have sought to highlight and delineate different eras in relation to corporate research and technology. It is important to note several points here. Firstly, that the periods denoted are from a developed industrialised economy perspective and that there will be significant variations between countries. Less developed economies still appear to lag substantially behind these phases (see, for example, Sikka, 1998). Secondly, most of the studies of corporate R&D and technical activity have taken a predominantly large firm perspective. In relation to external R&T small and medium-sized enterprises have always displayed a much higher propensity to use external R&T sources and therefore the trend in relation to using more or less external sources of technology are likely to be much less marked. Nonetheless large firms still play a major role in the R&T system and the chart highlights the considerable swings in relation external sourcing of research and technical activity over the twentieth century.

Table 13.2: Main Periods in Industrial Research and Technology

Period		Characteristics
1.	Up to late nineteenth Century (1890/1900)	Limited R&D; research and technical work mainly contracted out to universities, led by Germany
2.	1900-1914/18	Emergence of in-house R&D laboratories in a limited set of major companies, led by US
3.	Inter-War Period 1914/18-1939/45	Major diffusion and expansion of in-house R&D; relative decline in the use of external R&D expertise
4.	1939/45-1965/70	'Nadir' of relative use of external research and technical expertise; major growth period of corporate R&D; increasing government funding of R&D
5.	1965/70-1980/85	Gradual expansion of expansion of external R&T sources; continued growth of in-house capacity
6.	1985-?	More rapid take-off of external R&T; re-evaluation and restructuring of R&D within major firms
7.	?-	Growth of external R&T slows down; more balanced and holistic view of internal/external R&T resources by firms; R&T within more distributed innovation networks

Notes:
Compiled from the following sources: Sanderson (1972); Freeman (1982); Graham (1985); Liebenau (1985); Homburg (1992); Edgerton and Horrocks (1994) and Coombs (1996).

What were the factors underlying these fundamental shifts in the internal-external balance of R&D and technical work in and beyond the firm? Such explanations can be broken down to a number of key 'meta' parameters, namely changes in the:

- nature of R&D and technical work
- the emergence of a functioning R&T 'market' and new collaborative/institutional arrangements of innovation
- macro innovation environment associated with the life cycles of technologies and industries.

There have clearly been factors associated with the changing nature of the research process that have encouraged outsourcing of these activities. The most frequently cited factors are the increasing complexity of the research process and the cost and risks of R&D. As many of the simpler scientific and technical discoveries have now been made, companies are increasingly having to deal with much more difficult and intractable scientific problems. Products are also becoming more sophisticated catering for these new more complex problems and consumer demands. The number of technologies per product (TPP) is therefore increasing in many consumer and business products. An oft cited example here is the shift from mechanical to electro-mechanical systems in the automobile industry (Miller, 1994, p.30). Most companies simply do not have the necessary scientific resources to cope with additional burdens and seek outside support to overcome internal technical limitations (Haour, 1992). Many of these new (fundamental) research issues cover several scientific and technical disciplines further encouraging collaboration between organisations with strengths in different fields. This growth in the need for interdisciplinary research has been reflected in the shift from what Gibbons et al (1994) have termed Mode 1 to Mode 2 research operations.

The increasing complexity and inter-disciplinary nature of the R&D process in turn has increased the cost of research. Research may therefore become less attractive without partners to share the cost. More simply the firm may lack the financial resources to undertake research even if it remains an attractive proposition. Associated with this, the traditional barriers between scientific and technical disciplines are being broken down, as the interchange between basic research and development work grows. In the field of semiconductors basic research is undertaken with a view to future applications right from the outset. This pressure to improve the interface between basic research on the one hand and applied and developmental work on the other also stems from the pressure to reduce innovation cycle times. Scientific and institutional inertia may also play a key role. Pisano et al (1988, pp.191-3) have, for example, outlined the defensive response by many major pharmaceutical companies in the 1980s, which grounded in chemistry, sought links with new biotechnology firms and universities to gain research expertise in biotechnology.

The above have described mainly 'push' factors associated with firms seeking outside research and technical support, but there are also a number of key 'pull' factors encouraging such the trend towards further external R&D collaboration. These cover the relative attractiveness of external sources of expertise over internal firm resources which the firm may not possess or only very inadequately. Such factors also covers other motives such as the desire to enhance the scope and testing of in-house scientific and technical activities, as a general scanning mechanism of technological opportunities existing outside the company and as a means to network with other organisations and to enable in-house staff to be part of a wider 'invisible college' within the specific research community. Lastly, there are government policies and incentives to participate in collaborative research and technical projects. These push factors have been emphasised in Ringe's (1992, pp.28-9) survey of selected UK firms which found that the desire to gain specialist expertise was the most frequently cited motive for contracting out R&D, together with access to specialist techniques or equipment. The need to tap additional manpower was the third main factor, followed by the ability to gain tight control of R&D timescales and budgets, in order to get the job done.

These last points in turn relate to developments in the R&T 'market' (Howells, 1999a). Many companies are still reluctant to outsource 'critical' technologies to outside suppliers, but they are increasingly willing to contemplate the sub-contracting of more routine, low value-added research and technical activities. As the R&T 'market' continues to develop, less of this routine R&D and technical work will be undertaken 'in-house' and instead will be the responsibility of CRTOs and their employees either working on- or off-site. The R&T 'market' appears to be highly dynamic, not only in terms of the appearance of more dedicated CRTOs, but also of a much wider set of partial players, who may over time become more central participants (Table 13.3). Research and technology outsourcing is 'coming of age' and seen as a major industry in its own right. It is also becoming more complex in its form. Research and technology outsourcing is no longer just appearing in a 'pure' market form, but is also occurring in more hybridised, quasi-market and relational transactional forms.

Indeed this has been acknowledged by Williamson (1996) himself in his reworking of the transaction cost model more generally. He notes that there has been a plethora of new forms of firm collaboration emerging within the economy. There are therefore many more types of hybrid arrangement that firms can find themselves in aside from in-house work (hierarchy) or simple subcontracting (market-based) forms. This is evident more specifically within the field of research and technology relationships where these hybrid forms range from, for example: collaboration without equity participation; co-operation with equity participation; joint ventures with or without prior 'exit' strategies and varying degrees of equity stakes between the two partners; complex 'multi' ventures involving more than one firm; project specific collaborative networks with varying participants apportioned different intellectual property rights (IPRs) and commercialisation rights. Some of these relationships are time or project specific; others are more open-ended. Some collaborations may be centred around a technology or a specific innovation; other co-operative agreements have research and technology as a more indirect outcome of the collaboration which may be based on 'market' access, industrial or financial considerations. Even aside from the middle of the governance spectrum, both 'ends' also involve more complex arrangements. Thus the 'pure' hierarchy of internal R&D modes does in fact vary considerably from a centralised (U-form) of corporate structure to a more decentralised (M-form) structure which is nearer to the 'market' and more responsive to 'market' signalling (Croisier, 1998, p.291).

Table 13.3: Types of New Entrants in the Contract Research and Technology Market: UK Examples

	Types of New Entrants:	Examples:	
		Name of Firm/Organisation:	**Related/Affiliated Firm/Organisation:**
1.	*New Firm Types:*	Chemical Design	
		Oxford Molecular	
		Cambridge Combinatorial	Oxford Molecular 19.99% stake
2.	*Spin-Offs:*		
2.1	*Spin-Offs (from manufacturing and service firms)*	CRL	formerly Thorn EMI's central research laboratory
		International Research and Development	formerly part of Rolls Royce Industrial Power
2.2	*Spin-Offs (from universities)*	Aston Molecules	Aston University, Birmingham
		Bradford University Research	Bradford University, Bradford
3.	*Diversification:*	BSI	has over 250 research staff working on basic and applied R&D
4.	*Government Privatisation:*	ADAS	former government agency for agricultural R&D and laboratory and consultancy services
		AEA Technology	formerly the UK Atomic Energy Authority
		LGC	Formerly the Laboratory of the Government Chemist

Source: Howells (1999a, p.24)

Equally at the other end of the spectrum, supposed 'market' based contracts in relation to research and technology can differ substantially and have evolved over time. These can range from:

1) fee-for-service arrangements, such as

   a) fixed-price contracts; and
   b) cost-plus contracts; through to

2) performance-based service arrangements, ranging from

   a) fixed royalty payments; to
   b) variable royalty payments based on successful outcomes.

Even with performance-based service arrangements firms pursue different strategies based on one-off contract deals usually involving high royalty deals to more long term 'relational' low royalty deals with more familiar clients.

The boundaries within and between these different hierarchy-market based forms are therefore becoming more blurred as firms learn about collaboration and governance strategies. With the increasing range and diversity of these options firms are more likely to find the right 'fit' encouraging them to externalise their R&D and technical effort whereas before they may have decided that it was better to 'play safe' and retain the project or technology in-house. Moreover the more they experiment and learn from these external forms of technology sourcing the more comfortable they feel in making future external R&D relationships.

The above set of factors have focused on specific research, firm and market contexts, however there are also wider macro issues centred around technological regimes and the histories and life cycles of industries, based around Schumpeterian patterns of innovation (see, for example, Malerba and Orsenigo, 1995; 1997; Klepper, 1996). In brief, the Schumpeter Mark I model of innovation is associated with 'creative destruction' and the emergence of new firms related to low entry barriers, with a strong role played by entrepreneurs. By contrast, Schumpeter Mark II models are characterised by 'creative accumulation' and are dominated by large corporations and relatively high entry barriers for new innovators. The key issue here for R&T outsourcing is that under Schumpeter Mark I conditions and early in the history of an industry, the knowledge base of the industry is changing rapidly and consequently uncertainty is high. Moreover with low entry barriers, the number of new entrants or players in the market grows quickly, whilst the knowledge of each potential player about other potential players is also low. All these conditions suggest that outsourcing of research and technology would remain low when major new industries and innovations are emerging because of high levels of risk and uncertainty and associated lack of knowledge about other participants.

However against this argument, many of the participants will be new and small firms who simply do not have the resources to undertake much research and development in-house and who, therefore, are more likely to seek outside partners. Moreover these Schumpeterian patterns of industry dynamics become more problematic when one is moving from inferences about particular industries, in relation to the presence or absence of innovation discontinuities, and making more widespread inferences about the economy as a whole in terms of the propensity towards R&D internalising or externalising.

13.4 RESEARCH AND TECHNOLOGY OUTSOURCING: NEW INSIGHTS?

What can be concluded from these influences on industry to outsource research and technology or not and their longer term impact on internal/external R&T trends? The above has identified many of the factors driving this most recent change, although such factors do not make it possible to predict future changes. Has this current trend towards further relative shifts in the internal/external R&T boundary of the firm now reached its peak? There are suggestions that certain firms and certain industries have indeed reached such a peak and indeed gone too far; but there are few specific pointers to this impression. The conceptual frameworks provided by such models as the transaction cost approach or the capabilities model, outlined above, certainly provide little in the way of prediction or understanding of the dynamics of the process. From the transaction cost approach the only way a predictive element could be produced is if one could infer if the type of R&D projects in aggregate were to move in one direction (i.e. 'reverse engineering' the concept). Thus if in aggregate R&T projects were to become shorter term, more certain and predictable, more applied and developmental type work, then this would suggest more external, less hierarchical R&T activity. There may be some suggestion of this trend in some industries, but it only provides a very partial insight into such changes.

Innovation researchers however may also be too restricted in their analytical framework by generally taking a (single) firm perspective as the subject of study rather than taking a more object-based approach by investigating the innovation (Archibugi and Simonetti, 1998). If innovative activity is becoming more distributed amongst firms and organisations through R&T externalisation (Coombs and Metcalfe, 1998) it is perhaps less helpful to always take the single firm, 'subject' perspective. A more systemic view of R&T outsourcing may therefore be useful in providing insights and explanations to these changes which are not possible with taking solely a firm-based view.

Much of the literature on outsourcing in relation to (manufacturing) firms using contract (service) firms for research and technological inputs has assumed a highly contractual and formalised set of relationships. This is despite recent research on R&D collaboration between manufacturing firms which has shown the strong informal and non-contractual relationship ties existing between firms which are research partners (Dickson et al, 1993). As noted earlier, R&T outsourcing is often not undertaken in strict market governance situations, but rather under more relational and quasi-market conditions. Even where contracts are strictly laid down in more routine technical activities, these were 'put under the table' and were frequently not referred to again. Although the parties involved could therefore, in extremis, refer back to them if serious problems occurred and no agreement could be found, generally they were ignored and more informal discussions and bargaining took place to resolve difficulties. Formal, contractual documents bore very little relation to the usual and day-to-day contacts that were the 'real' research and technical links between parties. In part this misrepresentation may reflect much of the literature which has discussed more routine supply of service activities, where costs and efficiency criteria can be readily assessed under more predictable outcome conditions. It may also reflect the fact that a snapshot, on an individual basis, of a particular research and technical link between a firm and a contract research and technology organisation tends to ignore the wider 'envelope' within which the relationship may be part. The partners may have links stretching back decades, with close personal and informal links between them. The routines of the two organisations may also be very familiar to staff from both sides.

Often the standard, implicit model of manufacturing firms deciding, a priori, to outsource part of their research or technical activity and then selecting a service firm to deliver the service or 'solution' can often be misplaced. This is associated with the continued perception of service firms being 'passive' or 'reactive' in the innovation process. There are, however, examples of service firms (in this case CRTOs) who have initiated and developed a new product themselves and only 'putting out' the final more routine stages of the development (Howells, 1999c). This was associated with the tooling up and production of the new product by a single manufacturing firm or set of companies (where independent components or systems were required to be incorporated into the final good). In this context, manufacturing firms are reacting to the initiation of ideas and action by the service firms and supply limited technical capability and manufacturing as their input to the whole process. Here the focus is on the service firm, the CRO or technology intermediary firm, as the key actor in the innovation process and the manufacturing firm undertaking a more peripheral role. Indeed such firms may more centrally move into manufacturing to fully capture the benefits of the innovation process, following the 'soft' model of firm formation.

This particular construct therefore turns 'on its head' the simple linear, uni-directional concept which is typically envisaged in research and technical outsourcing process. This model sees outsourcing as involving a single manufacturing firm using a single service company (manufacturing⇒service links). Arising from this study, an alternative picture emerges which has at its centre a proactive service firm integrating routine production and technical inputs often from a set of manufacturing firms, sometimes involving a uni-directional linkage (service⇒manufacturing), but more often a two-way or multi-dimensional link (service⇔manufacturing). It should be stressed that outsourcing does not only imply manufacturing⇒/⇔service or service⇒/⇔manufacturing links. An increasing proportion of work forCRTOs is to provide research and technological inputs and resources to *other service* firms. Service⇒/⇔service innovation links need, therefore, to be recognised in the outsourcing and externalisation process.

13.5 RESEARCH AND TECHNOLOGY OUTSOURCING AND SYSTEMS OF INNOVATION

More specifically, the growth of research and technology outsourcing has a number of important issues and implications for innovation systems approaches and the wider understanding of the knowledge economy. The first is whether the whole activity of innovation outsourcing can be treated as a system in itself. On one level, it can be argued that the more restricted research and technological outsourcing 'market' outlined earlier has all the essential attributes of a system that is required by classical systems models (see, for example, Hall and Fagen, 1956), with CRTOs and their 'customer' firms and organisations forming the key elements of the system and the collaborative relations between as the links combining the different elements in the system.

However, defining the boundaries of such a system is not easy and there are clearly a whole set of systems associated with research and technological outsourcing and externalisation, not just a single system. These systems (or sub-systems) relate to specific geographical systems defined in terms of national (or even regional) research and technological outsourcing systems as well as more specific sectoral or technological outsourcing systems associated with specific technologies or industries, such as pharmaceuticals.

However, innovation outsourcing does not readily fit into either the systems of innovation or technological systems approaches as they cut across both national and technological boundaries. This is important in itself, but also highlights inadequacies about the existing approaches, namely that they assume that there are neat systems which have uniform structures and that tightly interlock with each other. The US national system of innovation is different from those of the French or Danish system of innovation; similarly the pharmaceutical technological system is different from that of the energy technological system. In one sense this is true and these systems can be treated as separate, but they largely ignore the issue of 'nesting' hierarchies (Blalock and Blalock, 1959) of systems and their overlap with each other (see Howells, 1999b, in relation to the issue of regional innovation systems versus national systems of innovation). A key element in defining a system is being able to draw a boundary around the set of elements (firms and organisations) and their chosen links within it. However, care should be taken in acknowledging that few systems have clear and mutually exclusive elements, links and boundaries, particularly in terms of a temporal perspective (unless it is static which is extremely unlikely in an innovation system). Thus there is no ready acceptance that firms or organisations can be members of multiple innovation systems. This represents a key danger of a systems approach in that it encourages the conceptual mindset of having separate and 'closed' (in terms of not having, or rather, not considering links and interrelationships external to the system) innovation systems. This is where considering innovation outsourcing challenges such a notion since research and technology outsourcing are usually involved in multiple sets of innovation and technological systems, both geographically and sectorally.

It is also revealing trying to define a system around the formal and informal innovation outsourcing market, which includes both service and manufacturing firms and organisations. Delimiting such a system obviously cuts across many technological systems and national systems of innovation which have already been highlighted in earlier studies. The key elements in such a system would be both CRTOs and their 'manufacturing' (but also service) customers (forming in turn a wider 'organizational population' of different actors and agencies; Reddy and Roa, 1990; see also Phillips, 1960). The boundary of the system could be defined in straightforward 'market' terms, but as has been indicated above, there are strong informal and non-traded elements in the R&T 'market' which involve high levels of reciprocity and trust and which parallels much of what Dore (1986) would term as 'relational' networks. In terms of this wider system, many of the links described would not be based on purely monetary transactions and formal contracts, but involve informal know-how trading and patterns of exchange (see, for example, Von Hippel, 1987; Kreiner and Schultz, 1993). A clear nesting of the wider system is also apparent hierarchically with sub-systems evident in relation to specific technological systems where the research and technology outsourcing market may operate, such as energy or environmental 'services', but also narrowing down to specific national or regional systems of research and technology outsourcing. This has outlined the basic framework of a system, but it has crucially left out the institutional, as well as more general cultural, social and economic, elements in such an innovation system. Without these elements the framework that has been described is more akin to a simplified innovation network (Håkansson, 1987) or linkage structure (Andersen and Lundvall, 1997, pp.248-51). Institutions are central to the notion of any kind of system of innovation (Edquist, 1997, p.24). They provide the dynamic context (the routines and 'guide posts'; Lundvall, 1992, p.10) through which systems evolve and change and linked with the notion of technological paradigms and trajectories.

Leading on from this is that, although the innovation systems literature has so frequently stressed the dynamic qualities of such systems and the approach, it then usually goes on to treat such systems in a static fashion. The wider innovation outsourcing system, or one of its sub-systems, in period n1 is going to be different in period n2. Key elements or actors will have disappeared between the two periods and have been replaced, leading to new and different linkage patterns and changes in the overall shape and configuration of the system (with some of these changes being measurable and therefore allowing systems to be compared).

Indeed much of systems theory development has been driven by its ability to describe and map different contexts in a dynamic fashion. What might be considered to be the appropriate 'system' in one time period or for one set of analyses is unlikely to be the same for a different period. Appropriate systems modelling should also allow the ability to move between different hierarchies or 'nests' so that particular phenomenon or processes can be focused down upon or conversely the analysis can be widened out to allow a broader view. In transport systems modelling the focus can be on the whole transport system covering all transport modes (car, rail, air and sea) on a national perspective, a single transport mode such as car traffic within for example a single urban system, or indeed the car itself can be treated as system (or indeed the electrical 'system' of the car can be viewed as a system). Much is lost from systems analysis by sticking rigidly to viewing and conceptualising one particular system and innovation systems, it is argued here, are no different from other systems. One of the chief merits of systems theory is its hierarchical nature and ability to flexibly move between these hierarchies over time, when conditions allow or when analysis requires this. Our view of what a systemis, depends on what phenomenon (or what attribute of a phenomenon) we wish to study. This determines what the 'element' (or basic unit) within the system will be.

Research and technology outsourcing in this way can be treated as a 'phenomenon-based' system of innovation (PBSI) which can be seen as nesting and overlapping with existing geographical and sectoral systems of innovation, but in a more time-specific and dynamic set-up. The transitory and evolutionary nature of such systems in this way is stressed but it accepts that other, more macro national or sectoral systems are operating at the same time and in part provide part of the institutional, economic and social context for these meta systems. Indeed this perspective has parallels with the notion of Lynn et al (1996, pp.99-101) on innovation community's which stresses the dynamic elements of change within such communities, such as entry and exits, as well as methodological issues to do with boundary permeability and connectedness of organisations within a group.

The growth of innovation outsourcing and externalisation also has more general implications for innovation and technological systems. In particular, that the growth of external innovation links by firms and organisations will lead to the overall increase in the connectivity within and openness between innovation systems. On a more abstract level, and in terms of systems theory, this may suggest innovation systems have increased susceptibility to volatility, dislocation and radical change. This, in turn, is related to the major effect such changes will have on the spatial and temporal diffusion processes in relation to innovation. With increased connectivity and interlinkage, innovation waves will diffuse in significantly different ways. Above all, innovations will diffuse more rapidly throughout the system associated with increased and more pervasive information flows between diffusion actors (although major barriers remain; see, for example,Turpin et al, 1996, p.271). From a policy perspective, Hicks and Katz (1966, pp.43-5), in relation more specifically to inter-firm collaboration, have reviewed some of the implications of an increasingly collaborative science system in industrialised economies.

The analysis of research and technology innovation in relation to innovation and technological systems also has important implications for how service firms, activities and indeed innovations are viewed in existing literature. Although Carlsson (1995, p.444) does place certain service firms in a more central role, within his technological system, as 'bridging organisations', the innovation system literature in general still tends to see service firms and innovators in a passive, peripheral and support role. Manufacturing firms still take the 'centre stage' and the systems are defined around them. Service firms and organisations (including here public agencies and research establishments and Higher Education Institutes) provide the 'supporting cast'. They at best represent what are admittedly seen as key elements in the innovation infrastructure (see, for example, Smith, 1997), but their role is still firmly placed as 'supporters' and 'facilitators' rather than those who lead the innovation process. Above all, the innovation and technological systems which have been outlined so far have been based on manufacturing firms and their innovative processes; around them the boundary of the system has been defined and in system terms they have formed the core 'elements'. By contrast, service firms have been viewed as being at the edges of the system and/or operating in a supporting or indirect role (as subsidiary or second tier 'elements') within the innovation system.

It is argued here that such a perspective is at best dated, partial and incomplete and at worst false and misconceived. Many service sectors and technologies now have a lead role in key areas of modern economies. Service activities and service firms are frequently the 'drivers' in the innovation process, not the other way around. The above analysis of research and technology outsourcing has also indicated that at the very least, 'manufacturing-centred' conceptual frameworks of innovation systems do not readily accommodate the realities of many CRTOs and the formal and informal R&T 'market'. As has been outlined earlier, CRTOs not infrequently take the 'lead' in the innovation process and undertake their own research and development work (see, for example, von Emloh, Pearson and Ball, 1994). They commission the research and undertake much of the basic, groundwork R&D before approaching likely manufacturing firms who may participate in delivery, support manufacturing and supply technical inputs which can help to produce a final innovation. Manufacturing firms represent the passive elements here, whilst the service firms are the proactive elements and represent the 'innovation integrators' and leaders.

13.6 CONCLUSIONS

This research has highlighted a number of issues which have not been adequately captured in previous studies that have explored external research relations and systems of innovation. The decisions-making process surrounding the decision to use external research and technology is much less rational and optimal than frequently assumed, with decisions generally made on an ad hoc, informal basis. The view of contract research and technology activity as being formal, routine, repetitive, cost-based and having short time horizons and levels of uncertainty is being challenged as contract research and technology organisations are moving towards longer, more relational types of research and technology partnerships. Different forms of innovation outsourcing are becoming less distinguishable and increasingly blurred over time.

Research and technology outsourcing and its changing dynamics also contribute to notions underlying the systems of innovation approach through widening its articulation and conceptualisation. The chapter has suggested that research and technology outsourcing can be treated as a PBSI which can be seen as nesting and overlapping with existing geographical and sectoral systems of innovation. A PBSI is more time-specific and dynamic than previous conceptualisations of systems of innovation and may be more fragile in this sense. These qualities arguably are missing in existing systems of innovations. However, PBSI is not meant to be seen as replacing existing systems of innovation approaches, but rather provide an additional mechanism to view innovation within a systemic way. Indeed many of the interesting facets derived from taking this approach is viewing the relation between systems. Thus the research and technology outsourcing system will be expected to be different between different countries because such research and technology outsourcing systems are operating against a different national system of innovation 'backcloth' which will crucially influence the underlying institutional economic arrangements between different national technology outsourcing systems.

Finally, this approach challenges the notion of the role of service firms and the support infrastructure in systems of innovation approaches. Service firms and (perceived) innovation intermediaries, such as contract research and technology organisations, should no longer be seen as passive agents or 'supporting cast' in relation to innovative activity. Increasingly they are taking on more proactive, central and pivotal roles in the innovation process. Indeed the service sector is becoming an important source of demand for external research and technology services in its own right. All these issues suggest that our current view of the innovation system in advanced industrial economies needs to be challenged, if not altered, on the basis of these prospective change elements.

References

Andersen, E.F. and Lundvall, D-A. 'National Innovation Systems and the Dynamics of the Division of Labor' in Edquist, C. (ed.), *Systems of Innovation: Technologies, Institutions and Organizations*, London: Pinter, 1997; pp.242-265.

Archibugi, D., Howells, J. and Michie, J. 'Innovation Systems and Policy in a Global Economy' in Archibugi, D., Howells, J. and Michie, J. (eds.), *Innovation Policy in a Global Economy*, Cambridge: Cambridge University Press, 1999, pp.1-16.

Archibugi, D. and Simonetti, R. 'Objects and Subjects in Technological Interdependence: Towards a Framework to Monitor Innovation', *International Journal of the Economics of Business*, 1998; 5, pp.295-309.

Arora, A. and Gambardella, A. 'Complimentarity and External Linkages: The Strategies of the Large Firms in Biotechnology', *Journal of Industrial Economics*, 1990; 38, pp.361-379.

Bergen, S.A. *R&D Management: Managing Projects and New Products*, Oxford: Basil Blackwell, 1990.

Blalock, H.M. and Blalock, A. 'Toward a Clarification of System Analysis in the Social Sciences', *Philosophy of Science*, 1959; 26, pp.84-92.

Boulding, K.E. *The World as a Total System*, Beverly Hills, CA: Sage, 1985.

Byatt, I.C.R. *The British Electrical Industry, 1875-1914*, Oxford: Oxford University Press, 1979.

Brockhoff, K. 'R&D Co-operation Between Firms - A Perceived Transaction Cost Perspective', *Management Science*, 1992; 38, pp.514-524.

Breschi, S. and Malerba, F. 'Sectoral Innovation Systems: Technological Regimes,Schumpeterian Dynamics, and Spatial Boundaries' in Edquist, C. (ed.), *Systems of Innovation: Technologies, Institutions and Organizations*, London: Pinter, 1997, pp.130-156.

Buckley, P.J. and Chapman, M. 'The Perception and Measurement of Transaction Costs', *Cambridge Journal of Economics*, 1997; 21, pp.127-145.

Carlsson, B. 'The Technological System for Factory Automation: An International Comparison' in Carlsson, B. (ed.), *Technological Systems and Economic Performance: The Case of Factory Automation*, Dordrecht: Kluwer, 1995, pp.441-475.

Carlsson, B. and Stankiewicz, R. 'On the Nature, Function and Composition of Technological Systems', *Journal of Evolutionary Economics*, 1991; 1 (2), pp.93-118.

Charles, D. and Howells, J. *Technology Transfer in Europe: Public and Private Networks*, London: Belhaven Press, 1992.

Chatterje, D. 'Accessing External Sources of Technology', *Research Technology Management*, 1996; 39 (2), pp.48-56.

Chen, S-H. 'Decision-Making in Research and Development Collaboration', *Research Policy*, 1997; 26, pp.121-135.

Chesbrough, H. and Teece, D.J. 'When is Virtual Virtuous?', *Harvard Business Review*, 1996; (Jan-Feb.), pp.65-73.

Chiesa, V. and Manzini, R. 'Managing Virtual R&D Organizations: Lessons from the Pharmaceutical Industry', *International Journal of Technology Management*, 1997; 13, pp.471-485.

Coase, R. 'The Nature of the Firm', *Economica*, 1937; 4, pp.386-405.

Coombs, R. 'Core Competencies and the Strategic Management of R&D', *R&D Management*, 1996; 26, pp.345-355.

Coombs, R. and Metcalfe, J.S. 'Distributed Capabilities and the Governance of the Firm', *CRIC Discussion Paper*, No. 16, ESRC Centre for Research in Innovation and Competition, University of Manchester, Manchester, 1998.

Cook, P.L. 'Research and Development Networks and Markets in a Complex Industry: The Example of Offshore Oil Equipment' in Saunders, C.T. (ed.), *Industrial Policies and Structural Change*, London: Macmillan, 1987; pp.105-117.

Cooper, A.C. 'Spin-Offs and Technical Entrepreneurship', *IEEE Transactions on Engineering Management*, 1971; 18 (1), pp.2-6.

Croisier, B. 'The Governance of External Research: Empirical Test of Some Transaction-Cost Related Factors', *R&D Management*, 1998; 28, pp.289-298.

Davis-Blake, A. and Uzzi, B. 'Determinants of Employment Externalization: A Study of Temporary Workers and Independent Contractors', *Administrative Science Quarterly*, 1993; 38, pp.195-223.

Dickson, K., Lawton Smith, H. and Smith, S. 'The Small Firm Perspective on Inter-Firm Collaboration for Innovation' in O'Doherty, D. (ed.), *The Co-operation Phenomenon: Prospects for Small Firms and the Small Economies*, London: Graham Trotham, 1990, pp.51-70.

Dore, R. *Flexible Rigidities: Industrial Policy and Structural Adjustment in the Japanese Economy: 1970-1980*, London: Athlone Press, 1986.

Edgerton, D.E.H. and Horrocks, S.M. 'British Industrial Research and Development Before 1945', *Economic History Review*, 1994; XLVII, pp.213-238.

Elfring, T. and Baven, G. 'Outsourcing Technical Services: Stages of Development', *Long Range Planning*, 1994; 27 (5), pp.42-51.

Elfring, T. and Baven, G. 'Spinning-Off Capabilities: Competence Development in Knowledge-Intensive Services' in Sanchez, R. Heene, A. and Thomas, H. (eds.), *Dynamics of Competence-Based Competition: Theory and Practice in the New Strategic Management*, Oxford: Pergamon, 1996; pp.209-225.

Englander, E.J. 'Technology and Oliver Williamson's Transaction Cost Economics', *Journal of Economic Behavior and Organization*, 1988; 10, pp.339-353.

Foray, D. 'Generation and Distribution of Technological Knowledge: Incentives, Norms and Institutions' in Edquist, C. (ed.), *Systems of Innovation: Technologies, Institutions and Organizations*, London: Pinter, 1997, pp.64-85.

Freeman, C. *The Economics of Industrial Innovation*, London: Pinter Publishers, Second Edition, 1982.

Freeman, C. *Technology Policy and Economic Performance: Lessons from Japan*, London: Frances Pinter, 1987.

Freeman, C. 'Japan: A New National System of Innovation?' in Dosi, G., Freeman, C., Nelson, R., Silverberg, G. and Soete, L. (eds.), *Technological Change and Economic Theory*, London: Pinter, 1988, pp.330-348.

Freeman, C. 'Networks of Innovators: A Synthesis of Research Issues',*Research Policy*, 1991; 20, pp.499-514.

Freeman, C. and Soete, L. *The Economics of Industrial Innovation*, London, Pinter, Third Edition, 1997.

Gibbons, M., Limoges, C., Nowotny, H., Schwartzman, S., Scott, P. and Troiw, M. *The New Production of Knowledge: The Dynamics of Science and Research in Contemporary Societies*, London: Sage, 1994.

Graham, M.B.W. 'Corporate Research and Development: The Latest Transformation',*Technology in Society*, 1985; 7 (2-3), pp.86-102.

Håkansson, H. (ed.). *Industrial Technological Development - A Network Approach*, London: Croom Helm, 1987.

Hall, R. 'The Strategic Analysis of Intangible Resources', *Strategic Management Review*, 1992; 13, pp.135-144

Hall, A.D. and Fagen, R.E. 'Definition of Wystem', *General Systems Yearbook*, 1956; 1, pp.18-28.

Haour, G. 'Stretching the Knowledge-Base of the Enterprise Through Contract Research',*R&D Management*, 1992; 22, pp.177-182.

Häusler, J., Hohn, H-W. and Lütz, S. 'Contingencies of Innovative Networks: A Case Study of Successful Interfirm R&D Collaboration', *Research Policy*, 1994; 23, pp.47-66.

Hicks, D. and Katz, J.S. 'Science Policy for a Highly Collaborative Science System',*Science and Public Policy*, 1996; 23, pp.39-44.

Homburg, E. 'The Emergence of Research Laboratories in the Dyestuffs Industry, 1870-1900', *British Journal of the History of Science*, 1992; 25, pp.91-111.

Howells, J. 'Externalisation and the Formation of New Industrial Operations',*Area*, 1989; 21, pp.289-299.

Howells, J. 'Research and Technology Outsourcing', *Technology Analysis & Strategic Management*, 1999a; 11, pp.591-603.

Howells, J. 'Regional Innovation Systems?' in Archibugi, D., Howells, J. and Michie, J. (eds.), *Innovation Policy in a Global Economy*, Cambridge: Cambridge University Press, 1999b, pp.67-93.

Howells, J. 'Research and Technology Outsourcing and Innovation Systems: An Exploratory Analysis', *Industry and Innovation*, 1999c, 6, pp.111-129.

Jones, O. 'From Post-It Notes to Post-Modernism: Organisations, Innovation and Virtual R&D', *Research Paper Series*, 1997; 9733, Aston Business School, Aston University, Birmingham.

Katz, J.S. and Martin, B.R. 'What is Research Collaboration?', *Research Policy*, 1997; 26, pp.1-18.

Klepper, S. 'Entry, Exit, Growth and Innovation over the Product Life Cycle',*American Economic Review*, 1996; 86, pp.562-583.

Kluth, M.F. and Andersen, J.B. 'Pooling the Technology Base: TheGlobalization of European Research and Technology Organizations' in Howells, J. and Michie, J. (eds.), *Technology, Innovation and Competitiveness*, Cheltenham, UK and Lyme, US: Edward Elgar, 1997; pp.65-88.

Kreiner, K. and Schultz, M. 'Informal Collaboration in R&D: The Formation of Networks Across Organizations', *Organization Studies*, 1993; 14, pp.189-209.

Liebenau, J.M. 'International R&D in Pharmaceutical Firms in the Early Twentieth Century', *Business History*, 1984; 26, pp.329-346.

Liebenau, J.M. 'Innovation in Pharmaceuticals: Industrial R&D in the Early Twentieth Century', *Research Policy*, 1985; 14, pp.179-87.

Lowe, J. and Taylor, P. 'R&D and Technology Purchase Through Licence Agreements: Complementary Strategies and Complementary Assets', *R&D Management*, 1998; 28, pp.263-278.

Lynn, L.H., Reddy, N.M. and Aram, J.D. 'Linking Technology and Institutions: The Innovation Community Framework', *Research Policy*, 1996; 25, pp.91-106.

Lundvall, B-Å. 'Introduction' in Lundvall, B-Å. (ed.), *National Systems of Innovation: Towards a Theory of Innovation and Interactive Learning*, London: Pinter, 1992; pp.1-19.

Lundvall, B-Å. 'Technology Policy in the Learning Economy' inArchibugi, D., Howells, J. and Michie, J. (eds.), *Innovation Policy in a Global Economy*, Cambridge: Cambridge University Press, 1999; pp.19-34.

MacNeil, I. 'Contracts: Adjustment of Long-Term Economic Relations Under Classical, Neoclassical, and Relational Contract Law', *Northwestern University Law Review*, 1978; 72, pp.854-906.

MacPherson, A. 'The Role of Producer Service Outsourcing in the Innovation Performance of New York State Manufacturing Firms', *Annals of the Association of American Geographers*, 1997; 87, pp.52-71.

Malerba, F. and Orsenigo, L. 'Schumpeterian Patterns of Innovation', *Cambridge Journal of Economics*, 1995; 19, pp.47-65.

Malerba, F. and Orsenigo, L. 'Technological Regimes and Sectoral Patterns of Innovative Activities', *Industrial and Corporate Change*, 1997; 6, pp.83-117.

Marsch, U. 'Strategies for Success: Research Organization in German Chemical Companies and IG Farben until 1936', *History and Technology*, 1994; 12, pp.25-77.

Mason, G. and Wagner, K. 'High Level Skills, Knowledge Transfer and Industrial Performance: Electronics in Britain and Germany', *Industry and Innovation*, 1999; 6, pp.85-109.

Meyer-Thurow, G. 'The Industrialization of Invention: A Case Study from the German Chemical Industry', *Isis*, 1982; 73, pp.363-381.

Miller, R. 'Global R&D Networks and Large-Scale Innovations: The Case of the Automobile Industry', *Research Policy*, 1994; 23, pp.27-46.

Mowery, D.C. 'The Relationship Between Intrafirm and Contractual Forms of Industrial Research in American Manufacturing, 1900-1940', *Explorations in Economic History*, 1983; 20, pp.351-374.

Mowery, D.C. 'Firm Structure, Government Policy, and the Organization of Industrial Research: Great Britain and the United States, 1900-1950', *Business History Review*, 1984; 58, pp.504-531.

Mowery, D.C. and Rosenberg, N. *Technology and the Pursuit of Economic Growth*, Cambridge: Cambridge University Press, 1989.

Nelson, R.R. 'National Innovation Systems: A Retrospective on a Study'*Industrial and Corporate Change*, 1992; 1 (2), pp.347-374.

Nelson, R.R. and Rosenberg, N. 'Technical Innovation and National Systems' in Nelson, R.R. (ed.), *National Innovation Systems: A Comparative Analysis*, New York: Oxford University Press, 1993; pp.3-21.

Oliver, N. and Blakeborough, M. 'The Multi Firm New Product Development Process' in Grieve-Smith, J. and Michie, J. (eds.), *Innovation, Co-operation and Growth*, Oxford: Oxford University Press, 1998; pp151-160.

Ouchi, W.G. 'Markets, Bureaucracies and Clans', *Administrative Science Quarterly*, 1980; 25, pp.129-141.

Ouchi, W.G. and Bolton, M.K. 'The Logic of Joint Research and Development', *California Management Review*, 1988; 30 (3), pp.9-19.

Penrose, E. *The Theory of the Growth of the Firm*, Oxford: Basil Blackwell, 1959.

Phillips, A. 'A Theory of Interfirm Organization', *Quarterly Journal of Economics*, 1960; 74, pp.602-613.

Pisano, G.P. 'The R&D Boundaries of the Firm: An Empirical Analysis', *Administrative Science Quarterly*, 1990; 35, pp.153-176.

Pisano, G.P. 'Knowledge, Integration and the Locus of Learning: An Empirical Analysis of Process Development', *Strategic Management Journal*, 1994; 15, pp.85-100.

Pisano, G.P., Shaw, W. and Teece, D.J. 'Joint Ventures and Collaboration in the Biotechnology Industry' in Mowery, D.C. (ed.), *International Collaborative Ventures in U.S. Manufacturing*, Cambridge, Mass.: Ballinger, 1988; pp.182-222.

Powell, W.W. 'Neither Market Nor Hierarchy: Network Forms of Organization', *Research in Organizational Behaviour*, 1990; 12, pp.295-336.

Prahalad, C. and Hamel, G. 'The Core Competence of the Corporation', *Harvard Business Review*, 1990; 68, pp.79-91.

Reddy, N.M. and Rao, M.V.H. 'The Industrial Market as an Interfirm Organization', *Journal of Management Studies*, 1990; 27, pp.42-59.

Richardson, G.B. 'The Organization of Industry', *Economic Journal*, 1972; 82, pp.372-383.

Ringe, M.J. *The Contract Research Business in the United Kingdom: The European Dimension*, Luxembourg: Report to the Directorate General for Telecommunications, Information Industries and Innovation, Commission of the European Communities, 1992.

Robins, J.A. 'Organizational Economics: Notes on the Use of Transaction-Cost Theory in the Study of Organizations', *Administrative Science Quarterly*, 1987; 32, pp.68-86.

Rose, A. 'Transfers of Funds for Research and Development in Canadian Industry, 1993', *Science and Technology Redesign Project Working Paper*, 1997; ST-97-05, Statistics Canada, Ottawa.

Rothwell, R. 'External Networking and Innovation in Small and Medium-Sized Manufacturing Firms in Europe', *Technovation*, 1991; 11, pp.93-112.

Sanderson, M. 'Research and the Firm in British Industry, 1919-39', *Science Studies*, 1972; 2, pp.107-151.

Sen, F. and Rubenstein, A.H. 'An Exploration of Factors Affecting the Integration ofIn-House R&D with External Technology Acquisition Strategies of a Firm', *IEEE Transactions on Engineering Management*, 1990; 37 (4), pp.246-258.

Sikka, P. 'Analysis of In-House R&D Centres of Innovative Firms in India', *Research Policy*, 1998; 27, pp.429-433.

Simon, H. *Models of Man*, New York: John Wiley & Son, 1957.

Smith, K. 'Economic Infrastructures and Innovation Systems' in Edquist, C. (ed.), *Systems of Innovation: Technologies, Institutions and Organizations*, London: Pinter, 1997; pp.86-106.

Swann, J.P. *Academic Scientists and the Pharmaceutical Industry: Co-operative Research in Twentieth-Century America*, Boston, Mass.: John Hopkins University Press, 1989.

Thomas, L.G. 'Multifirm Strategies in the U.S. Pharmaceutical Industry' in Mowery, D.C. (ed.), *International Collaborative Ventures in U.S. Manufacturing*, Cambridge, Mass.: Ballinger, 1988; pp.147-181.

Turpin, T., Garrett-Jones, S. and Rankin, N. 'Bricoleurs and Boundary Riders: Managing Basic Research and Innovation Knowledge Networks', *R&D Management*, 1996; 26, pp.267-282.

Tyler, B.B. and Steensma, H.K. 'Evaluating Technological Collaborative Opportunities: A Cognitive Modelling Perspective', *Strategic Management Journal*, 1995; 16, pp.43-70.

Varma, R. 'Restructuring Corporate R&D: From an Autonomous to a Linkage Model', *Technology Analysis & Strategic Management*, 1993; 7, pp.231-245.

Veugelers, R. 'Internal R&D Expenditures and External TechnologySourcing', *Research Policy*, 1997; 26, pp.303-315.

von Emloh, D.A., Pearson, A.W. and Ball, D.F. 'The Use of R&D in International Process Plant Contracting', *International Journal of Technology Management*, 1994; 9 (1), pp.61-76.

Von Hippel, E. 'Co-operation Between Rivals: Informal Know-How Trading', *Research Policy*, 1987; 16, pp.291-302.

Whittington, R. 'The Changing Structure of R&D: From Centralization to Fragmentation' in Loveridge, R. and Pitt, M. (eds.), *The Strategic Management of Technological Innovation*, London: Wiley, 1990; pp.183-203.

Winter, S. 'Knowledge and Competence as Strategic Assets' in Teece, D. (ed.), *The Competitive Challenge: Strategies for Industrial Innovation and Renewal*, Cambridge, Mass: Ballinger, 1987; pp.159-183.

Williamson, O.E. 'Markets and Hierarchies: Some Elementary Considerations', *American Economic Review*, 1973; 63, pp.316-325.

Williamson, O.E. *Markets and Hierarchies: Analysis and Antitrust Implications*, New York: Free Press, 1975.

Williamson, O.E. 'Transaction Cost Economics: The Governance of Contractual Relations', *Journal of Law and Economics*, 1979; 22, pp.233-261.

Williamson, O.E. *Economic Organisation: Firms, Markets and Policy Control*, New York: Harvester Wheatsheaf, 1986.

Williamson, O.E. 'Comparative Economic Organisations: The Analysis of Discrete Structural Alternatives', *Administrative Science Quarterly*, 1991; 36, pp.269-296.

Williamson, O.E. *The Mechanisms of Governance*, Oxford: Oxford University Press, 1996.

---

[1] This chapter is based on earlier paper published in Industry and Innovation (seeHowells, 1999c). The research is supported by the UK Economic and Social Research Council. I am grateful to comments made by Bo Carlsson, Rod Coombs, Stan Metcalfe and Ian Miles on earlier drafts of the chapter.

[2] The word 'market' is hyphenated here, because as will be shown many of the links are not formal, market-based relationships but are informal and involve a high degree of non-monetary reciprocity, i.e. what may be generally termed 'non-market' relations.

Chapter 14
Horndal at Heathrow? Incremental Innovation through Procedural Change at a Congested Airport[1]

Bruce S Tether and J Stanley Metcalfe

"*To many of those involved in getting the maximum out of our airports and runways, the methods used in calculating a runway's capacity remain largely a 'black art*'"
Keith Williams, NATS Director of Operations, December 1993 [CAP 627]

14.1 INTRODUCTION

Heathrow airport is busy. Indeed, it is Europe's busiest and most congested airport (CAA, 1995). Last year (1998) over 60 million passengers used Heathrow, on 441,000 flights[2], but what is remarkable is not that Heathrow is busy, but that it keeps getting busier, despite the lack of obvious investments, especially new runways and terminal buildings. In this chapter, we explore the development of Heathrow's capacity to handle an ever-increasing number of flights, despite retaining the same basic (runway) infrastructure. Our primary interest is therefore in the development of runway capacity at Heathrow, and how this has been increased through procedural and organisational changes - so called 'soft innovations' - rather than through investments in infrastructure or hardware innovations. This said, our purpose is not to deny the role of hardware innovations, such as new technical equipment or new pieces of infrastructure, such as taxiways. Instead, we will highlight the important role of 'soft innovations' which are complementary to 'hardware' innovations but which are based on procedural knowledge and co-operation between agents.

We begin the chapter with a brief historical account of the development of air traffic activity at Heathrow, before introducing the notion of capacity and showing how this has changed over time. Later, we discuss how the development of the airport's capacity depends on co-operation between different groups of actors at the airport. Because the development of the airport's capacity is based on institutional practices and requires co-operation and co-ordination, we describe the process of innovation behind the development of capacity at Heathrow as systemic.

14.2 HEATHROW AIRPORT – A BRIEF HISTORICAL REVIEW

Heathrow airport opened in 1946, and in that year handled 63,000 passengers and 2,046 commercial flights (or 'air transport movements' (ATMs), in the industry jargon)[3]. By 1998 the airport was handling more than two and a half times that number of passengers in a single 'average' day, and two-hundred times as many flights in a year. This growth in traffic, together with that of London's three other major commercial airports[4], is shown in Figure 14.1 (for passengers) and Figure 14.2 (for ATMs).

Figure 14.1: Terminal Passengers at London's Airports

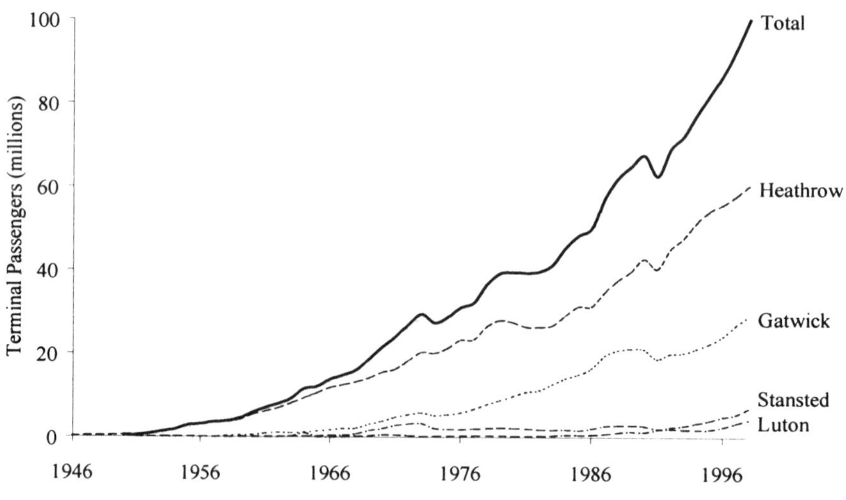

Figure 14.2: Air Transport Movements at London's Airports

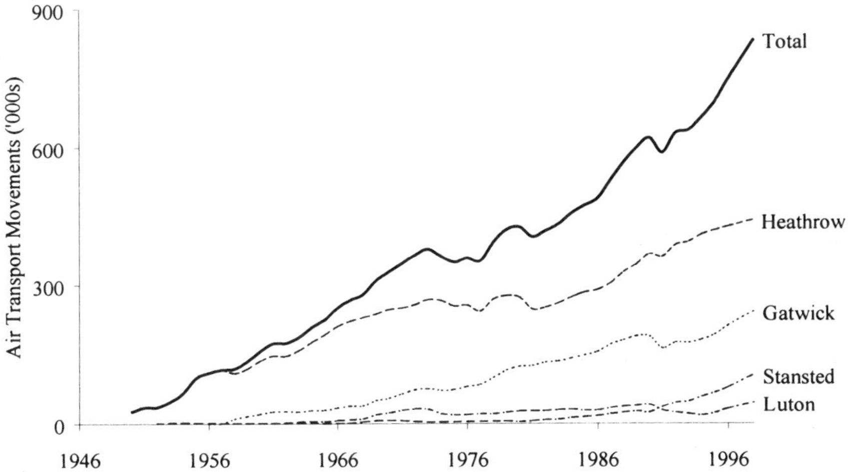

The average annual rate of growth in the number of passengers using Heathrow and London's other airports has been high, as Table 14.1 shows, but the rate of growth has been slowing since the 1960s. Because of the use of larger aircraft and higher average 'seat loadings'[5], growth in the number of ATMs has not kept pace with the growth in the number of passengers, yet the number of ATMs per year also continues to increase, at roughly 2% per annum in recent years.

Table 14.1: Traffic Growth at Heathrow and London's Airports, 1960-1998

	Heathrow Airport			London's Airports		
	Pax. (a)	ATMs (b)	a - b	Pax. (a)	ATMs (b)	a - b
1960 – 1970	11.3%	6.2%	5.1%	14.0%	7.7%	6.3%
1970 – 1980	5.9%	1.0%*	4.9%	6.3%	2.6%	3.7%
1980 – 1990	4.5%	3.3%*	1.2%	5.5%	3.8%	1.7%
1990 – 1995	4.9%	2.8%*	2.1%	4.0%	2.4%	1.6%
1995 – 1998	3.7%	1.8%*	1.9%	6.9%	6.0%	0.9%

Figures are for the mean annual growth over the periods stated. Pax. – terminal passengers; ATMs – air transport movements. *ATMs for Heathrow after 1970 include only passenger flights.

For our purposes the growth in traffic at Heathrow is not particularly interesting in and of itself, but is interesting because Heathrow's capacity to handle traffic is constrained by the (supply side) infrastructure of the airport. This leads us into a discussion of the definition of capacity at an airport, but before turning to that we briefly consider the demand side.

14.3 THE DEMAND SIDE

14.3.1 A Case of Exogenous Growth

In this analysis, we will treat demand, and the growth of demand, as exogenous. For Heathrow, we are justified in so doing for a number of reasons. Firstly, econometric studies have shown that the development of demand for air transport services in the UK has historically been closely related to the growth in GDP, and that, from a statistical perspective at least, growth in GDP largely 'explains' growth in demand for air transport services[6]. Given that the London airports have dominated air traffic activity in the UK, and that Heathrow is by far the largest airport in the London system, it follows that growth in demand for air transport activities at Heathrow is closely related to growth in GDP. Figure 14.3 shows a range of forecasts for the growth of air passenger activity at the London airports until 2015. These forecasts predict the number of air passengers will roughly double in the twenty years between 1995 and 2015.

Figure 14.3: Forecasts to 2015 of Demand in Terms of Terminal Passengers at London's Airports

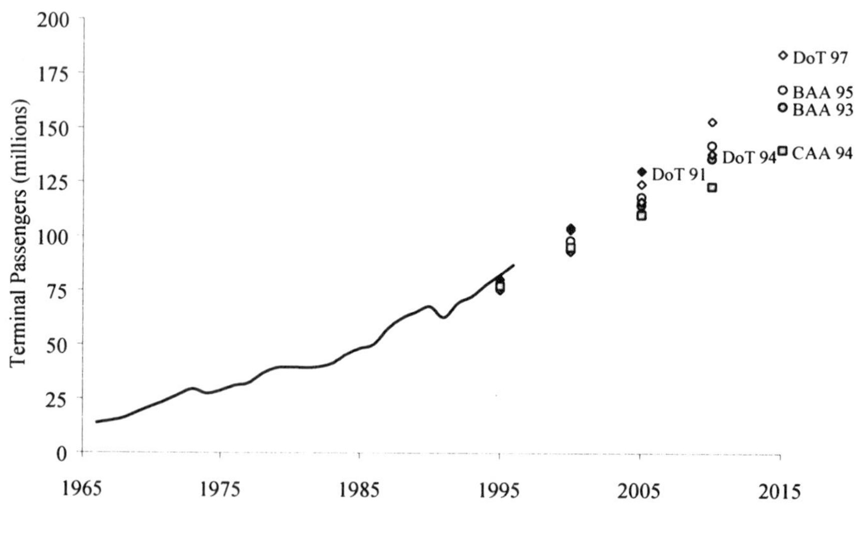

Secondly, Heathrow is the airport of preference for most airlines serving, and wishing to serve, London. Consequently, most traffic, given the choice, would prefer to use Heathrow than Gatwick, Stansted or Luton airports. As Heathrow's capacity has not developed - and has not been allowed to develop - in line with demand, the airport has become saturated, and many airlines wishing to serve London have been unable to gain adequate access to Heathrow so they have had to use Gatwick, Stansted, Luton or the City Airport instead, or not to serve London at all.

Evidence of Heathrow's saturation is found in the demand for slots exceeding the available supply, and this situation has existed for many years, particularly at peak times[7]. For example, in 1994: 'The initial demand from airlines over the summer season as a whole was for some 292,000 slots, about 10% more than the total number of available slots in the daytime hours. In the busiest hours demand exceeded the available slots by more than 30%' (CAA, 1995, para 10)[8].

Because of Heathrow's saturation, particularly at busy times of the day, much of the traffic which uses the other London airports can be considered overflow traffic from Heathrow[9]. Figures 14.1 and 14.2 show that a significant proportion of London's air traffic has been using these other airports, and particularly Gatwick, since the mid-1960s. On these bases, we treat total demand as growing, but exogenous, and in the case of Heathrow significantly in excess of the available supply. We consider the growing demand is the primary stimulus for innovation on the supply side of the airport.

14.3.2 Demand – Some Further Notes on Traffic Mix and the Size of Aircraft Used

We have just argued that demand for air traffic activities at Heathrow significantly exceeds supply, yet one interesting aspect of demand at Heathrow is that the average number of passengers per flight tends to be quite low, relative to that which is theoretically possible. Figure 14.4 shows that even in August, the busiest month, the average aircraft carried about 150 passengers, yet there are aircraft capable of carrying several hundred passengers. So why is the average so low, and why are larger aircraft not more widely used?

There are a number of reasons for the traffic mix at Heathrow. The first reason is that access to the airport is determined by the possession of airport slots. A pair of slots essentially provides the airline which holds them with the right to land and take-off using the airport's runways, and to use the ground facilities of the airport whilst the aircraft is on the ground. Most slots are retained by the airlines over time by a system of 'grandfather rights'. This institution means that the airline which held a given slot in one year has the right to retain it for the next year. Moreover, before a legal judgement earlier this year, slots could not be bought or sold openly (although they could be swapped). These institutions mean that many slots are used by airlines which, in a global sense, are not making the most efficient possible use of them.

Figure 14.4: Mean Number of Passengers per Passenger ATM at Heathrow

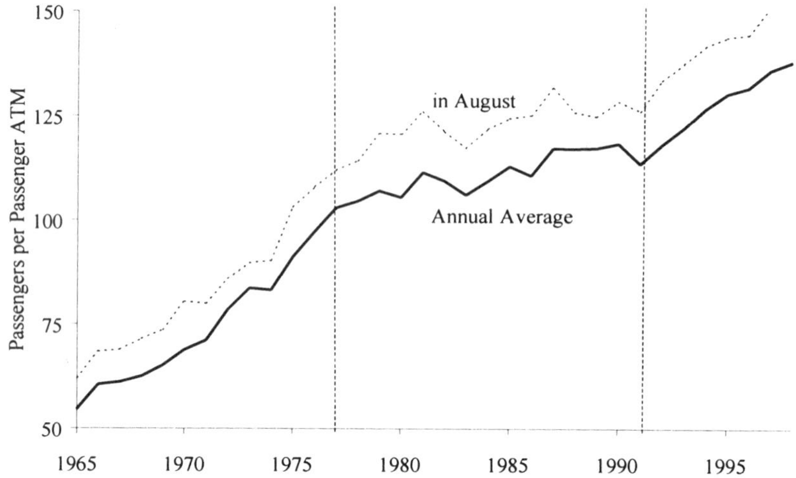

Secondly, airlines tend to use smaller aircraft on short haul feeder services that supply passengers to their long haul services for which they use large aircraft. In some cases, these feeder routes are too thin, in terms of market demand, to justify the use of larger aircraft, but smaller aircraft are also used when the frequency of a service is high. It is well known in the airline industry that frequency of service is one of the most important aspects of the service considered by passengers, particularly high paying business class passengers. Consequently, when airlines compete on a route they typically try to maintain as many flights a day (or week) as their rivals, even if this means using smaller aircraft (CAA, 1995). For example, in 1997 British Midland introduced a rival service to that of British Airways between Manchester and London Heathrow, matching British Airways in both the number of flights per day and the timing of those flights. This resulted in a loss of passengers to British Airways, and both airlines now use smaller aircraft than British Airways employed when it had a monopoly on the route. Recently, the competition authorities in Europe have been encouraging the introduction of a second or even a third airline on routes which were previously monopolies or duopolies, and this has encouraged the use of smaller aircraft.

It is also widely known in the airline industry that, as demand for a route grows over time, airlines will, if they are able, first increase the number of flights then the size of the aircraft serving the route, rather than using larger aircraft then adding more flights. This behaviour reflects passengers' preference for more frequent rather than lowest cost services, especially on short haul routes.

Consequently, whilst the total demand for air traffic activities at Heathrow can be considered an exogenous stimulus to the creation of more capacity on the supply side, the actual structure of demand is not wholly unrelated to the supply side capacity constraints. Figure 14.4 shows the steady growth in passengers per ATM during the 1970s, the period during which large wide-bodied aircraft and especially the Boeing 747 were introduced. This growth in passengers per ATM slowed during the 1980s, but has since increased again in the 1990s as Heathrow has become increasingly constrained in terms of its available capacity[10].

14.4 INTRODUCTION TO THE SUPPLY SIDE

14.4.1 Supply Side Strategies in Relation to Capacity Creation

Faced with growing demand, the airport authorities have, theoretically, a number of choices as to how to react in terms of the provision of facilities on the supply side. Firstly, they can do nothing, essentially resisting the increased demand and let the airport stagnate. Whilst this is possible, the value of each additional flight at Heathrow to both the airline which operates it and to the airport is such that the airport has every incentive to increase the capacity of the airport and to run its existing facilities as efficiently as possible, within the constraints posed by the regulatory authorities and by the practices of the airlines[11]. For this reason, we disregard the 'do nothing' option in this analysis.

The second, and perhaps most obvious response, is to build more infrastructure – a new runway, terminal building or parking area – as required. This approach has been followed in France, particularly at Charles de Gaulle airport to the north of Paris, because the airport is located in a sparsely populated area and has a large area within which to expand, and the French approach to planning transport infrastructure has been much more strategic than that in the UK.[12] Heathrow, unlike Charles de Gaulle airport in Paris, is located in a highly urbanised area of west London, and is locationally constrained by the small area of the airport. Heathrow occupies a site of 1,200 hectares, which is considerably smaller than that of several other major European airports, including Charles de Gaulle airport which occupies a site of 3,200 hectares.

Because of its relatively small site, the reluctance of the British planning system to allow the air transport industry to develop strategically and unhindered, and the lack of determination on the part of British governments of all political hues to support the industry, Heathrow's basic infrastructure has remained largely unchanged over the years. Figure 14.5 shows a plan of the airport, with the location of Terminals 1, 2 and 3 between the main runways, Terminal 4 and the cargo area to the south, the maintenance area to theeast, and the location of the proposed Terminal 5 to the west. Most importantly from our perspective, the airport has two main runways (and a rarely used cross-runway) – this is the same basic runway infrastructure that it has had since the 1960s. Indeed, previously the airport had six runways, but three of these were built over during the expansion of the central terminal area (Graves, 1998). Consequently, the emphasis at Heathrow has been on making the most of existing facilities, adding incrementally to these facilities, and 'innovating' largely on the basis of procedural changes which improve the efficiency with which the existing facilities are used. The aim of this chapter is to highlight and discuss the important role of this sort of 'soft innovation'.

Figure 14.5: Plan of Heathrow Airport

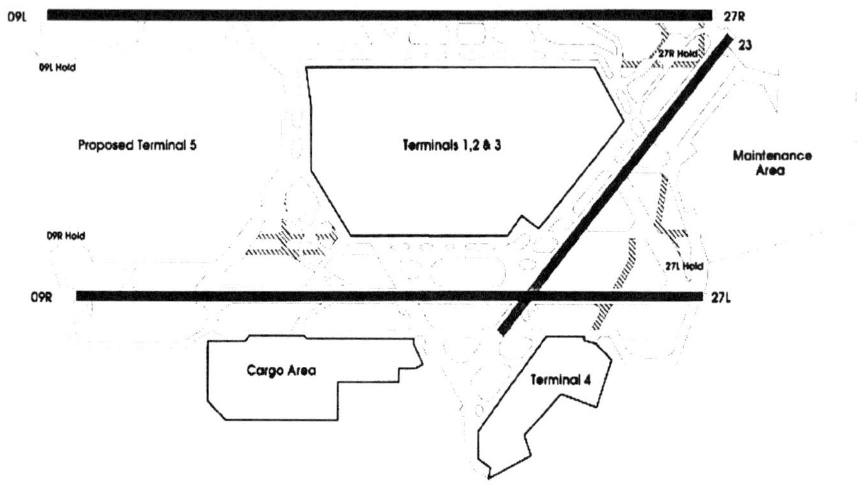

14.4.2 The Supply Side – Defining Airport and Runway Capacity

The first point to note in discussing airport capacity is that airports do not have one, but multiple constraints that impinge upon their capacity. These constraints relate to, amongst other things, the capacity of the runways to handle aircraft, the capacity of the stands to accommodate parked aircraft, the capacity of the terminals to handle passengers and the capacity of the transport infrastructure which links the airport to the outside world. Any of these may be, at a given time, the binding constraint on the capacity of the airport.

Our interest is in runway capacity, and the constraint this imposes on the level of traffic at the airport. We note, however, that runway capacity has not always been the primary, or the binding constraint at Heathrow. Indeed, presently the BAA, the airport's owner, argue that it is terminal and stand capacity are the primary constraints at Heathrow; Terminal 5, is their proposed solution to these problems. Nevertheless, runway capacity is certainly a central element in determining the overall capacity of an airport, particularly in terms of air transport movements.

The maximum number of aircraft that a runway can handle, for a given set of conditions, such as the weather and traffic mix, is known as the runway's 'service rate'. The 'service rate' does not equate with capacity, because it takes no account of delays. Instead, runway capacity is 'the number of aircraft movements that may be scheduled to use a runway such that their average delay during the airport's busy period does not exceed a specified value' (CAA, 1993). Until 1990, the agreed delay standard was an average of 5 minutes, but in 1990 this was increased to 10 minutes.

Before assessing further how runway capacity is determined, we consider the 'environmental constraints' which have a bearing on runway capacity. These environmental constraints are, to a greater or lesser extent, exogenous to the determination of runway capacity, but are none the less important in influencing the way in which the runways are operated (in a sub-optimal way).

14.4.3 Runway Capacity – Environmental Constraints

Although a private company, Heathrow is not permitted to operate in the most efficient possible way, because of legal restrictions and by agreement with the airlines and the communities around the airport.

14.4.3.1 Safety Related Constraints

Aircraft are subject to separation rules, which ensure aircraft are kept a safe distance apart. For aircraft landing and taking-off, separations are enforced for three main reasons:[13]

- Firstly to ensure that the leading aircraft has successfully completed a manoeuvre before the following aircraft is committed to undertake that manoeuvre.
- Secondly, aircraft are separated for wake vortex reasons. Large aircraft especially have powerful engines that create turbulence in the aircraft's wake. Generally the larger the aircraft the greater the turbulence it can endure, so when a small aircraft follows a large aircraft it has to follow at a greater distance than when a large (or 'Heavy') aircraft follows another large (or 'Heavy') aircraft. Aircraft are categorised by wake vortex separation criteria for which separation minima are specified on departure and on arrival (CAA, 1993).
- Thirdly, aircraft operate at different speeds, with turbo-prop aircraft being 'slow' compared with 'fast' jets, and it is necessary to enforce longer separations to allow slow aircraft to manoeuvre in an environment in which fast aircraft dominate.

For our purposes, the wake vortex separations are the most significant separations. As we shall show shortly, the cumulative significance of these can be minimised by clever procedures. However, it is notable that wake vortex separations 'are based more on experience than scientific research' (CAA, 1993). They are, however, now under-review, and a more scientific approach is being taken, which implies a shift from knowledge based on experience to knowledge based on scientific principles in the determination of these separations[14].

14.4.3.2 Noise Related Constraints

As Figure 14.5 shows, Heathrow has two main runways, and these can be operated in different modes. In *segregated mode*, one of the runways is used for landing aircraft, whilst the other is used for taking-off. In *mixed mode* both runways are used for aircraft landing and taking-off, with departing aircraft interleaved between arrivals. It is known that more aircraft are able to use the runways within a given time period in mixed mode than in segregated mode, and that mixed mode is therefore more efficient. A study in 1993 estimated that Heathrow could handle 92 aircraft movements (arrivals and departures) in an hour if mixed mode were used, compared with 82 movements per hour in segregated mode (NATS et al, 1994). 'The major disadvantage of segregated mode is that – because of aircraft vortex wake - streams of arriving and departing aircraft have to be comparatively widely spaced and therefore potential runway capacity is wasted. Moreover, its lack of flexibility means that capacity may be unused on one runway whilst excess arrivals or departures on the other runway suffer delays' (NATS et al, 1994, para 2.2).

Heathrow operates in segregated mode because this imposes less noise pollution on the surrounding communities[15]. Also for noise reasons, aircraft departing Heathrow must follow a restricted number of paths when leaving the airport, called Noise Preferential Routes (or NPRs) under 4,000 feet. Because of the need to separate traffic, NPRs also restrict the possible number of departures in a given period, but again clever procedures can make the best of this situation. Later in this chapter, we will discuss how the airport is able to squeeze out extra capacity within these restrictions.

14.4.3.3 Constraints related to User Requirements

Apart from constraints imposed by safety considerations, and by the need to placate the neighbouring communities with respect to the noise nuisance, airport operations are also compromised by the need to satisfy the users. On the airside, that means the airlines.

For airlines, delays are a significant problem, not just because they annoy passengers, but also because they waste fuel and make idle valuable assets. However, if no delays are allowed at a busy airport, this would severely reduce the operating efficiency of the airport, for there is a well understood relationship between the average delay and the number of aircraft movements the runways are able to handle (see Figure 14.6):

> "As the volume of traffic scheduled to use the runway increases, delays remain low until a critical point is reached. As a guide, this critical point occurs when aircraft scheduling reaches about 80% of the runway's 'service rate'. At this stage, delays begin to increase rapidly for relatively small increases in movements. This is because the number of aircraft in the arrival or departure queue is building gradually to the point where long queues develop, and with more aircraft waiting for longer and longer periods, the average delay soars. The point at which this dramatic rise in delays starts – and the rate at which delays then continue to increase – depends on the length of the busy period.[16] The longer the busy period, the earlier and steeper is the increase in average delays and the greater the 'peak delay'" (NATS, 1993).

Figure 14.6: The Relationship between Runway Capacity, Delays and Length of Busy Period

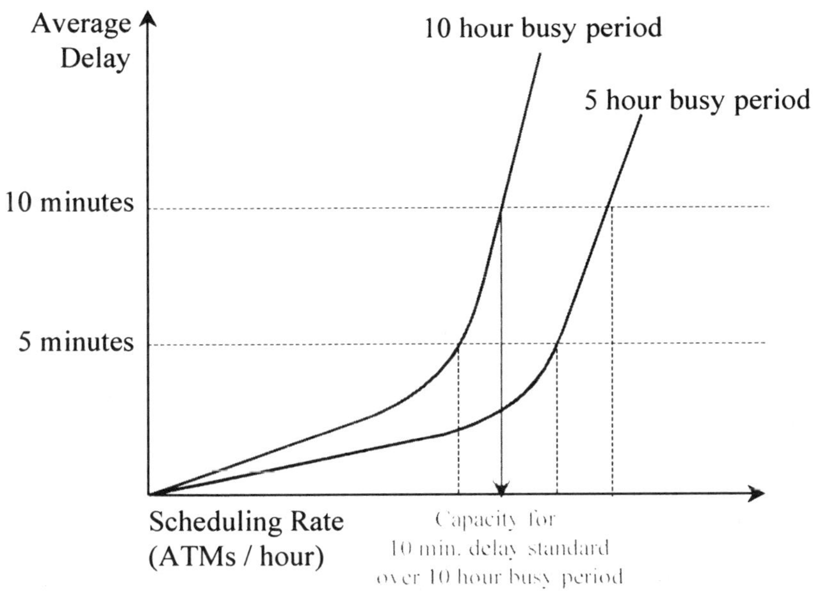

The effect of increasing activity on average delays was noted by a study in 1988: “[A]n increase at Heathrow by one movement over the declared capacity for arrivals or departures would produce a rise in the average delay from 5 to 7 - 8 minutes” (CAA, 1988, para 4.1).

A compromise has therefore to be reached between the scheduling rate – the number of movements the runways can handle in an hour – and the average delay the airlines will tolerate. Between the mid-1970s and 1990, the agreement between the airlines, the airport and air traffic control authorities was that the average delay criteria should be 5 minutes, but in 1990 this standard became 10 minutes.

14.4.4 The Supply Side – Defining Runway Capacity

After consideration of the geometry of the airport, and the safety and environmental constraints within which air traffic can operate, runway capacity is primarily determined by the length of the delays regarded as acceptable by the airlines. Thus:

> “The capacity of a runway is defined as that level of average demand, over a busy period of defined length which on being serviced gives rise to average delays of a specified amount. The capacity, consequently, depends not only upon the ‘hardware’ of the runway, the queue discipline and the air traffic control procedures and separation standards, but also on the length of the busy period and on the level of delay regarded as tolerable by the users of the system, that is, predominantly, commercial airline operators” (CAA, 1983).

We will discuss the importance of ‘queue discipline’ and ‘air traffic control procedures’ later in the chapter, as it is innovation in these that have given rise to a considerable amount of increased capacity through ‘soft innovation’, but also of note from the above quote is the significance of capacity being defined over a ‘busy period’. The busy period is the period over which aircraft are arriving and/or departing in a continuous stream, so there is always at least one aircraft demanding to use the runway(s).

A notable feature of this definition is that it shows that runway capacity is largely based on experience, and experience which only arises when the airport reaches a level of activity at which demand exceeds the ability of the airport to handle that demand. It was only in the 1970s that the level of air traffic activity at Heathrow and Gatwick became sufficiently high that capacity studies began. Thus capacity is not fully known in the abstract, but is instead ‘discovered’ through a learning process over time, which is closely related to the practical experience of operating the airport. We discuss this further below (in Section 14.4.4.1). The techniques for assessing runway capacity have also developed over time, for example with the introduction of increasingly sophisticated computer simulation.

In the UK, the determination of runway capacity is a distributed process, in which air traffic control, the airport and the airlines all play a role.

> "Assessment has a number of stages. First there is the monitoring of actual operations to record exactly what is happening, and the analysis of that data. The level of traffic, delays and time intervals for aircraft operations are monitored by observations at Heathrow and Gatwick by DORA personnel[17]. The next step is the use of fast-time computer simulation – using that data – to calculate the levels of delays that would occur for a given schedule. The result is a count of the movements that used the runway in each hour together with delay estimates which translate into 'capacity profiles'.
>
> Following an assessment of the operational implications by NATS ATC [Air Traffic Control] managers, capacity 'profiles' are presented and agreed with the airport's scheduling committees [which are composed of airline representatives] who then use them to plan their schedules. In considering the various 'profiles' available, the scheduling committee and airport operator have the opportunity to consult with NATS and airlines and take decisions on the patterns of their schedules, on the traffic mix and on the level of acceptable delay in order to meet the optimum capacity profile which will meet the needs of all parties" (CAA, 1993).

Our interest in runway capacity is twofold, although these interests are strongly inter-linked:

1. Incremental increases in capacity, despite fears of saturation
2. The distributed nature of capacity creation

14.4.4.1 *Incremental Increases in Capacity, in the Face of Fears of Saturation*

Firstly, we observe that the capacity of the runways at Heathrow (and Gatwick) has increased steadily over time (although throughout the period of our study there has been surplus demand – particularly at Heathrow). Figure 14.7 and Table 14.2 show this development in the airport's capacity over time. In 1981 the airport was able to handle 67 air transport movements in each hour, whilst by 1998 this figure, whilst now varying across the day, reached 84 movements per hour. The table shows that across the fourteen hours of the day between 0600 and 1959 GMT, the number of available slots increased by 20% over the 18 year period. Although there has been some deterioration in service – particularly with the acceptance of a longer delay criteria – this growth in capacity is quite remarkable, given the context of continually expressed fears that the capacity of London's airports, and Heathrow in particular, was becoming severely limited.

This concern is reflected in a report by the CAA and a Parliamentary Inquiry into Airport Capacity in the South East of England. In 1986, for example, a high level committee was established to investigate runway utilisation at Heathrow and Gatwick. This committee was chaired by the Chairman of the CAA, and included airline industry 'heavy-weights', such as Micheal Bishop, chairman of British Midland.

> "The Committee saw its task as being to demonstrate conclusively (or to put in hand studies which would ultimately demonstrate conclusively) whether or not it was possible to increase the capacities of Heathrow and Gatwick beyond the levels currently forseen" (CAA, 1988, para 3.1).

Figure 14.7: Heathrow Summer Scheduling Limits 1985-1998

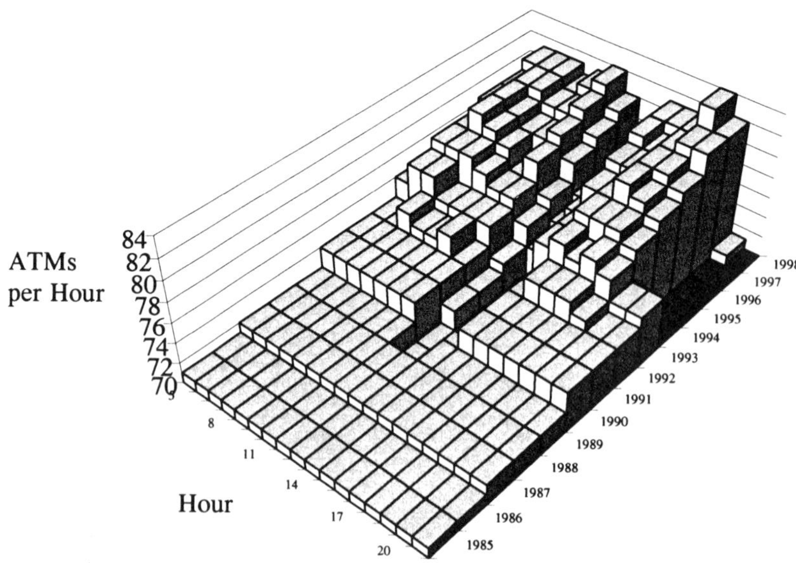

Table 14.2: Development of Capacity at Heathrow – 14 hour day Summer Slots Available, 1981 – 1998

Year	1981	1982	1983	1984	1985	1986	1987	1988	1989
Slots: 14 hour total	938	966	980	994	994	994	1,008	1,008	1,000
Index 1981 = 100	100	103	104	106	106	106	107	107	107
Year	1990	1991	1992	1993	1994	1995	1996	1997	1998
Slots: 14 hour total	1,020	1,024	1,037	1,061	1,084	1,099	1,110	1,119	1,127
Index, 1981 = 100	109	109	111	113	116	117	118	119	120

After a year of research, the Commission reported various means by which capacity could be increased, but by only one arrival and one departure movement per hour, and that only in part of the day[18]. Beyond this, and although the Committee recognised that the capacity of Heathrow had increased over the years, the report stated firmly that there was little room for further improvement, and that consequently the government should investigate, as a matter of urgency, the issue of providing further air traffic capacity in the south east of England:

> "Declared hourly capacities have shown a gradual increase at both Heathrow and Gatwick over the last decade or so. This has been the result of changes to ATC procedures, including changes in the vortex wake classification scheme and a general tightening of achieved separations, and, to a lesser extent, improvements in the ground layout at both airports which BAA has made" (para 4.4).

> "After a detailed and critical examination of the technical issues, the Committee has determined that, within its terms of reference, the scope for increasing runway capacities of Heathrow and Gatwick beyond that now envisaged is necessarily small" (para 8.1).

> "The pressure of demand on runway capacity at Heathrow and Gatwick combined with current forecasts of passenger demand in the South East (of England) strongly suggest that unless Stansted airport is allowed to operate to its full potential as a single runway airport, capacity in the London area will be insufficient to meet likely traffic growth beyond the early to mid 1990s. Even with the full utilisation of Stansted's single runway there is a strong possibility that runway capacity will be insufficient by the turn of the century" (para 8.5).

After this, a number of further studies were undertaken, including a major study on 'Runway Capacity to Serve the South East (RUCATSE) (DoT, 1993), a runway capacity enhancement study at Heathrow (NATS et al, 1994), and a Parliamentary Select Committee Inquiry into UK Airport Capacity (HMSO, 1996). Two important points emerge from these studies:

1. Firstly, the capacity of the runways at Heathrow and Gatwick are not known in an abstact theoretical sense, instead they are understood in relation to the practical experience of operating the runways. An example helps make the point. In response to fears that runway capacity in the London area would soon become inadequate, the airport's owner (the BAA), together with the air traffic control service (NATS) and the civil aviation authority (CAA) were asked to assess the potential capacity of a second 'close parallel' runway at Gatwick. A 'close parallel' is a runway which is too close to the first runway to be operated with complete independence from the first runway, and it is 'known' that the capacity of a close parallel is less than a fully independent runway. However, what was not known was what that capacity would be. The BAA, NATs and the CAA conducted trials over a three year period to determine the capacity of a 'close parallel' runway at Gatwick. The conducting of these trials shows that the capacity of a runway is not simply known by experts in advance.

2. The second important point arising from these studies is that whilst it is generally stressed that scope for further capacity formation at Heathrow and Gatwick (with their existing infrastructures and operating restrictions) is extremely limited, and the airports are close to saturation, the studies have also normally found some ways through which a small amount of increased capacity can be provided through incremental change, which depend rather more on changes to operating procedures than to the adoption of new technologies. For example, the Heathrow Airport Runway Capacity Enhancement Study (NATS et al, 1994) found:

   > "Our examination of segregated mode shows that measures could be introduced to increase both arrival and departure capacity from 39 to 41 movements per hour (82 movements per hour in total), subject to changes to NPRs [Noise Preferential Routes] being approved for new SIDs [Standard Instrument Departure routes]" (para 2.10).
   >
   > "We concluded that there is unlikely to be any further strategic improvement in peak hour runway capacity at Heathrow in segregated mode beyond 82 movements per hour" (para 2.22).

This combination of incremental improvements to capacity using existing infrastructures and operating procedures with an unknown ultimate capacity makes planning difficult – or at least makes it easier for politicians to defer the difficult decision as to when and where new runway capacity should be built. Of course because new capacity is not provided through the construction of new runways, further incremental change using the existing facilities is encouraged.

14.4.4.2 *The Distributed Process of Capacity Creation*

Secondly, we are interested in the process of capacity creation, and in particular the distributed nature of this process, which is illustrated by the fact that although the air traffic control (ATC) authority bears the ultimate responsibility for deciding the capacity levels that can safely be accommodated (because it bears a legal responsibility for safely operating the procedures affecting runway capacity), 'the declaration process involves considerable iterative discussion between airport ATC, the airport operator and the airport scheduling committee' (CAA, 1993). Thus the operation of airports, particularly on the air-side, depends on the interaction and co-operation of the airport operator, the airlines and the safety authorities, particularly air traffic control. Over time, institutions develop by which these different agents understand their role and what is expected of them. Sometimes these expectations are incorporated into legal requirements, but others are norms which are not enshrined in law.

The case of slots is an important example, for airport slots are not really owned but they are controlled by the airlines which have the right to keep their slots from one year to the next under the principle of 'grandfather rights'. A slots is an important institution, which provides the airline which holds it with precedence in the use of the runways within a limited time period around the slot-time. Aircraft which miss their slots times lose this privilege, and must wait for another slot to become available. The existence of slots, and associated rights, in some ways limits the scope for efficiency savings for the airport operators and air traffic control.

We shall show later that if aircraft are processed in strict 'first come first served' order the efficiency of the runways will be less than if the airport operator/air traffic control have some scope for re-ordering the traffic into a more efficient sequence. Principles, such as 'first come first served', can be hard to break, and individual airlines must be persuaded that they will not lose out by such changes, possibly through the introduction of new rules that limit the extent of the re-ordering (such as limiting the number of times an individual aircraft can be overtaken in the queue to land or depart).

We emphasise that while incremental innovation through procedural change often depends on the creation of new rules or institutions, new rules are difficult to negotiate because of the vested (and different) interests of the parties involved. Air transport is a global industry, and airlines can be reluctant to change their 'standard operating practices' (or SOPs) to suit the needs of individual airports[19] – particularly as there is no direct link between good performance and rewards in terms of lower airport changes or the award of more slots as these are made available. The performance of the airlines is observed and compared, and discussions are held with those that are less efficient in their use of the runways. Interestingly, however, there is no attempt to single out individual pilots, and appeals are made through the Chief Pilots on the basis of professionalism and good practice, rather than on the basis of penalties and rewards, because of the rule that everyone must be treated equally[20].

The existence of SOPs is often reinforced by industry wide standards which are agreed at the International Civil Aviation Organisation (ICAO) and other international bodies, but compliance with ICAO standard operating procedures is important for insurance purposes. Airlines are often reluctant to deviate from ICAO standards because their insurance and licensing can depend on compliance with these standards.

Heathrow, and a limited number of other airports around the world, have sought to improve their efficiency which has meant trying to find the most efficient way of operating within the ICAO standard operating procedures. This has necessitated the negotiation of a new set of rules, called 'High Intensity Runway Operations' (HIRO). These provide a number of special operating procedures. For example, an aircraft on final approach may be given clearance to land whilst the preceding aircraft is still on the runway, so long as the first aircraft is at least 2,500 meters from the threshold when the landing aircraft crosses the threshold (Duke, 1996). This contrasts with normal operating procedures whereby the first aircraft must have cleared the runway before the second is cleared to land. As most airports do not have the same need for efficiency related measures as Heathrow, there is a reluctance on the part of many airport operators and many unaffected airlines to renegotiate the existing practices.

These changes in practice must also be negotiated because they may have cost implications, and an example helps make this point. Runways operate most efficiently when the occupancy time of each aircraft is minimised, and the gap between aircraft operating the runway is minimised. Clearly, one way to minimise runway occupancy is to request that aircraft brake severely on landing, so that they slow down rapidly and can leave the runway quickly. Apart from not wanting to terrify the passengers, such a request may be refused by the airlines because of the costs it imposes on them – through increased brake-wear, the cost of new aircraft brakes being considerable[21].

The key point of this section is that changes in operating procedures upon which efficiency savings largely depend are themselves the result of negotiations between the different agents involved. Many lines of change are resisted, because vested interests do not make them acceptable to one or more of the interested parties. Thus not only is it important to know what might potentially be done to improve the efficiency of runway operations, but it is also important to know whether a change is likely to be acceptable, and if not whether there are ways of circumventing the objection. In the following section, we will discuss various 'innovations' which have gradually been introduced in the last twenty or so years which are largely responsible for improving the efficiency with which Heathrow's runways are operated, but in general these 'innovations' do not depend on new ideas, instead they depend on changing procedures and institutions which tend to be very sticky to change. For example, rapid exit taxiways (or RETS) were 'invented' through basic research at the University of Californian in the late 1950 (Hornonjeff et al, 1958), but were only adopted slowly, and little was done about measuring the effectiveness and optimal positioning of RETs until the late 1970s and 1980s (Warskow, 1970).

Clearly, there is a relationship between the highly institutionalised way in which airports, and runways in particular, are operated, and the lack of definite knowledge about an individual runway's ultimate capacity – for capacity depends on these institutions and the extent to which they may be changed.

14.5 'INNOVATIONS' RELATED TO RUNWAY UTILISATION

In this section, we will finally turn to some specific 'innovations' which have been introduced and which have contributed to the incremental increases in runway capacity at Heathrow over the last twenty or so years, but we begin by showing evidence of improvements in runway occupancy times and inter-aircraft separation times which are the keys to improving runway capacity.

14.5.1 Reducing Separations and Runway Occupancy – the Keys to Efficiency Savings

The key to maximising capacity – in terms of air traffic movements within a given time (usually an hour) - is maximising throughput of aircraft in a given period of time, and the keys to this are twofold: minimising separations between aircraft using the runways, and minimising the time taken by each aircraft on the runways.

14.5.1.1 *Reducing Separations between Aircraft*

The time spacings between aircraft when arrivals follow each other directly is partially determined by the wake vortex criteria, such that when a Heavy aircraft leads a Light aircraft the light aircraft must stay further back than when a Heavy follows a Heavy, for example, or when a Heavy follows a Light. Table 14.3 provides separation matrices, derived through visual observations, for each combination of leading and following aircraft. These matrices are important for bunching, an 'innovation' which we shall discuss shortly. Our immediate interest is in how the separations have reduced, albeit slightly, over the years. For example, between 1980 and 1988/89 a saving of between 7 and 8% was found in the inter-arrival time between two Medium aircraft, but significant savings of over 10% were common. The only increases were with Light aircraft following Mediums in 1988 and Light aircraft following other Light aircraft in 1989, but these may be 'rogue' results, because these particular means are based on very few observations.

Table 14.3: Mean Separation Times between Westerly Arrivals at Heathrow – 1980, 1988 and 1989

Separation Matrix – 1980

LEAD \ FOLLOWING	Heavy	Medium	Light
Heavy	104	142	178
Medium	99	94	128
Light	91	92	68

- Times are in Seconds

Separation Matrix - 1988

LEAD \ FOLLOWING	Heavy	Medium	Light
Heavy	100.7	132.1	None
Medium	85.8	87.5	146.6
Light	82.4	81.4	None

Improvement cf. 1980

LEAD \ FOLLOWING	Heavy	Medium	Light
Heavy	-3%	-7%	n.a.
Medium	-13%	-7%	+15%
Light	-9%	-12%	n.a.

Separation Matrix - 1989

LEAD \ FOLLOWING	Heavy	Medium	Light
Heavy	98.8	122.7	170.0
Medium	83.8	86.9	122.5
Light	74.8	81.4	71.0

Improvement cf. 1980

LEAD \ FOLLOWING	Heavy	Medium	Light
Heavy	-5%	-14%	-4%
Medium	-15%	-8%	-4%
Light	-18%	-12%	+4%

The key point we wish to stress is the apparent improvement in spacings between aircraft over time. Whilst it is true that the improvements are relatively small, this is a business in which small changes can yield significant benefits. Moreover, these small improvements are consistent with our story of incremental innovation, based largely on procedural change. As the CAA points out, the accuracy of spacings is very important to the efficient use of the runways, and although air traffic control bears ultimate responsibility for maintaining separations[22], the achieved spacings also depend on the co-operation and skills of the pilots, and on technology, as well as on the skills of the air traffic controllers:

> "The accuracy with which the required spacings can be achieved in practice also affects capacity substantially. Ideally, the actual spacings achieved between aircraft using the runway should match the required safe separations as closely as possible. This depends on a combination of controller skill, pilot skill and the use of ground-based and aircraft technology to achieve the required spacings" (CAA, 1993 – our emphasis).

In relation to this, there is evidence to suggests that controllers learn to separate aircraft more efficiently over time, although this is an unconscious learning-by-doing process, made more difficult by the fact that at the time controllers set up the second arrival behind the first they do not know exactly how long the first arrival will occupy the runway as this is affected by different aircraft operating characteristics and pilot performance (CAA, 1993)[23].

14.5.1.2 *Reducing Runway Occupancy Times*

The separations just discussed also depend on runway occupancy times – if runway occupancy times can be reduced, separation times can be reduced, and again, small improvements matter. According to an official report: 'saving an average of 5 seconds on every aircraft's runway occupancy would add another 1 - 1½ movements per hour' (CAA, 1993), however, industry experts have told us the gain can be double this, at nearly 3 additional movements per hour.

Table 14.4 below shows the runway occupancy times for runway 27R in 1980, 1984 and sometime around 1994[24]. This Table appears to provide evidence for a gradual reduction in runway occupancy times, with aircraft making greater use of the early exits rather than progressing further down the runway (note the decreased use of the exits in Block 9 and 11, and the increased use exits in Block 13). Moreover, occupancy times for aircraft exiting in most of the blocks appear to be declining.

We should note that direct comparisons are difficult, due to changes in the traffic mix – i.e. the types of aircraft using the runways – larger aircraft tend to spend longer on the runways than smaller aircraft – as the lower part of the Table shows. Although we are still gathering information on this, we would expect that changes in the traffic mix would, ceteris paribus, increase runway occupancy times, as more larger aircraft are using the runways. Against this, modern aircraft have better brakes allowing them to slow down sooner.

Finally, the bottom part of the Table is interesting because for each category of aircraft a large proportion are identified as having 'scope for improvement' in their use of the exits and consequently runway occupancy times. This illustrates the interest of the airport operator in gathering such information in order to improve the efficiency of the system, particularly because pilot skills and technique rather than technology is seen as the key to runway occupancy times (CAA, 1993). Since 1993, the airport has run a 'runway occupancy awareness campaign' to make pilots and air traffic controllers aware of the significance of minimising runway occupancy.

Table 14.4: Runway Occupancy on Heathrow 27R – developments between 1980 and c.1994

Runway 27R – Exit Blocks	9	11	13	14	15	16
Proportion per Exit, c. 1994	17%	22%	37%	18%	5%	1%
Proportion per Exit, 1984	22%	37%	14%	18%	6%	3%
Proportion per Exit, 1980	23%	34%	26%	14%	3%	n.a.
Seconds on Runway c. 1994	70 - 73	50 - 60	46 - 52	43 - 49	32 – 41	33 -39
Seconds on Runway (mean), 1984	77	58	49	48	41	36
Seconds on Runway (mean), 1980	78	61	50	50	45	n.a.
By Aircraft Type – c. 1994 data:						
B747, Concorde	85%*	15%				
A300/A310, B767, MD11	36%*	38%*	26%			
B737, MD80, A320, B727	5%*	26%*	47%*	19%	3%	
B757		9%*	43%*	40%	8%	
BA46, FK28, DH7			55%*	32%	9%	4%
FK27, FK50, ATP				28%*	36%	28%

- scope for improvement in runway occupancy

Table drawn from Duke, 1996, except 1980 data (CAA Paper 81014) and 1984 data (CAA Paper 85010)

14.5.2 Specific 'Innovations' in Runway Occupancy

In this sub-section, we discuss some specific 'innovations' which have been introduced and which have contributed to the incremental increases in runway capacity at Heathrow over the last twenty or so years. We stress that at present we do not have a full picture of the relative importance of these innovations. We hope to develop a more complete understanding of their relative importance in the course of future research.

14.5.2.1 *Rapid Exit (and Access) Taxiways and Procedures*

Traditionally, exit taxiways are perpendicular to the runway, but it has long been recognised that one way of reducing runway occupancy times, and thereby increasing runway capacity, is to design runway exit taxiways at 30° to the runway which allow aircraft to exit at a much higher speed than with conventional exits (Hornonjeff et al, 1958; Warskow, 1970). An expert at the BAA claims the optimal positioning of Rapid Exit Taxiways (RETs) is worth 2 to 3 additional movements per hour.

But the introduction and use of RETs appears to be a classic example of how procedural change is more important than the hardware component of innovation, for RETs can be built but pilots (and airlines) have to be persuaded to use them effectively. Gradually, RETs have become standardised, and indeed the ICAO publishes a book of standard designs for RETs.[25] Standardisation has encouraged their widespread use. The introduction of indicator lighting systems is also encouraging the use of the earliest possible rapid exit.

Efforts are also being made to introduce Rapid Access procedures, to minimise runway occupancy on departure. These appear to be part of the High Intensity Runway Operation rules in force at Heathrow, whereby aircraft line-up on the runway as soon as the preceding aircraft has commenced its take-off roll. Moreover, pilots are encouraged to have completed all cockpit checks prior to line-up, and to minimise the number of checks requiring completion on the runway. The aim is to ensure pilots commence the take-off roll immediately take-off clearance is issued.

14.5.2.2 *The Re-ordering, or 'Bunching', of Arrivals*

It has long been recognised that whilst a runway can be regarded as a single server queuing system, it need not have a first-come first serve queue discipline, and by skilful choice of the next movement to be handled, the runway controller can increase the number of movements handled in a given period of time (i.e. the rate of service) (CAA, 1983)[26].

The controller has the opportunity to re-order, or re-sequence, the aircraft, rather than treat each aircraft on a random 'first come first served basis', and this can provide some significant benefits in terms of the efficiency with which the runways are used. The key to re-sequencing is the fact that separation differences between different types of aircraft are not the same. (Note that if there were only one type of aircraft using the airport there would be no benefit from re-sequencing). The controllers aim is therefore to minimise the distance between (in this case) arrivals. Re-ordering is not simple, partly because it is based on rules, but also because the strategy used can vary with the traffic mix and with the separation matrix.

However, to illustrate the benefits of re-sequencing, we developed a simple simulation model which re-ordered arriving aircraft in a simple way. This model considered just three types of aircraft (Heavy, Medium and Small), and assumed all aircraft were in the same speed category; thus time can substitute directly for distance. We assumed a traffic mix of 27% Heavy, 67% Medium, and 6% Small, roughly similar to that at Heathrow in 1993. We also assumed the time-based inter-arrival separation matrix shown below in Table 14.5. According to this matrix, the separations were greatest when a Small Aircraft followed a Heavy aircraft – due to wake vortex problems - (140 seconds) and were smallest when any type of aircraft followed a Small aircraft (80 seconds). The strategy of the model was therefore simple – as far as possible, the controller (i.e. computer) should avoid Small aircraft following Heavy aircraft.

Table 14.5: Arrivals Sequencing Simulation – a. Assumed Separation Matrix

		Following Aircraft		
		Heavy	Medium	Small
Leading Aircraft	Heavy	100	120	150
	Medium	85	85	100
	Small	80	80	80

b. Summary of Results of Simulation Exercise over 50 Runs

	Arrivals in Random Order (a) (Cumulated Time)	Sequenced Arrival Order (Cumulated Time)	Runway Time Saved (b)	(b) as a percentage of (a)
Mean	9377.6	9027.3	350.3	3.7%
S.D.	47.0	20.4		
Maximum	9495	9070	425.0	4.5%
Minimum	9290	8970	320.0	3.4%

The first step was to allow the controller (i.e. computer) to 'look ahead' a certain number of aircraft. We chose five. Thus instead of having to land the next arrival in strict 'first come – first served' order, the controller is able to assess the next five flights due to land. Initially, the controller selects the first arrival amongst the largest category by size of aircraft available. This lands and is replaced in the pool of five possibilities by the next arrival in the random sequence. The controller then lands any other aircraft of the same size category in the pool of five, each time replacing the aircraft which lands with the next to arrive in the random sequence. When the supply of aircraft of the size being picked is exhausted, the controller moves to the next category of aircraft smaller than that from which it has been selecting (i.e. Medium following Heavy, or Small following Medium). If, however, it has been selecting Small aircraft, there is no smaller category, and it instead has to select the first arrival amongst the largest available category of aircraft by size.

This simple procedure reduces the number of times a Small aircraft follows a Heavy aircraft in the landing sequence, and it is remarkable that such a simple procedure can yield quite significant time savings. Using the inter-arrival separation matrix in Table 14.5a, and allowing the controller to look ahead just five aircraft, we assessed the total landing time of 100 aircraft (in a traffic mix 27% Heavy, 67% Medium and 6% Small) arriving in a random order and compared this with the total time taken by these aircraft if they were re-ordered, or 'bunched', by the procedure described above. Table 14.5b provides a summary of the results after 50 simulation runs.

This shows that over the 100 arrivals, an average time saving of 3.7% was achieved by 'bunching' the aircraft compared with the time the aircraft would have taken had they landed in a random, 'first come-first served', order. Moreover, apart from this significant time saving, the other notable feature of the simulation runs was that the standard deviation between the total times taken was substantially reduced, and the difference between the largest amount of time taken and the smallest amount of time taken was reduced by half (from 205 seconds to 100 seconds). Thus, not only was time saved (thus allowing greater capacity to be handled within the time available), but the pattern of activity also became more regular than with the simple random ordering of arrivals.

We know that air traffic controllers have been using bunching techniques for some time, and the evidence we have collected so far suggests that the extent of bunching is increasing. The actual re-sequencing strategies may vary, and we do not yet know how significant the strategies used are in terms of reducing cumulated landing times. However, it is likely that they provide significant time savings.

An important element of bunching is that it depends on the principle of 'first come first served' being abandoned. This means that safeguard rules must be introduced to prevent individual aircraft loosing out badly in the re-sequencing of traffic[27].

14.5.2.3 *Bunching Departures*

The bunching of departures is done for the same reason as the bunching of arrivals – time can be saved by reordering departures – but it is an even more complex problem. The controller's objective is to re-order the aircraft at the holding point so as to select the appropriate departure sequence which minimises departure intervals – ideally 1 minute between departures wherever possible. One minute intervals are achieved when the following aircraft departs on a different route from the lead aircraft. The minimum time between departures using the same route is two minutes, but can be longer if wake vortex separations apply (Graves, 1998). The departure rate is increased by switching departures between those using the left, right and straight-ahead tracks.

But there is more to it than that. Apart from their wake vortex categories, aircraft are grouped according to size and speed and when an aircraft from a faster group follows a slower one extra time must be added. Thus there is a lot of mental arithmetic involved, and 'Achieving the optimum departure sequence is by no means easy. The scale of difficulty is illustrated by the fact that at Heathrow the departure controller is expected to memorise rules for over 500 route combinations and then apply the appropriate speed allowances before deciding whether wake vortex separations should take precedence. With say ten holding aircraft, he or she must then determine which of 3½ million possible sequences matches flow control requirements to give the least delay and achieve maximum use of the runway.' (CAA, 1993). He or she must also take account of the slot times of the aircraft waiting at the hold, as aircraft with slot times have to be fitted in within a limited time-period, because scheduling is based on these slot times (Graves, 1998).

The efficiency of departure sequencing has been further improved by the redesign of the departure routes (known as Standard Instrument Departures – SIDs), so that these do not interfere with each other.

14.6 DISCUSSIONS AND CONCLUSIONS

The title of this chapter - 'Horndal at Heathrow?' - refers to the Horndal iron works in Sweden, which Lundberg (1961) discovered increased its productivity[28] by an average of 2% per annum despite receiving no new investment for 15 years. Lundberg thus identified 'the Horndal effect' – an increase in output based on learning, which itself is 'the product of experience' (Arrow, 1962), rather than by investments in new capital equipment.

Other examples of productivity due to learning by experience are provided by Arrow (1962) and Rosenberg (1982), and include Wright's (1936) study which found the number of labour-hours required to produce an airframe was a decreasing function of the total number of airframes of the same type previously produced.

Arrow emphasises the importance of 'learning by doing', which, according to Rosenberg (1982, p. 121), 'is a form of learning that takes place at the manufacturing stage after the product has been designed. ... Learning at this stage ... consists of developing increased skills in production. This has the effect of reducing real labour costs per unit of output.' Rosenberg goes on to identify a related form of learning by doing, that has not been highlighted in the literature, which concerns small improvements to the production process which arise out of a direct involvement in the production process. For this type of learning by doing:

> "The learning involved requires participation in the production process. Such participation is obviously not sufficient, because a perception of possible improvements depends not only upon the opportunity to make certain observations but also upon prior training and experience. The point is that productive activities always involve specialised kinds of knowledge, much of which may be unique to a specific industrial process. There is typically a range of possible improvements that require intimate familiarity with the minutiae of the productive sequence" (Rosenberg, 1982, p.122).

Rosenberg (1982) then goes on to identify another sort of learning – learning by using – which is based on an increasing familiarity amongst users with a particular product or technology. Thus 'learning by using refers to a very different locus of learning than does learning by doing' (p.122), and although the experiences of users with a technology can feedback to inform producers about possible improvements to the next generation of that technology, learning by using essentially takes place on the technology as given.

In this chapter, we have been interested in the growth of air traffic activity at one of the world's busiest airports – Heathrow, and have observed that air traffic activity has grown substantially despite the fact that the airport retains the same basic infrastructure today – in terms of the number of runways - as it has had for more than 30 years. We must stress that we are not saying Heathrow has not had any investment over this period – indeed, investments have been substantial. What we are claiming is that much of the improvement in the airport's capacity is due to procedural changes in which new technological hardware plays a relatively minor role.

Our study is novel because it examines a service activity, whereas most studies of learning in industry focus on manufacturers. This is not a trivial difference, because an important feature of services is that they are consumed as they are produced. This is sometimes referred to as 'co-terminality'. The 'learning-by-doing' identified by Arrow (1962) and Rosenberg (1982) is an activity which provides productivity improvements to producers, whilst 'learning-by-using', as identified by Rosenberg (1982), is an activity which provides productivity improvements to users. To the extent that such learning is conscious and subject to encouragement and choice[29], within the context of a manufactured product these activities are essentially separate and the producer is free to make changes to the process by which the technology is produced, whilst the user is free to make changes to the pattern of use of the technology. This is different in services, because the 'product' does not exists as an entity independent of the producer, or indeed the user, and the service is consumed as it is produced. Thus the producer's freedom of action is likely to be more constrained by the operational requirements of the user. In the case of airports, this situation is complicated further by the importance of the regulatory authorities whose primary interest is safety.

We have been seeking to open up the 'black art' by which runway capacity has developed over time. We have stressed the importance of an evolving understanding of how the system works, together with an iterative process by which a relatively (but not optimally) efficient compromise situation emerges between the provider (the airport operator), the user (the airlines) and the regulatory agencies (air traffic control and the civil aviation authority) so that the system may be made to work more efficiently. Improvements are not automatically or easily imposed. For example, airlines have Standard Operating Procedures which they are reluctant to change for individual airports, not least because of the insurance cost implications. So the (conscious) learning that is significant is not just about what are the possible changes that might be made (the possibility space), but extends to understanding the needs of the other actors in the system, so the number of real options is less than that offered by the possibility space. In this sense, innovation in services through procedural change has a close relation to the design function for manufactured goods.

Another interesting feature of this inter-dependence is that because procedural change at an airport requires the close collaboration of the airport operator, air traffic control (which may be independent of the airport), the airlines and the safety regulator, and because the issue of capacity is so important politically, a lot of information has been documented revealing the known options and conscious choices taken in relation to capacity creation. Thus we can ascribe a considerable part of the increased capacity to choices made on the basis of this conscious understanding (although further improvements will also arise through unconscious learning by doing/using amongst pilots and air traffic controllers). Compare that with the situation at a factory, which increased its output significantly, but neither customers nor regulators had any direct interest in this process, and the management were free impose any changes they saw fit without the need for extensive documentation. We might well ascribe the productivity improvement of this factory to an unconscious and automatic 'learning-by-doing' process, because we have no other explanations.

References

Arrow, K.J. 'The Economic Implications of Learning by Doing', *Review of Economic Studies*, 1962; June, pp.155-173.

Brooks, P.W. 'A Short History of London's Airports', *The Journal of Transport History*, 1957; Vol. 3, pp. 12-22.

CAA. *Runway Capacity and Aircraft Delays at Gatwick Airport*, CAA Paper 81002, London: Civil Aviation Authority, 1981.

CAA. *The Methodology of Runway Capacity Assessment – A Summary Paper*, CAA Paper 83019, London: Civil Aviation Authority, 1983.

CAA. *Arrival Runway Capacity and Arrival Delays – Heathrow Airport, Summer 1984*, CAA Paper 85010, London: Civil Aviation Authority, 1985.

CAA. *Report of the Committee on Runway Utilisation at Heathrow and Gatwick*, CAP 534, London: Civil Aviation Authority, 1988.

CAA. *Heathrow Demand and Delays: Monitoring of Arrivals, Summer 1988*, CAA Paper 89008, London: Civil Aviation Authority, 1989.

CAA. *Heathrow Demand and Delays: Monitoring of Arrivals, Summer 1989*, CAA Paper 90004, London: Civil Aviation Authority, 1990.

CAA. *A Guide to Runway Capacity – For ATC, Airport and Aircraft Operators*, CAP 627, London: Civil Aviation Authority, 1993.

CAA. *Slot Allocation: A Proposal for Europe's Airports*, CAP 644, London: Civil Aviation Authority, 1995.

DoT. *Runway Capacity to Serve the South East (RUCATSE) – Report of the Working Party*, London: Department of Transport, June, 1993.

Drew, A.E. and Logan, H.K. *Arrival Runway Capacity and Arrival Delays at London Heathrow Airport, Summer 1980*, CAA Paper 81014, London: Civil Aviation Authority, 1981.

Drew, A.E. and Logan, H.K. *The Effect of Certain Facilities on the Runway Capacity and Aircraft Delays at Gatwick Airport, 1981*, CAA Paper 82019, London: Civil Aviation Authority, 1982.

Duke, G. *Air Traffic Control*, 6th edition, Shepperton, Surrey: Ian Allan, 1996.

Graves, D. *UK Air Traffic Control: A Layman's Guide*, 3rd Edition, Shrewsbury, England: Airlife, 1998.

Higham, R. 'A Matter of the Utmost Urgency: The Search for a Third London Airport, 1918-1992' in Leary, W.M. (ed.), *From Airships to Airbus – the History of Civil and Commercial Aviation – Volume 1, Infrastructure and Environment*, Washington and London: Smithsonian Institution Press, 1995.

HMSO. *UK Airport Capacity*, Second Report in the Session 1995-96 of the Transport Committee of the House of Commons, London, 1996.

Horonjeff, R. (and others). *Exit Taxiway Location and Design*, Institute of Transportation and Traffic Engineering, University of California, 1958.

Malerba, F. 'Learning by Firms and Incremental Technical Change', *Economic Journal*, 1992; 102 (July), pp. 845-859.

NATS, IATA and BAA. *Report on the Heathrow Airport Runway Capacity Enhancement Study*, Cheltenham, England: National Air Traffic Services, Civil Aviation Authority, 1994.

Rosenberg, N. *Inside the Black Box – Technology and Economics*, Cambridge: Cambridge University Press, (see particularly Chapter 6 on 'Learning by Using'), 1982.

Warskow, M.A. *Means of Increasing Airport Capacity*, in "World Airports the Way Ahead", Proceedings of the Conference held at the Queen Elizabeth Hall, London, 23-25 September, 1969, Institute of Civil Engineers, London, 1970.

---

[1] This chapter develops ideas first presented in our paper 'Productivity Growth and Procedural Knowledge - Analysing the Development of London's Airports from a "Systems of Innovation' Perspective', presented at the CRIC, Workshop on 'Systems and Services', on 17th – 18th March, 1998, in Manchester. We are grateful to the workshop participants for their comments on the work. We are also grateful to the BAA's research department at Gatwick for their help in providing data and discussing 'innovations' for this project and to Mr Keith Robinson, air traffic control manager at Manchester airport for our discussions with him and his comments on an earlier version of this paper. As usual, responsibility for any errors is ours alone.

[2] Data from BAA plc, the owner of Heathrow airport. See: http://www.baa.co.uk

[3] For a short account of the early history of London's airports see Brooks (1957)

[4] The London City Airport is London's fifth airport, but traffic levels are small relative to the other four.

[5] Seat loading is the term for passengers on a flight as a proportion of the total number of seats available.

[6] Although air fares, exchange rates and trade variables have also been found to be significant.

[7] Some slots do remain unused in certain off-peak periods – particularly late at night. These are unused largely because of airlines are not able to gain adequate access to the airport to maintain a service, or because they are not commercially viable.

[8] Note that because demand is difficult to quantify, there is no definitive proof that Heathrow is saturated, but this remains the belief of the authorities. 'We believe that, unconstrained, demand to use Heathrow would significantly exceed capacity ... although the extent is difficult to quantify' (NATS et al, 1994). 'The potential demand for slots for particular uses cannot be measured precisely given the complex of influences involved and the sometimes unpredictable nature of individual management decisions' (CAA, 1995 – CAP 644, para 43).

[9] London's airports have a certain hierarchy, related to their size, suchthat airlines normally prefer to serve Heathrow, and failing that Gatwick, and failing that Stansted or Luton.

[10] In fact, the principal constraints at Heathrow are currently terminal and stand capacity. Stand capacity is where aircraft park whilst at the airport. It is these constraints that the proposed Terminal 5 at Heathrow is intended to overcome.

[11] The monetary value of a pair of runway slots is not fully known as a market for slots has only recently become legal, but the value is thought to run into several million pounds, particularly for slots during busy parts of the day. In 1987 a high level study (CAA, 1988 – CAP 534) noted: 'Each additional five slots, i.e. one in each peak hour, was thus associated with an annual average earnings potential of £180 million per annum at Heathrow and £70 millions at Gatwick'.

[12] French planning law has been favourable to the development of large infrastructural projects considered to be in the national interest. British planning law is very different, as the case of the proposed Terminal 5 at Heathrow makes clear. In fact, air transport in the UK has had repeated battles over planning permission over the years. There were major inquiries into the construction of a third London airport, initially at Maplin, then later at Stansted, and there have been inquiries into the development of major facilities, such as Terminal 4 at Heathrow, which, incidentally, the airport operator initially opposed. (For the history and debate about UK planning in relation to airports and especially London's third airport – see Higham, 1995.)

[13] We recognise that this is a simplification of the reality.

[14] Also of interest is how the categories were established. The International Civil Aviation Organisation (ICAO) divides aircraft into three categories based on maximum take off weight in kilograms. These are Light (7,000 or less), Medium (7,001 to 135,999) and Heavy (136,000 or more). In the UK, however, the Medium Category is divided into Small (7,001 to 40,000) and Medium (40,001 to 135,999). In an industry in which insurance plays a key role, and in which insurance often depends on compliance with international (ICAO) standards, it would be interesting to know how the UK came to introduce the distinction between Small and Medium traffic in the early 1980s. Moreover, around the same time, aircraft such as the Boeing 707 and Douglas DC-8, although in the Heavy category for their take-off weights, were reclassified as Medium category aircraft. Beyond this, there is some suggestion that if the Medium category, which dominates traffic at Heathrow but which includes aircraft of quite widely differing sizes, were divided up, some savings on runway time might be made. It would be interesting to know whether any attempts have been made to seek a further reclassification, and if so whether these attempts have met with resistance.

[15] Further details of the operational procedures undertaken to reduce the noise nuisance are (NATS et al, 1994): 1. An 'alteration procedure' applies to the use of the runways during westerly operations. This means that one runway is used for landing between 7 am and 3pm, after which the other runway is used until 11pm. This is done to equalise the noise distribution and give periods of respite to the communities under the respective approach flight paths. During westerly operations there are no restrictions to using both runways for departures. This means that some departures can be interleaved between arrivals on the 'landing runway'. 2. With easterly operations, the close proximity of the village ofCranford to the end of the runway severely restricts the use of 09L for departures. The 'Cranford Agreement' means that aircraft can only take-off from 09R, but both runways can be used for landing. 3. 'Preferred Runway Selection' is theprocedure whereby, even in light easterly winds, westerly operations are sustained in preference to easterly operations, provided that tailwind and crosswind limits are not exceeded. This is to reduce the number of times that aircraft on 'take-off power' overfly the built up areas to the east of Heathrow. 4. Departures from Heathrow must follow prescribed noise preferential routes (NPRs) below 4,000 feet.

[16] The 'busy period' is the period over which there is a constant stream of traffic seeking to use the runways.

[17] DORA = Department of Operational Research and Analysis, in the Chief Scientist's Division, of National Air Traffic Services (NATS), of the Civil Aviation Authority.

[18] This report began the process of declaring a variable capacity – taking into account the traffic mix and how that differed throughout the day – for example, in the afternoon the proportion of heavy traffic is less, allowing an increase of 0.5 movements per hour during this period.

[19] For example, US aircraft do not power up before turning onto the runway because of what might be behind – i.e. a light aircraft. This, however, slows the manoeuvre and extends runway occupancy, reducing the efficiency of the runways.

[20] For example, it was discovered that one US airline took longer in using the runways on departure than the average. Discussion revealed that this was because the airline's standard practice was to use a satellite downlink to set the flight computers just prior to departure. This led to a delay in the airline's use of the runways.

[21] Moreover, modern aircraft tend to be highly automated, and land on 'auto-brake' which is designed to ensure comfort and to protect the brakes from excessive wear.

[22] In the US, under certain conditions responsibility for separations can pass to the pilots, and this normally results in smaller average separations. Such a change would require a change in the law in the UK. It is also thought that because Heathrow has so many different airlines, many of which have low frequency flights into the airport, the lack of familiarity amongst the pilots with the airport might increase the average separations rather than decrease them if responsibility for separations were given to the pilots.

[23] Another example of learning-by-doing amongst air traffic controllers is found in their improved handling of the LINK helicopter which linked Heathrow and Gatwick between 1978 and 1986 after which it was withdrawn due to the completion of the M25 motorway. Following observations in 1981, it was reported that 'The effect of the LINK helicopter is to reduce runway capacity (at Gatwick) by 0.7 movements (per hour), an improvement of 30% on the 1979 observations. This seems to be a consequence of the greater experience of controllers in handling the helicopter in a manner which minimises its effect on the other traffic.' (CAA, 1982 - Paper 82019).

[24] The actual date was not provided in the reference.
[25] Note that these standards often have to be adapted to local circumstances, for at Heathrow there is very little room between the runways and the outer-taxiways.
[26] Consequently, as demand increases queues tend to become longer, which in turn provide the controller with a greater choice as to the next movement. Because of this, the service rate is not independent of the level of demand.
[27] Indeed, this was found with our sequencing model, for potentially an aircraft had to wait some considerable time before being allowed to land. We therefore introduced a new rule, such that no aircraft could be passed in the queue by more than 10 other aircraft. This solved the problem, but reduced the efficiency of the re-sequencing, such that the mean saving in runway time was reduced from 3.7% to 2.1% - that is more than 40% of the efficiency gain. Most probably a more sophisticated approach would not lead to such a severe reduction in the time saving.
[28] Output per manhour.
[29] There is a difficulty with 'learning-by-doing' and 'learning-by-using' in the sense that it is not clear whether these types of learning are confined to sub-conscious processes (which are automatic and costless – Malerba, 1992), or whether they extend to conscious and deliberate changes, based on choice, to the organisation of the production process or the pattern of use of the technology, where the opportunities for organisational change arise out of learning by experience. With reference to the experience of the Horndal iron works, Arrow (1962) presumes that because there was no new investment there was also no 'significant change' in the methods of production. This raises the question as to what constitutes a significant change, but our point is that whilst some of the productivity improvements at Horndal may have been due to unconscious and automatic 'learning-by-doing', another component may have been due to conscious decisions by management and workforce about how to better organise the production process. If we ascribe productivity increases to 'learning-by-doing' in the narrow sense of unconscious automatic processes, then to what extent is it really a measure of the importance of this, and to what extent is it a measure of our ignorance about the importance of deliberate organisational and procedural change.

List of Invited Participants to CRIC Workshops

Workshop on "Innovation Systems and Services" Workshop, 17th-18th March 1998

Professor Esben Sloth Andersen, DRUID, University of Aalborg, Denmark
Professor Cristiano Antonelli, University of Turin, Italy
Professor Bo Carlsson, Case Western Reserve University, Cleveland, USA
Dr Ken Ducatel, IPTS, Seville and PREST, University of Manchester, UK
Dr Rinaldo Evangelisto, ISRDS – CNR, Rome, Italy
Professor Faiz Gallouj, IFRESI, University of Lille, France
Dr Geoff Mason, National Institute of Economic and Social Research, London, UK
Professor Richard Nelson, Colombia University, New York, USA
Dr Paul Nightingale, Research Fellow, CoPS, SPRU, University of Sussex, UK
Dr Brigitte Preissl, German Institute for Economic Research (DIW), Berlin, Germany
Dr Joanne Roberts, University of Northumbria, UK
Dr Jon Sundbo, Department of Social Sciences, Roskilde University, Denmark

Workshop on "Conceptualizing and Measuring Service Innovation Systems", 20th-21st May 1998

Professor Gilberto Antonelli, IDSE, Milan and University of Bologna, Italy
Mr John Barber, Department of Trade and Industry, UK
Mr G Cainelli, IDSE, Milan, Italy
Dr Rinaldo Evangelista, ISRDS – CNR, Rome, Italy
Dr Dominique Guellec, OECD, France
Daood Hamdani, Statistics Canada
Dr Johan Hauknes, STEP Group, Oslo, Norway
Ms Christiana Hipp, Mannesmann Pilotentwicklung, Munich, Germany
Professor Alan Hughes, ESRC Centre for Business Research, University of Cambridge, UK
Professor Alfred Kleinknecht, Delft University of Technology, The Netherlands
Dr Georg Licht, Fraunhofer Institute for Innovations and Systems Research, Karlsruhe, Germany
Professor Morris Teubal, The Jerusalem Institute for Israel Studies, Jerusalem, Israel
Eric Wood, ESRC Centre for Business Research, Cambridge, UK
Professor Peter Wood, University College London, UK

Index

K

S

Economics of Science, Technology and Innovation

1. A. Phillips, A.P. Phillips and T.R. Phillips:
Biz Jets. Technology and Market Structure in the Corporate Jet Aircraft Industry. 1994 ISBN 0-7923-2660-1
2. M.P. Feldman:
The Geography of Innovation. 1994 ISBN 0-7923-2698-9
3. C. Antonelli:
The Economics of Localized Technological Change and Industrial Dynamics. 1995 ISBN 0-7923-2910-4
4. G. Becher and S. Kuhlmann (eds.):
Evaluation of Technology Policy Programmes in Germany. 1995 ISBN 0-7923-3115-X
5. B. Carlsson (ed.): *Technological Systems and Economic Performance: The Case of Factory Automation.* 1995 ISBN 0-7923-3512-0
6. G.E. Flueckiger: *Control, Information, and Technological Change.* 1995 ISBN 0-7923-3667-4
7. M. Teubal, D. Foray, M. Justman and E. Zuscovitch (eds.):
Technological Infrastructure Policy. An International Perspective. 1996 ISBN 0-7923-3835-9
8. G. Eliasson:
Firm Objectives, Controls and Organization. The Use of Information and the Transfer of Knowledge within the Firm. 1996 ISBN 0-7923-3870-7
9. X. Vence-Deza and J.S. Metcalfe (eds.):
Wealth from Diversity. Innovation, Structural Change and Finance for Regional Development in Europe. 1996 ISBN 0-7923-4115-5
10. B. Carlsson (ed.):
Technological Systems and Industrial Dynamics. 1997 ISBN 0-7923-9940-4
11. N.S. Vonortas:
Cooperation in Research and Development. 1997 ISBN 0-7923-8042-8
12. P. Braunerhjelm and K. Ekholm (eds.):
The Geography of Multinational Firms. 1998 ISBN 0-7923-8133-5
13. A. Varga:
University Research and Regional Innovation: A Spatial Econometric Analysis of Academic Technology Transfers. 1998 ISBN 0-7923-8248-X
14. J. de la Mothe and G. Paquet (eds.):
Local and Regional Systems of Innovation ISBN 0-7923-8287-0
15. D. Gerbarg (ed.):
The Economics, Technology and Content of Digital T V ISBN 0-7923-8325-7
16. C. Edquist, L. Hommen and L. Tsipouri
Public Technology Procurement and Innovation ISBN 0-7923-8685-X
17. J. de la Mothe and G. Paquet (eds.):
Information, Innovation and Impacts ISBN 0-7923-8692-2
18. J. S. Metcalfe and I. Miles (eds.):
Innovation Systems in the Service Economy: Measurement and Case Study Analysis ISBN 0-7923-7730-3

KLUWER ACADEMIC PUBLISHERS — BOSTON / DORDRECHT / LONDON